Various Deep Learning Algorithms in Computational Intelligence

Various Deep Learning Algorithms in Computational Intelligence

Editor

Oscar Humberto Montiel Ross

MDPI • Basel • Beijing • Wuhan • Barcelona • Belgrade • Manchester • Tokyo • Cluj • Tianjin

Editor
Oscar Humberto Montiel Ross
Instituto Politécnico Nacional
Tijuana
Mexico

Editorial Office
MDPI
St. Alban-Anlage 66
4052 Basel, Switzerland

This is a reprint of articles from the Special Issue published online in the open access journal *Axioms* (ISSN 2075-1680) (available at: https://www.mdpi.com/journal/axioms/special_issues/Deep_Learning_Algorithms).

For citation purposes, cite each article independently as indicated on the article page online and as indicated below:

LastName, A.A.; LastName, B.B.; LastName, C.C. Article Title. *Journal Name* **Year**, *Volume Number*, Page Range.

ISBN 978-3-0365-8138-5 (Hbk)
ISBN 978-3-0365-8139-2 (PDF)

Contents

About the Editor

Oscar Humberto Montiel Ross

He is currently a researcher at the Centro de Investigación y Desarrollo de Tecnología Digital (CITEDI-IPN). He received a D.Sc. in Computer Science from Universidad Autónoma de Baja California (UABC) in 2006, an M.Sc. in Computer Science from Instituto Tecnológico de Tijuana in 2000, an M.Sc. in Digital Systems from Instituto Politécnico Nacional (IPN) in 1999, and a B.S. in Electrical Engineering in Electronics from UABC in 1985; all of these degrees were obtained in México. He has published articles, books, and book chapters about quantum computing, evolutionary computation, mobile robotics, mediative fuzzy logic, ant colonies, type-2 fuzzy systems, and embedded systems. Prof. Montiel has led 20 research projects and participated in about 20 others, most of them in the Computational Intelligence field. He is a senior member of the Institute of Electrical and Electronics Engineers, a member of the International Association of Engineers (IANG), and a member of the Mexican Science Foundation CONAHCYT (Consejo Nacional de Ciencia Humanidades y Tecnología). He has experience editing and publishing books. He is member of the Editorial Board of Axioms and Associate Editor of *Expert Systems with Applications*; as Guest Editor, he has published Special Issues for the journals *Advances in Fuzzy Systems*, *Computación y Sistemas*, *Axioms*, and *Soft Computing*. He received the Research Award in 2016 from IPN. He is a Co-Founder and an active member of the HAFSA (Hispanic American Fuzzy Systems Association) and the Mexican Chapter of the Computational Intelligence Society (IEEE).

Preface to "Various Deep Learning Algorithms in Computational Intelligence"

We are very proud of the meticulous editing of this reprint, titled "Various Deep Learning Algorithms in Computational Intelligence," which compiles journal articles and showcases a diverse range of contributions from authors representing various fields across the globe. Our gratification stems from achieving our objective of gathering different perspectives, enabling us to comprehensively explore Deep Learning (DL) from multiple disciplines.

The above is highly important, since DL has garnered significant attention in science, industry, and academia due to its essential nature, which draws inspiration from the functioning of the human brain and the concept of learning. Unlike traditional and machine learning methods, deep learning techniques emulate the human brain's neural networks at a lower scale, allowing them to process and analyze substantial quantities of unstructured data. The remarkable proficiency of deep learning in unveiling intricate structures within extensive datasets genuinely resembles the extraordinary aptitude of the brain to recognize patterns and form complex connections. This unique characteristic allows DL to excel in modeling and solving complex problems across various scientific and technological fields. Just as the brain learns from experience, DL architectures learn through algorithms from data by adjusting numerous parameters during training to optimize their performance and accuracy. This concept of learning and adaptation is fundamental to DL's success.

The publication of this reprint serves as an excellent opportunity to disseminate current knowledge beyond academic boundaries, reaching a diverse audience encompassing academics, professionals, and the general public. This wide readership fosters the potential for meaningful connections to established projects and the cultivation of collaboration for future research endeavors.

Finally, I express my gratitude to everyone involved in the publication of this reprint and the Special Issue. I extend my appreciation to the reviewers that helped to improve all the contributions, the editors, and the assistant staff for their valuable help in upholding the highest quality standards for the published scientific papers. I want to give special recognition to the Managing Editor for their exceptional efforts in overseeing this Special Issue and the reprint.

Oscar Humberto Montiel Ross
Editor

MDPI

Editorial

Various Deep Learning Algorithms in Computational Intelligence

Oscar Humberto Montiel Ross

Centro de Investigación y Desarrollo de Tecnología Digital, Instituto Politécnico Nacional, Mexico City 07738, Mexico; oross@ipn.mx

Citation: Ross, O.H.M. Various Deep Learning Algorithms in Computational Intelligence. *Axioms* **2023**, *12*, 495. https://doi.org/10.3390/axioms12050495

Received: 16 May 2023
Accepted: 18 May 2023
Published: 19 May 2023

Deep Learning (DL) is an essential topic of increasing interest in science, industry, and academia. Unlike traditional and machine learning methods, DL methods can process large volumes of unstructured data discovering intricate structures in large data sets. It is rapidly becoming a tool for modeling and solving complex and difficult problems in different fields of science and technology. For example, in medicine for breast cancer, COVID-19 detection and diabetes detection and prediction; in autonomous vehicles in various tasks such as perception, mapping and localization; in astronomy, for classifying and detecting stars and galaxies; in the design of future wireless networks, and others. Although DL has been successfully applied in many fields and there are some theoretical developments, there are still many challenging problems in theory and applications that need to be solved for improving these techniques. For instance, it is important to find new methods to train the massive number of parameters required by DL architectures, solve overfitting and transfer learning problems, and others.

This Special Issue was dedicated to contributing to the state-of-the-art progress on DL with theoretical, practical, and creative insights that provided vanguard solutions to challenging problems or could demonstrate competitive performance. The participation of the community in sending works to this Special Issue was high and important, and after a rigorous peer-reviewing process only fourteen papers were accepted to be published. Next, we mention in chronological publication order the accepted works.

The first contribution was the "Optimization of Convolutional Neural Networks Architectures Using PSO for Sign Language Recognition" paper by Jonathan Fregoso, Claudia I. Gonzalez, and Gabriela E. Martínez [1]. Here, the authors presented an approach to designing convolutional neural network architectures using the particle swarm optimization algorithm (PSO), where they face the challenge of finding the optimal parameters of convolutional neural networks, particularly the number of layers, the filter size, the number of convolutional filters, and the batch size. They evaluated two different approaches to perform the optimization. In the first one, the parameters obtained by PSO were kept under the same conditions in each convolutional layer, and the classification rate gave the objective function evaluated by PSO. In the second one, the PSO generated different parameters per layer. The objective function was composed of recognition rate in conjunction with the Akaike information criterion; the latter helps to find the best network performance but with the minimum parameters. Finally, they tested the optimized architectures with three study cases: sign language databases where the Mexican Sign Language alphabet was included, the American Sign Language MNIST, and the American Sign Language alphabet. The authors found that the proposed methodologies achieved favorable results with a recognition rate higher than 99%, showing competitive results compared to other state-of-the-art approaches.

The paper "Multicriteria Evaluation of Deep Neural Networks for Semantic Segmentation of Mammographies" by Yoshio Rubio and Oscar Montiel [2] studied the critical problem of breast segmentation for automatic and accurate analysis of mammograms to increments the probability of a correct diagnostic while reducing the computational cost, which was done through an extensive evaluation of deep learning architectures for semantic segmentation of mammograms, including segmentation metrics, memory requirements, and average inference time. They used different combinations of two-stage segmentation architectures composed of a feature extraction net (VGG16 and ResNet50) and a segmentation net (FCN-8, U-Net, and PSPNet). The author used examples from the mini–Mammographic Image Analysis Society (MIAS) database for the learning phase. The experimental results showed that the best net scored a Sørensen–Dice similarity coefficient of 99.37% for breast boundary segmentation and 95.45% for pectoral muscle segmentation.

In the paper "RainPredRNN: A New Approach for Precipitation Nowcasting with Weather Radar Echo Images Based on Deep Learning" by Do Ngoc Tuyen, Tran Manh Tuan, Xuan-Hien Le, Nguyen Thanh Tung, Tran Kim Chau, Pham Van Hai, Vassilis C. Gerogiannis, and Le Hoang Son [3], the authors proposed a novel approach named RainPredRNN, which combines the UNet segmentation model and the PredRNN_v2 deep learning model for precipitation nowcasting with weather radar echo images. They found that it is possible to reduce the number of operations of the RainPredRNN model by leveraging the abilities of the contracting–expansive path of the UNet model. Consequently, the result offers the benefit of reducing the processing time of the overall model while maintaining reasonable errors in the predicted images. They validated their proposed mode through experiments on real reflectivity fields collected from the Phadin weather radar station in Dien Bien province in Vietnam. Some credible quality metrics, such as the mean absolute error (MAE), the structural similarity index measure (SSIM), and the critical success index (CSI), were used for analyzing the performance of the model. It was certified that the proposed model had produced improved performance, about 0.43, 0.95, and 0.94 of MAE, SSIM, and CSI, respectively, with only 30% of training time compared to the other methods.

The paper "Cubical Homology-Based Machine Learning: An Application in Image Classification" by Seungho Choe and Sheela Ramanna [4] contributed with a cubical homology-based algorithm for extracting topological features from 2D images to generate their topological signatures; they propose a score metric to measure the significance of the subcomplex calculated from the persistence diagram (topological signatures). Additionally, they used gray-level co-occurrence matrix (GLCM) and contrast limited adapting histogram equalization (CLAHE) to obtain additional image features to improve the classification performance; finally, they discussed the results of our supervised learning experiments of eight well-known machine learning models trained on six different published image datasets using the extracted topological features.

In the study, "Towards Predictive Vietnamese Human Resource Migration by Machine Learning: A Case Study in Northeast Asian Countries", the authors Nguyen Hong Giang, Tien-Thinh Nguyen, Chac Cau Tay, Le Anh Phuong, and Thanh-Tuan Dang [5] presented a forecast of Vietnamese labor migration using the kNN, RFR, and BPNN models. The data collected in the study included 29 observations from 1992 to 2020. With the support of the collected data, the paper analyzed the role of labor exports in Vietnam's socio-economic development and compared this export labor with that of other Asian countries. After that, they compared the three algorithms based on the results of their statistical accuracy indicators. This research highlighted the likely future contexts of labor migration between Vietnam and East Asia, including Korea, the Republic of China (Taiwan), and Japan. Furthermore, this research could assist the government of Vietnam in enacting new regulations for Vietnamese migrant workers to boost the socio-economic situation.

In "Lexicon-Enhanced Multi-Task Convolutional Neural Network for Emotion Distribution Learning", by Yuchang Dong and Xueqiang Zeng [6], the authors dealt with the problem of emotion analysis behind massive human behavior such as text, pictures, movies, and music. To address this problem, they proposed a text emotion distribution learning model based on a lexicon-enhanced multi-task convolutional neural network (LMT-CNN). The overall architecture of the LMT-CNN model has three major modules: semantic information, emotion knowledge, and multi-task prediction. They constructed the input of the multi-task prediction module from the outputs of the first two modules. Then, they predicted the final emotion distribution through a fully connected layer. They presented extensive comparative experiments on nine commonly used emotional text datasets, showing that the proposed LMT-CNN model is superior to the compared EDL methods for both emotion distribution prediction and emotion recognition tasks.

In "A Schelling Extended Model in Networks—Characterization of Ghettos in Washington D.C", Diego Ortega and Elka Korutcheva [7] proposed a method to bridge the gap between Sociophysics models and the measure of segregation by geographic information systems (GIS) techniques. Their procedure relies on capturing the ghettos' location with a segregation model and basic data from the city in situations where detailed data cannot be completely reliable. This is a novel approach to the segregation field. They started by using basic GIS information from the region of interest to define a network and then ran an extended Schelling model on this framework. Their enhanced version considered the economic contribution to segregation, including terms for the housing market and the financial gap between the population. The result was the location of ghettos on the network, which could be mapped into their corresponding city areas. Finally, the predicted ostracized regions were compared to the ones characterized as ghettos by spatial analysis (SA) and machine learning (ML) algorithms. For the real case study of Washington D.C., the obtained accuracy was $80 \pm 7\%$.

In "Hybrid Deep Learning Algorithm for Forecasting SARS-CoV-2 Daily Infections and Death Cases" by Fehaid Alqahtani, Mostafa Abotaleb, Ammar Kadi, Tatiana Makarovskikh, Irina Potoroko, Khder Alakkari, and Amr Badr [8], the authors used a hybrid deep learning algorithm to predict new cases of infection which is crucial for authorities to get ready for early handling of the virus spread. The hybrid deep learning method was used to improve the parameters of long short-term memory (LSTM). To evaluate the effectiveness of the proposed methodology, the authors collected a dataset based on the recorded cases in the Russian Federation and Chelyabinsk region between 22 January 2020 and 23 August 2022. In addition, five regression models were included in the conducted experiments to show the effectiveness and superiority of their proposed approach. The achieved results showed that their proposal could reduce the mean square error (RMSE), relative root mean square error (RRMSE), mean absolute error (MAE), coefficient of determination (R Square), coefficient of correlation (R), and mean bias error (MBE) when compared with the five base models. The results confirmed the effectiveness, superiority, and significance of the proposed approach in predicting the infection cases of SARS-CoV-2.

In "Score-Guided Generative Adversarial Networks" by Minhyeok Lee and Junhee Seok [9], a generative adversarial network (GAN) that introduces an evaluator module using pre-trained networks was proposed, and it was called a score-guided GAN (Score-GAN). The module was trained using an evaluation metric for GANs, i.e., the Inception score, as a rough guide for the training of the generator. Using another pre-trained network instead of the Inception network, ScoreGAN circumvented the overfitting of the Inception network such that the generated samples do not correspond to adversarial examples of the Inception network. In addition, they used evaluation metrics only in an auxiliary role to prevent overfitting. When evaluated using the CIFAR-10 dataset, ScoreGAN achieved an Inception score of 10.36 ± 0.15, corresponding to state-of-the-art performance. To generalize the effectiveness of ScoreGAN, the model was evaluated further using another dataset, CIFAR-100. ScoreGAN outperformed existing methods.

In the paper "Improved Method for Oriented Waste Detection", Weizhi Yang, Yi Xie, and Peng Gao [10] investigated automated waste classification and image-related problems when using robotic arms. Here, the authors, to solve the problem of low-accuracy image detection caused by irregular placement angles of robotic arms, proposed an improved oriented waste-detection method based on YOLOv5 by optimizing the detection head of the YOLOv5 model. This method generated an oriented detection box for a waste object at any angle. Based on the proposed scheme, they further improved three aspects of the performance of YOLOv5 in the detection of waste objects: the angular loss function was derived based on dynamic smoothing to enhance the model's angular prediction ability, the backbone network was optimized with enhanced shallow features and attention mechanisms, and the feature aggregation network was improved to enhance the effects of feature multi-scale fusion. The experimental results showed that the detection performance of the proposed method for waste targets was better than other deep learning methods. Its average accuracy and recall were 93.9% and 94.8%, 11.6% and 7.6% higher than the original network, respectively.

In "Two Novel Models for Traffic Sign Detection Based on YOLOv5s" by Wei Bai, Jingyi Zhao, Chenxu Dai, Haiyang Zhang, Li Zhao, Zhanlin Ji, and Ivan Ganchev [11], the problem of object detection and image recognition for unmanned driving technology was studied, particularly, the detection and recognition of traffic signs which are affected by diverse factors such as light, the presence of small objects, and complicated backgrounds. It is well known that traditional traffic sign detection technology does not produce satisfactory results when facing the above problems. To solve this problem, the authors proposed two novel traffic sign detection models called YOLOv5-DH and YOLOv5-TDHSA. Experiments conducted on two public datasets showed that both proposed models perform better than the original YOLOv5s model and three other state-of-the-art models.

In "Barrier Options and Greeks: Modeling with Neural Networks", the authors, Nneka Umeorah, Phillip Mashele, Onyecherelam Agbaeze, and Jules Clement Mba [12], proposed a non-parametric technique of option valuation and hedging. Using the fully connected feed-forward neural network, they replicated the extended Black–Scholes pricing model for the exotic barrier options and their corresponding Greeks. Their methodology involved benchmarking experiments resulting in an optimal neural network hyperparameter that effectively prices the barrier options and facilitates their option Greeks extraction. They compared the results from the optimal NN model to those produced by other machine learning models, such as the random forest and the polynomial regression; the output highlighted the proposed methodology's accuracy and efficiency for the pricing problem. The results showed that the artificial neural network could effectively and accurately learn the extended Black–Scholes model from a given simulated dataset. This concept can similarly be applied to the valuation of complex financial derivatives without analytical solutions.

In "Developing a Deep Learning-Based Defect Detection System for Ski Goggles Lenses" by Dinh-Thuan Dang, and Jing-Wein Wang [13], the authors presented a study that developed a deep learning-based defect detection system for ski goggles lenses. In the study, the first step was to design an image acquisition model that combines cameras and light sources; this step aimed to capture clear and high-resolution images on the entire surface of the lenses. Next, defect categories were identified, including scratches, watermarks, spotlight, stains, dust-line, and dust-spot. Then, they were labeled to create the ski goggles lens defect dataset. Finally, the defects were detected automatically by fine-tuning the mobile-friendly object detection model, the MobileNetV3 backbone used in a feature pyramid network (FPN), and the Faster-RCNN detector. The experiments demonstrate the effectiveness of defect detection at faster inference speeds. The defect detection accuracy achieved a mean average precision (mAP) of 55%. The work automatically integrated all steps, from capturing images to defect detection.

In "A Unified Learning Approach for Malicious Domain Name Detection" by Atif Ali Wagan, Qianmu Li, Zubair Zaland, Shah Marjan, Dadan Khan Bozdar, Aamir Hussain, Aamir Mehmood Mirza, and Mehmood Baryalai [14]. The authors presented a novel unified learning approach that uses both numerical and textual features of the domain name to classify whether a domain name pair is malicious or not. They conducted experiments on a benchmark domain names dataset consisting of 90,000 domain names. The experimental results show that the proposed approach performs significantly better than the six comparative methods in terms of accuracy, precision, recall, and F1-Score.

The diversity of fields of publications covers a wide range of interesting topics for researchers, scholars, and students interested in vanguard scientific contributions and practical applications to solve real-life problems.

Funding: This research was funded by Instituto Politécnico Nacional—CITEDI grant numbers SIP20220079 and SIP20230104.

Acknowledgments: I thank all of you that have participated in the publication of the Special Issue and this book. I thank the Editors and assistant personnel that have contributed to maintaining high-quality standards of the published scientific paper. Special thanks to the Managing Editor of this Special Issue and book.

Conflicts of Interest: The author declares no conflict of interest.

References

1. Fregoso, J.; Gonzalez, C.I.; Martinez, G.E. Optimization of Convolutional Neural Networks Architectures Using PSO for Sign Language Recognition. *Axioms* **2021**, *10*, 139. [CrossRef]
2. Rubio, Y.; Montiel, O. Multicriteria Evaluation of Deep Neural Networks for Semantic Segmentation of Mammographies. *Axioms* **2021**, *10*, 180. [CrossRef]
3. Tuyen, D.N.; Tuan, T.M.; Le, X.-H.; Tung, N.T.; Chau, T.K.; Van Hai, P.; Gerogiannis, V.C.; Son, L.H. RainPredRNN: A New Approach for Precipitation Nowcasting with Weather Radar Echo Images Based on Deep Learning. *Axioms* **2022**, *11*, 107. [CrossRef]
4. Choe, S.; Ramanna, S. Cubical Homology-Based Machine Learning: An Application in Image Classification. *Axioms* **2022**, *11*, 112. [CrossRef]
5. Giang, N.H.; Nguyen, T.-T.; Tay, C.C.; Phuong, L.A.; Dang, T.-T. Towards Predictive Vietnamese Human Resource Migration by Machine Learning: A Case Study in Northeast Asian Countries. *Axioms* **2022**, *11*, 151. [CrossRef]
6. Dong, Y.; Zeng, X. Lexicon-Enhanced Multi-Task Convolutional Neural Network for Emotion Distribution Learning. *Axioms* **2022**, *11*, 181. [CrossRef]
7. Ortega, D.; Korutcheva, E. A Schelling Extended Model in Networks—Characterization of Ghettos in Washington D.C. *Axioms* **2022**, *11*, 457. [CrossRef]
8. Alqahtani, F.; Abotaleb, M.; Kadi, A.; Makarovskikh, T.; Potoroko, I.; Alakkari, K.; Badr, A. Hybrid Deep Learning Algorithm for Forecasting SARS-CoV-2 Daily Infections and Death Cases. *Axioms* **2022**, *11*, 620. [CrossRef]
9. Lee, M.; Seok, J. Score-Guided Generative Adversarial Networks. *Axioms* **2022**, *11*, 701. [CrossRef]
10. Yang, W.; Xie, Y.; Gao, P. Improved Method for Oriented Waste Detection. *Axioms* **2023**, *12*, 18. [CrossRef]
11. Bai, W.; Zhao, J.; Dai, C.; Zhang, H.; Zhao, L.; Ji, Z.; Ganchev, I. Two Novel Models for Traffic Sign Detection Based on YOLOv5s. *Axioms* **2023**, *12*, 160. [CrossRef]
12. Umeorah, N.; Mashele, P.; Agbaeze, O.; Mba, J.C. Barrier Options and Greeks: Modeling with Neural Networks. *Axioms* **2023**, *12*, 384. [CrossRef]
13. Dang, D.-T.; Wang, J.-W. Developing a Deep Learning-Based Defect Detection System for Ski Goggles Lenses. *Axioms* **2023**, *12*, 386. [CrossRef]
14. Wagan, A.A.; Li, Q.; Zaland, Z.; Marjan, S.; Bozdar, D.K.; Hussain, A.; Mirza, A.M.; Baryalai, M. A Unified Learning Approach for Malicious Domain Name Detection. *Axioms* **2023**, *12*, 458. [CrossRef]

Article

Optimization of Convolutional Neural Networks Architectures Using PSO for Sign Language Recognition

Jonathan Fregoso, Claudia I. Gonzalez * and Gabriela E. Martinez

Division of Graduate Studies and Research, Tijuana Institute of Technology, Tijuana 22414, Mexico; jonathan.fregoso@tectijuana.edu.mx (J.F.); gmartinez@tectijuana.mx (G.E.M.)
* Correspondence: cgonzalez@tectijuana.mx

Abstract: This paper presents an approach to design convolutional neural network architectures, using the particle swarm optimization algorithm. The adjustment of the hyper-parameters and finding the optimal network architecture of convolutional neural networks represents an important challenge. Network performance and achieving efficient learning models for a particular problem depends on setting hyper-parameter values and this implies exploring a huge and complex search space. The use of heuristic-based searches supports these types of problems; therefore, the main contribution of this research work is to apply the PSO algorithm to find the optimal parameters of the convolutional neural networks which include the number of convolutional layers, the filter size used in the convolutional process, the number of convolutional filters, and the batch size. This work describes two optimization approaches; the first, the parameters obtained by PSO are kept under the same conditions in each convolutional layer, and the objective function evaluated by PSO is given by the classification rate; in the second, the PSO generates different parameters per layer, and the objective function is composed of the recognition rate in conjunction with the Akaike information criterion, the latter helps to find the best network performance but with the minimum parameters. The optimized architectures are implemented in three study cases of sign language databases, in which are included the Mexican Sign Language alphabet, the American Sign Language MNIST, and the American Sign Language alphabet. According to the results, the proposed methodologies achieved favorable results with a recognition rate higher than 99%, showing competitive results compared to other state-of-the-art approaches.

Keywords: PSO; sign language recognition; optimization of convolutional neural networks

Citation: Fregoso, J.; Gonzalez, C.I.; Martinez, G.E. Optimization of Convolutional Neural Networks Architectures Using PSO for Sign Language Recognition. *Axioms* **2021**, *10*, 139. https://doi.org/10.3390/axioms10030139

Academic Editor: Oscar Humberto Montiel Ross

Received: 29 May 2021
Accepted: 25 June 2021
Published: 29 June 2021

Publisher's Note: MDPI stays neutral with regard to jurisdictional claims in published maps and institutional affiliations.

1. Introduction

Deep neural networks have demonstrated their capacity to solve classification problems using a hierarchical model, millions of parameters, and learning with big databases. Convolutional neural networks (CNN) are a special class of deep neural networks that consist of several convolutions, pooling, and fully connected layers; this has proven to be a robust method for image or video processing, classification, and pattern recognition. In recent years CNN has attracted attention for achieving superior results in various applications in the computer vision domain, such as medicine, aerospace, natural language processing and robotics [1,2].

CNN are widely used in the field of industry, however, when designing CNN architectures, we face some challenges which include the high computational costs for information processing and finding the optimal CNN parameters (architecture) for each problem [3]. CNN architectures are made up of numerous parameters and, depending on their configuration, can generate a variety of classification results when applied to solve the same tasks; the setting of the hyper-parameter values is usually based on a random search, performing several tests or adjusting manually and this represents a complex search process. To solve this challenge, various researchers have proposed the implementation of evolutionary computation approaches to automatically design the optimal CNN architectures and to increase its performance [4,5]. In Sun et al. [6,7], an evolutionary approach is implemented to automatically obtain CNN architectures, achieving good results against the state-of-the-art architectures. In Ma et al. [8], the authors present an analysis of different methodologies based on evolutionary computing techniques to optimize CNN architectures, these were tested on benchmark data sets and achieved competitive results. Baldominos et al. [9] implement an approach to automatically design CNN architectures, using genetic algorithms (GA) in conjunction with grammatical evolution.

In the state of the art, we can find a variety of meta-heuristics that are applied to optimize CNN hyper-parameters, including the FGSA [10–12], harmonic search (HS) [13], differential evolution (DE) [14], microcanonical optimization algorithm [15], Whale optimization algorithm [16] and tree growth algorithm framework [17] to mention a few.

In other research works, the PSO algorithm is used to optimize CNN architectures, obtaining favorable results in the solution of different applications. In Sun et al. [18], Singh et al. [19] and Wang et al. [20], PSO is applied to automatically design CNN architectures; these approaches are tested on known benchmark datasets, and the results obtained are competitive against the state-of-the-art architectures. Besides this, PSO has been implemented in other fields of machine learning, including the optimization of different types of artificial neural network architectures, given favorable solutions for a plethora of problems [21,22]. In [23], PSO optimizes models of modular neural networks and is applied to obtain the blood pressure trend. In [24], a hybrid ANN-PSO method is applied to model the electricity price forecasting for the Indian energy exchange. As well as in [25], the PSO variants are applied to generate optimal modular neural network architectures obtaining competitive results for human recognition. In [26], the PSO algorithm is used to optimize deep neural network architectures and is tested in image classification tasks. In [27] a new paradigm of hybrid classification based on PSO is presented, which is applied for the prediction of medical diagnoses and prognoses. Furthermore, in [28] the artificial bee colony (ABC) [29] and PSO are used to optimize multilayer perceptron neural networks (MLP); the approach is applied to estimate the heating load and cooling load of energy efficient buildings; and the authors report that PSO outperforms ABC, improving the MLP performance. In the listed works, we can note the advantages that PSO offers in the optimization process, increasing performance in different tasks.

In research related to CNN approaches applied to the recognition of sign language, we find the work presented in [30] where a CNN model with stochastic pooling is implemented in the recognition of the Chinese sign language spelling, achieving a rate of 89.32 $\pm$ 1.07% recognition. In [31] a CNN method for Arabic sign language (ArSL) recognition was applied, where the authors report a value of 90.02% precision. In [32] a 3D-CNN approach is applied to sign language recognition for extensive vocabulary, images are captured through a Kinect, the authors report effectiveness of 88.7%.

The contribution of this research work focuses on implementing a hybrid methodology, where the PSO algorithm is applied to find the optimal design of parameters for CNN architectures. This work presents two optimization approaches; in both, the parameters considered are the number of convolution layers, the filter size used in each convolutional layer, the convolution filters number and, the batch size. In the first approach, the consistency of the parameters between each layer is maintained in the same conditions and the objective function is given by the recognition rate. In the second approach, the aim is to find more random searches in the architectures that the PSO produces; in this case, the values for each convolution layer are completely different, and the objective function is given by the highest recognition rate, and the lowest Akaike information criterion (AIC); the latter helps to obtain more robust performance of the network with the minimum parameters as the AIC allows penalizing the number of parameters used in each training. The optimized architectures are tested with three sign language databases, including the Mexican Sign Language (MSL) alphabet, the American Sign Language (ASL) alphabet [33], and the American Sign Language MNIST (ASL MNIST) [34]. This research aims to impulse the investigation in the soft computing area for the development of tools to help the deaf community for a more inclusive society [35].

The structure of this work is organized as follows. Section 2 presents the general theory about convolutional neural networks. Section 3 introduces PSO theory, including definitions, functionality, and the main equations. Section 4 details the methodology for developing the two PSO-CNN optimization approaches. Section 5 describes an analysis of the experimental results achieved after the optimized architectures are implemented for the three databases. Additionally, Section 5 presents a statistical test to compare the two optimization proposals, and we also show a comparative analysis against other CNN approaches focused on sign language recognition. Finally, Section 6 gives important conclusions and future works.

2. Convolutional Neural Networks

Biologically inspired computational models are capable of far outperforming previous forms of common artificial intelligence of machine learning. One of the most impressive forms of ANN (artificial neural network) architecture is that of CNN, which is mainly implemented to solve difficult image-based pattern recognition tasks.

CNNs are a specialized type of ANN with supervised learning, which process their layers by emulating the visual cortex of the human eye. This procedure allows the recognition of characteristic patterns in the input data, which makes it possible to identify objects through a set of hidden layers, which have a hierarchy and are specialized. The first layers are capable of detecting curves and lines and to the extent that you work with deeper layers, it is possible to achieve the recognition of more complex shapes, such as a silhouette or peoples' faces.

These types of networks are designed to operate specifically with image processing. The design of its architecture emulates the behavior of the visual cortex of the brain when processing and recognizing images [36]. Its main function is to locate and learn the information characteristic patterns, such as curves, lines, color tones, etc., through the application of convolution layers, which facilitate the process of identification and classification of objects [37,38].

The basic CNN architecture is presented in Figure 1, which consists of five layers: the input, convolution, non-linearity (ReLu), pooling, and classification layer [39,40], these are described in the following subsections.

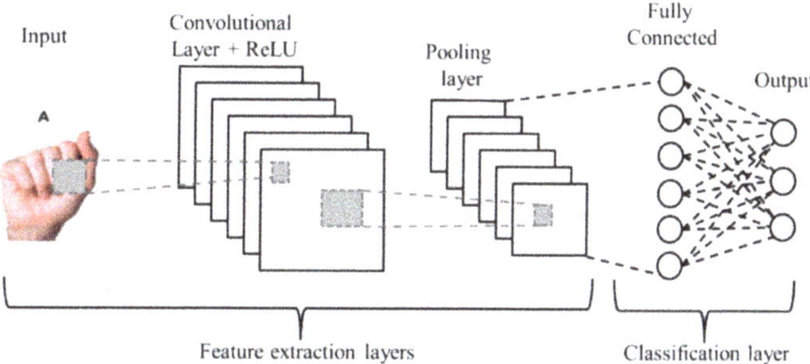

Figure 1. The minimal architecture of a CNN.

CNNs are widely implemented in applications that need the use of artificial vision techniques. Although the results that have been obtained are very promising, the reality is that they incur high computational costs; therefore it is essential to implement techniques that allow your performance to be increased. For this reason, an optimization of the CNN parameters is presented to improve the recognition percentage and reduce computational cost. In Figure 2, we can appreciate some parameters that can be optimized in each CNN layer [41].

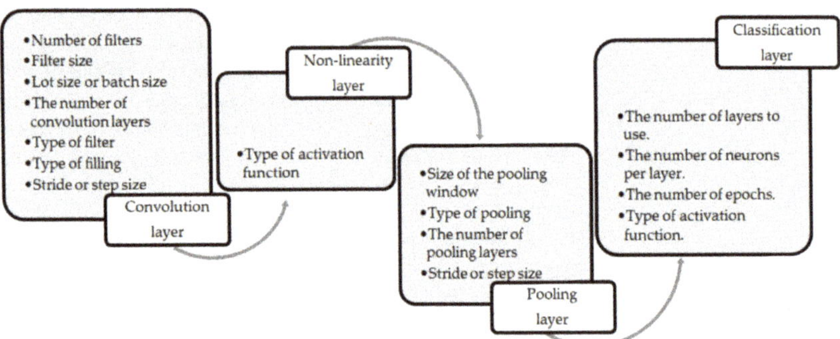

Figure 2. Layers and the parameters per layer of a CNN.

2.1. Input Layer

It is the first layer of a CNN, here the images or videos are entered that are going to be processed by the neural network to extract their characteristics. All information is stored in two-dimensional matrices. To increase the effectiveness of the algorithms and reduce the computational cost, it is recommended to carry out a previous preprocessing of the images to be trained, such as segmentation, normalization of pixel values, extraction of characteristics of the objects or the background to keep the most relevant information, working them in grayscale, etc.

2.2. Convolution Layer

One of the most distinctive processes of this type of network is convolutions. It consists of taking a group of pixels from the input image and making a dot product with a kernel to produce numerous images that are the feature maps; these maps are distinct and depend on the type and size of the convolution filter implemented in the image.

Among the important characteristics that it gives to the kernel, is to detect lines, edges, focus, blur, curves, colors, among others. This is achieved by performing the convolution between the image and the kernel, multiplying the filter values pixel by pixel with those of the image, by traveling the filter from left to right; this representation can be appreciated in Figure 3 [42], where * stands the convolution operation.

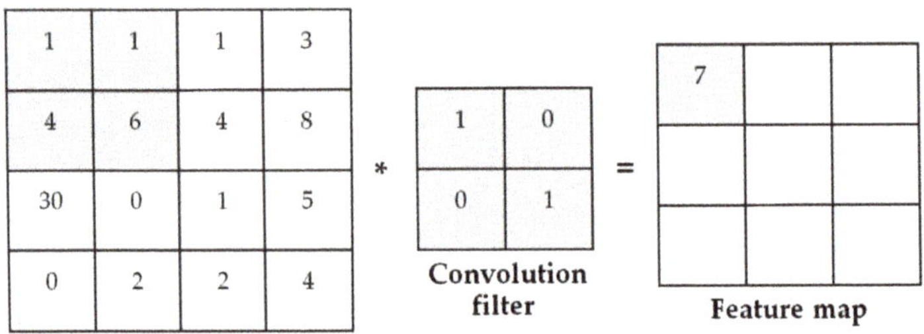

Figure 3. Feature maps generated by the convolution process.

2.3. Non-Linearity Layer

The activation function in the convolutional layer has the same proposal that the activation used in any neural network, commonly a non-linearity function is used to normalize the images. There exist different activation functions; one of the most used in this type of models is the rectified linear unit (ReLU) function which brings back a value of zero if it receives a value less than zero as input, nevertheless for any value greater than zero the same parameter comes back [41,42].

2.4. Pooling Layer

The pooling task is used to reduce the dimensionality of the network, in other words, allows the reduction of the number of parameters, which shortens training time and combats over-fitting [41]. Among the most used types of grouping, we can mention the following: (1) mean, select the arithmetic mean of the values, (2) max pooling, select the pixel with the largest value in the feature map and (3) sum, take the sum of all the elements present in the feature map.

The pooling operation is usually done using a 2 × 2 filter, assuming that we have a 4 × 4 future map (obtained after the convolution layer), and this operation is carried out; first, the future map is divided into 4 segments with the size of the filer (2 × 2), second, in each segment an pixel value is selected according to pooling type (mean, max, sum). An example is illustrated in Figure 4.

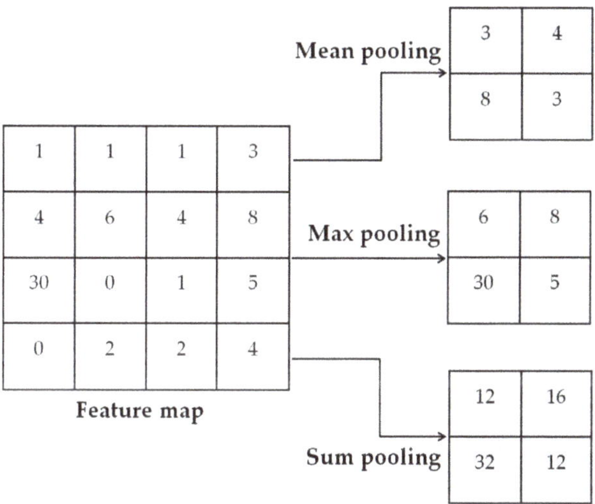

Figure 4. Examples of pooling using the mean, max and sum operation.

2.5. Classifier Layer

This layer appears in the CNN architecture after total convolutional and pooling layers; this is a fully connected layer that interprets the feature representations obtained by the previous layers and performs the high-level reasoning function. It has a similar principle to the conventional multilayer perceptron neural system, and in this layer, the CNN recognizes and classifies the images that are part of the output. In a multiclass classification problem, this fully connected layer has the same number of outputs as the classes defined in the study case to be solved. The Softmax function has become one of the most popular options for the classification task, due to its effectiveness [42].

3. Particle Swarm Optimization

It is a stochastic algorithm established on the intelligence of the swarm and inspired by the way birds forage for food; each bird is represented using particles which "move" in a multidimensional search space and "adjust" based on the experience of neighbors and your own.

The possible solution to the problem is depicted by the particle, which can be considered as "an individual element in a flock" [43]. PSO uses local and global information to find the best solution using a fitness function and the speeds at which the particles are moving.

PSO is very prone to premature convergence and falls into local optimum, so since its introduction in 1995 by Kennedy and Eberhart [44], various optimization variants have been proposed [45–48].

Algorithm 1. The PSO algorithm

Initialize the parameter of the problem (a random population).
while (completion criteria are not met)
begin
For each particle i do
begin
Update the position p_i using (1).
Update the velocity x_i using (2).
Evaluate the fitness value of the particle
If is necessary using (3)(4)
Update pbest$_i$(t) and gbest$_i$(t).
end
end

Algorithm 1 describes the process carried out by the PSO. This algorithm is defined by the equations that allow updating of the velocity with Equation (2) and the position with Equation (1).

$$p_i(t+1) = p_i(t) + x_i(t+1), \tag{1}$$

In Equation (1), $p_i(t)$ is the position of particle i in a time t, within the search space. By adding a velocity $x_i(t)$ it is possible to change the position of the particle [45].

$$x_i(t+1) = x_i(t)\omega + c_1 r_1 [y_i - p_i(t)] + c_2 r_2 [\hat{y} - p_i(t)], \tag{2}$$

In Equation (2), x represents the velocity and i the particle. The parameters c_1 and c_2 define the cognitive and social factors, respectively. The random values in the interval [0,1] are depicted by r_1 and r_2, ω is an inertia weight and the best position of the particle (*pbest$_i$*) is determined by y_i and the best global position (*gbest*) by $\hat{y}$.

The swarm is assumed to consist of n particles, so an objective function f is implemented to perform the computation of particle fitness with a maximization task. The personal and global best values are updated using Equations (3) and (4), respectively, at a time t [48].

Thus, $i \in 1 \ldots n$

$$pbest_i(t+1) = \begin{cases} pbest_i(t) \ if \ f(pbest_i(t)) \leq f(p(t+1)) \\ p_i(t+1) \ if \ f(pbest_i(t)) > f(p_i(t+1)) \end{cases} \tag{3}$$

$$gbest(t+1) = max\{f(y), \ f(gbest(t))\}$$
$$where, \quad y \ \in \ pbest_0(t), \ pbest_1(t), \ \ldots, \ pbest_n(t) \tag{4}$$

According to Equations (1) and (2), the movements of the particle in the search space are illustrated in Figure 5.

The red and yellow circles represent the movement that a particle makes when the parameters c_1 and c_2 are updated. When $c_1 > c_2$, the particle moves in the direction of the yellow circle. When this condition is met, it means that the swarm performs the exploration process, so they "fly" in the search space to find the area that allows it to find the global optimum.

This movement allows the particles to perform long displacements, thus covering the whole search space. In the case of $c_2 > c_1$ then, the particle motion will be towards the red circle. It is here that the exploitation process takes place; it consists of the swarm "flying" in the best area of the search space, making small motions, which allow an intensive search [49].

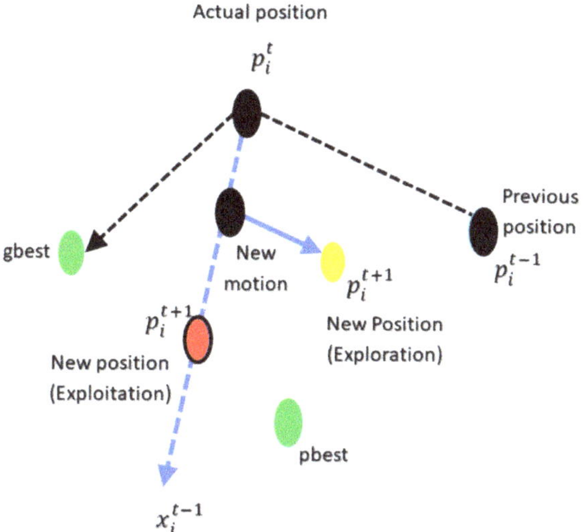

Figure 5. Representation of the movement of the particle.

4. Convolutional Neural Network Architecture Optimized by PSO

This Section presents two optimization approaches where the PSO algorithm is applied to optimize the parameters of CNN architectures, these approaches are denoted as PSO-CNN-I and PSO-CNN-II. The first objective is to select the most relevant parameters that have influence to obtain good performance of CNN and then implement the PSO algorithm to find these optimal parameters.

The parameters to be optimized were selected after evaluating the performance of a CNN with an experimental study, where the parameters were changed manually. As mentioned above, different CNN parameter values produce a variety of results for the same task; for this reason, the aim is to find the optimal architectures. The parameters listed below were chosen to be optimized in this work.

- The number of convolutional layers;
- The filter size or filter dimension used in each convolutional;
- The number of filters to extract the future maps (the convolution filter number);
- The batch size number: this value represents the number of images that are entered into CNN in each training block.

The general methodology of the proposal is presented in Figure 6, as the "training and optimization" block is the most important part of the whole process, where the CNN is initialized to integrate the parameter optimization by applying the PSO algorithm. In this process, the PSO is initialized according to the parameter given for the execution (the parameters are explained below) and this generates the particles. Each particle is a possible solution and its position has the parameter to be optimized, so each solution represents a complete CNN training.

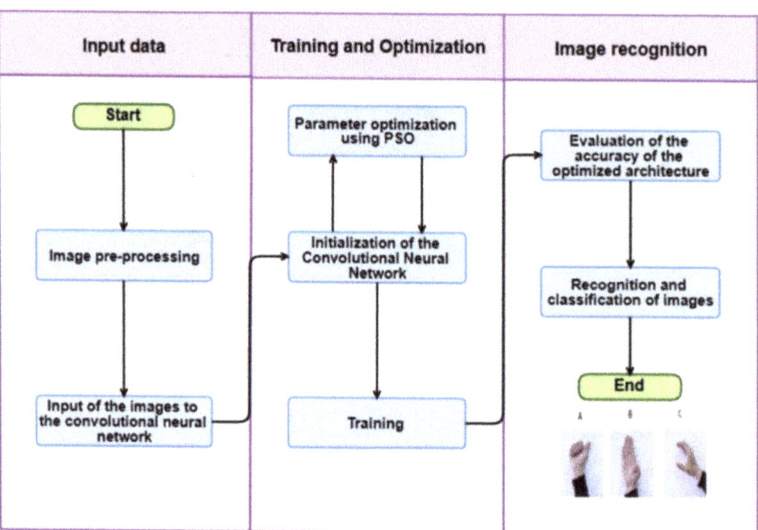

Figure 6. General CNN optimization process using the PSO algorithm. The letters A, B, and C stand the illustrated sign language in image form.

The training process is an iterative cycle that ends when all the particles generated by the PSO are evaluated for each generation. The computational cost is higher and, it depends on the database size, the size of particles, the number of iterations of the PSO and, the number of particles in each iteration. That is to say, if the PSO is executed with 10 particles and 10 iterations, the CNN training process is executed 100 times. The steps to optimize the CNN by the PSO algorithm are illustrated in Figure 7 and explained as follows.

1. Input database to train the CNN. This step consists of selecting the database to be processed and classified for the CNN (ASL alphabet, ASL MINIST and MSL alphabet). Is important to mention that all the elements of each database need to keep a similar structure or characteristics. In other words, images with the same scale and color gamma (grayscale, RGB, CMYK); additionally, with the same dimensions of pixels and a similar format of file (JPGE, PNG, TIFF, BMP, etc.).

2. Generate the particle population for the PSO algorithm. The PSO parameters are set to include the number of iterations, the number of particles, inertial weight, cognitive constant (W1), and social constant (W2); the parameters used in the experimentation are presented in Table 8. This step involves the design of the particles; the structures of these are presented in Tables 1 and 3 according to the two optimization architecture proposals in this paper.

3. Initialize the CNN architecture, with the parameter obtained by the PSO (convolution layers number, the filter size, number of convolution filters, and the batch size) the CNN is initialized and in conjunction with the additional parameter specified in Table 8, the CNN is ready to train the input database.

4. CNN training and validation. The CNN reads and processes the input databases taking the images for training, validation, and testing; this step produces a recognition rate and the AIC value. These values return to the PSO as part of the objective function.

5. Evaluate the objective function. The PSO algorithm evaluates the objective function to determine the best value. As in this research, we are considering two approaches, in the first, the objective function is only the recognition rate (Equation (5)) and in the second, the objective function consists of the recognition rate and the AIC value (Equation (6)).

6. Update PSO parameters. At each iteration, each particle updates its velocity and position depending on its own best-known position (Pbest) in the search-space and the best-known position in the whole swarm (Gbest).
7. The process is repeated, evaluating all the particles until the stop criteria are found (in this case, it is the number of iterations).
8. Finally, the optimal solution is selected. In this process, the particle represented by Gbest is the optimal one for the CNN model.

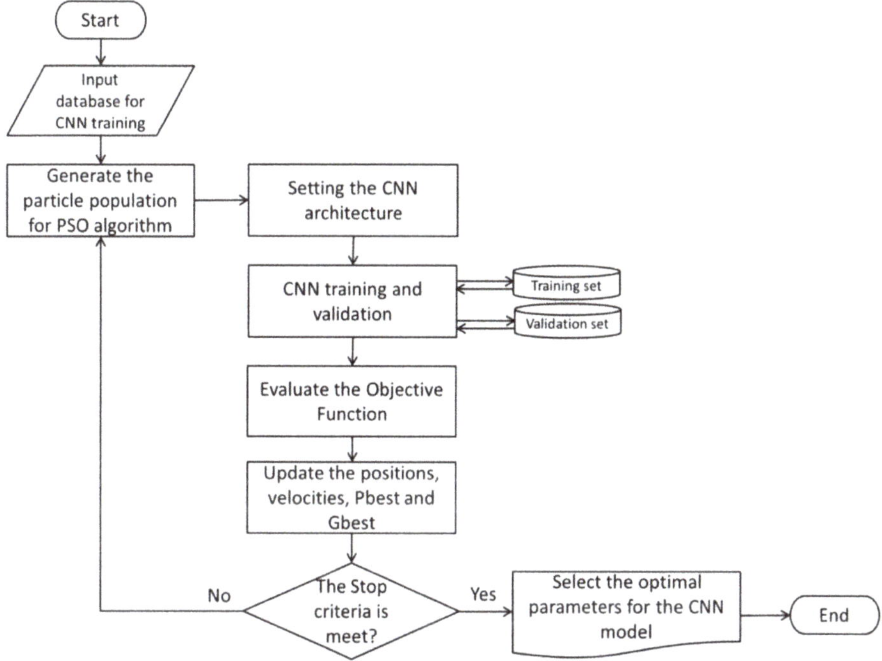

Figure 7. Flowchart of CNN optimization process using PSO.

4.1. PSO-CNN Optimization Process (PSO-CNN-I)

This first approach, which we are going to identify as PSO-CNN-I, consists of implementing a particle with four positions, one position for each parameter to be optimized (Figure 8). Table 1 presents the detail of the particle composition where the position x_1 corresponds to the number of layers with a search space from 1 to n, that is to say, that method can produce architectures with a minimum of one layer and maximum n, for the purposes of this work, we are using n = 3. The x_2 position represents the number of convolution filters used to extract the characteristics, with a search space of 32 to 18 filters. Position x_3 is the filter size; the search space is from 1 to 4 where this values represents a position, the value reached is mapped with the values of Table 2 to obtain the filter size (i.e., if the particle generates a value of 1 this represents a filter size of [3 × 3], to get a value of 2 the filter size will be [5 × 5] and so on, respectively, for each value. The last position represents the batch size (x_4), this is initialized considering the search space ranges from 32 to 256. In this optimization process, the consistency of the parameters between the layers is maintained in the same conditions, that is, if after the PSO execution it generates a particle with 3 convolutional layers (x_1), 50 filters (x_2), a filter dimension of 3 × 3 (x_3) and batch size of 50 (x_4). The same values of filter numbers (x_2) and filter size (x_3) will apply to the three convolution layers of the CNN.

Figure 8. Structure of the particle used in the PSO-CNN-I approach.

Table 1. Search spaces used to define the particle in the PSO-CNN-I approach.

Particle Coordinate	Hyper-Parameter	Search Space
x_1	Number of convolutional layers	[1, 3]
x_2	Filter number	[32, 128]
x_3	Filter size	[1, 4]
x_4	Batch size in the training	[32, 256]

Table 2. Convolutional filter dimensions for the x_3 position.

x_3 Value	Search Space
1	[3, 3]
2	[5, 5]
3	[7, 7]
4	[9, 9]

In this process, the objective function defined by Equation (5) is given by the recognition rate (precision) that the CNN returns after it is trained with the parameters generated by the PSO.

$$Objective\ function = Recognition\ Rate, \tag{5}$$

4.2. PSO-CNN Optimization Process (PSO-CNN-II)

In this second proposal, identifying as PSO-CNN-II, the particle structure consists of eight positions whose structure is presented in Figure 9, where each position represents the parameter to be optimized. The difference from the previous approach (PSO-CNN-I in Section 4.1) is finding more random searches in the architectures that the PSO produces; because in this case, the values for each convolution layer are completely different. Table 3 presents the detail of the particle composition, the description of each position, and the search space used. As we can see in Table 3, the positions x_3, x_5 and x_7 represent an index with an integer value between 1 to 4, and depending on the value taken by the PSO, a mapping is made with values presented in Table 2.

Figure 9. Structure of the particle used in the PSO-CNN-II approach.

Table 3. Search spaces used to define the particle in the PSO-CNN-II approach.

Particle Coordinate	Hyper-Parameter	Search Space
x_1	Convolutional layer number	[1, 3]
x_2	Filter number (layer 1)	[32, 128]
x_3	Filter size (layer 1)	[1, 4]
x_4	Filter number (layer 2)	[32, 128]
x_5	Filter size (layer 2)	[1, 4]
x_6	Filter number (layer 3)	[32, 128]
x_7	Filter size (layer 3)	[1, 4]
x_8	Batch size in the training	[32, 256]

According to the values to optimize in this new approach, the x_1 position is used to control the number of convolution layers and the activation of the positions x_2 to x_7. If PSO generates a particle with a value of one for x_1, only the position x_2 and x_3 will be activated to generate the number of filters of the convolutional layer 1 and the filter size to use in this layer. In other words, if PSO produces a particle with a value of three in the x_1 position, the positions from x_2 to x_7 will be activated to generate the number of filters to use in the convolutional layer 1 (x_2), the filter size of layer 1 (x_3), the number of filters of layer 2 (x_4), the filter size for layer 2 (x_5), the number of filters of layer 3 (x_6), and the filter size for layer 3 (x_7), respectively; these values are completely different from each other, therefore this methodology helps to produce more heterogeneous CNN architectures.

Another difference between this proposal and the previous one (PSO-CNN-I) is that the objective function changes, for this we are using the recognition rate together with the Akaike information criteria (AIC). The AIC penalizes the architectures according to the number of parameters used; that is to say, the model is penalized when it needs more parameters. The objective function is considered the highest recognition rate and the lowest AIC. The AIC is defined in Equation (6).

$$AIC = 2k - 2ln(L), \tag{6}$$

According to our problem, in Equation (6) k is the number of parameters of the model (number of layers and filter number) and L is the maximum value of the recognition that the CNN can reach; in this case, the value is 100. Figure 10 illustrates an example of a particle generated by PSO.

| 3 | 100 | 1 | 85 | 2 | 50 | 3 | 32 |

Figure 10. Example of a particle generated by PSO.

Based on the structure of Figure 10, we have a three-layer convolutional architecture, where the first layer consists of 100 convolution filters with a filter size of 3×3. The second layer has 85 convolution filters with a filter size of 5×5 and the third convolution layer has 50 convolutional filters with a filter size of 7×7. Finally, the batch size is 32. The CNN is training with these values, and the recognition rate is calculated, additionally the AIC is obtained based on the parameters of the positions x_1, x_2, x_4 and x_6 which represents the number of convolution layers and the number of filters for each convolutional layer. After applying Equation (6), this architecture produces the AIC defined in Equation (7).

$$AIC = 2(3 + 100 + 85 + 50) - 2ln\,(100),$$
$$AIC = 466.7897, \tag{7}$$

Assuming that there are two architectures with the same recognition rate but with different AICs (Table 4), the model will take the architecture with the lowest AIC, as this would help penalize the parameters that are needed to train the network and thus produce optimized and simpler architectures.

Table 4. Objective function values based on the recognition rate and the AIC value.

Architecture Number	Recognition Rate (%)	AIC Value
1	98.50	466.78
2	98.50	350.85

5. Experiments and Results

This section describes the three databases implemented in the case studies (ASL alphabet, ASL MNIST, and MSL alphabet), the static parameters used to set the PSO algorithm and the CNN process, the experimental results obtained in the two optimization approaches that were performed (PSO-CNN-I and PSO-CNN-II), as well as the comparison analysis against other approaches.

5.1. Sign Language Databases Used in the Study Cases

The characteristics of the sign databases are described below.

5.1.1. American Sign Language (ASL Alphabet)

The ASL alphabet consists of 87,000 images in color format, with a dimension of 200×200 pixels. This database contains 29 classes, these are labeled in a range of 0 to 28, with a one-to-one assignment for each letter of the American alphabet A–Z (0 to 25 for the alphabet; that is, 0 = A and 25 = Z) the other three classes correspond to the space symbols, delete, and null (26 to 28; i.e., 26 = space, 27 = delete and, 28 = null). Table 5 presents the general description of the ASL alphabet database and Figure 11 illustrates a sample of the images.

Table 5. ASL Alphabet database description.

Name	ASL Alphabet Detail
Total images	87,000
Images for training	82,650
Images for test	4350
Images size	32×32
Database format	JPGE

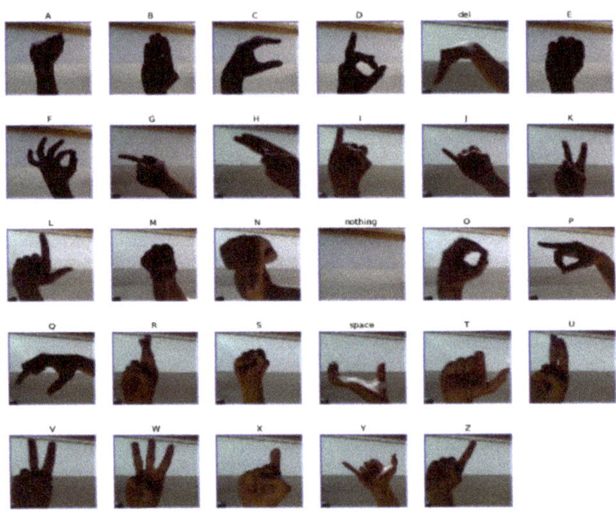

Figure 11. A sample of the ASL alphabet database.

5.1.2. American Sign Language (ASL MNIST)

ASL MNIST consists of a collection of 34,627 grayscale images with a dimension of 28×28 pixels. This database has 24 labeled classes in a range from 0 to 25 with assignment for each letter of the alphabet A–Z (the class 9 = J and 25 = Z, were excluded due to gestural movements). Table 6 presents a description of this database and Figure 12 illustrates a sample of the sign images.

Table 6. ASL MNIST database description.

Name	ASL MNIST Detail
Total images	34,627
Images for training	24,239
Images for test	10,388
Images size	28×28
Database format	CSV

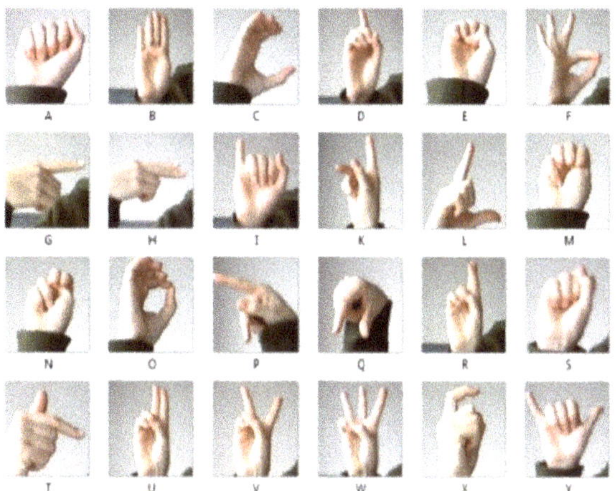

Figure 12. A sample of the ASL MNIST database.

5.1.3. Mexican Sign Language (MSL Alphabet)

The MSL alphabet database was obtained from a group of 18 people, including deaf students and sign language translation teachers. Students are part of an inclusive group in a high school in Mexico. This database consists of 21 classes with the alphabet of the MSL without movement as illustrated in Figure 13. Ten images were captured for each letter, achieving a total of 3780 grayscale images with a dimension of 32 by 32. Table 7 displays a general overview of the MSL alphabet database.

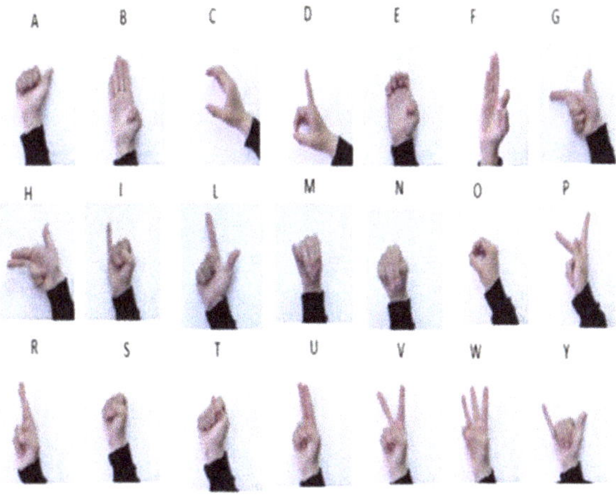

Figure 13. Sample of the MSL alphabet database.

Table 7. MSL alphabet database description.

Name	MSL Alphabet Detail
Total images	3780
Images for training	2646
Images for test	1134
Images size	32×32
Database format	JPG

5.2. Parameters Used in the Experimentation

In the CNN parameter settings, some static parameters were used, including the learning function, the activation function in the classifying layer, the non-linearity activation function, and the epoch number. The fixed parameters considered in the PSO configuration are the number of particles, the iterations number, the inertial weight, and the social and cognitive constants. The static parameters used for PSO and CNN are presented in Table 8. The dynamic parameters optimized by PSO are the number of convolutional layers, the size of the filters used in each convolutional layer, the number of convolutional filters, and the batch size (Tables 1 and 3).

Table 8. Static parameters for CNN and PSO.

Parameters of CNN	
Learning function	Adam
Activation function (classifying layer)	Softmax
Non-linearity activation function	ReLU
Epochs	5
Parameters of PSO	
Particles	10
Iterations	10
Inertial weight (W)	0.85
Social constant (W2)	2
Cognitive constant (W1)	2

5.3. Optimization Results Obtained by the PSO-CNN-I Approach

This section presents the simulation results produced after the CNN architecture is optimized considering the approach described in Section 4.1. The experimentation consists of 30 executions carried out on the three databases; the aim is to obtain the optimal CNN architecture, that is, minimum parameters necessary to maximize the recognition rate.

The first experiment was applied in the ASL alphabet (Table 5), using a distribution of 80% of the total images for training and 20% for testing. Table 9 shows the values achieved after 30 executions, where the higher recognition rate was a value of 99.87% and the mean was 99.58%. Based on the results, we can see that the optimal architecture achieved by the PSO was as follows: three convolutional layers, 128 filters per layer with a filter size of 7×7, and the batch size with a value of 256.

In another test, the PSO-CNN-I approach was applied to the ASL MNIST database; Table 10 presents the results achieved by the CNN where the best recognition rate was a value of 98.82% and the mean of 99.53%. According to this analysis, the optimal architecture for this study case is two convolutional layers, with 117 convolutional filters in both layers with a filter size of 7×7 and the batch size with a value of 129.

Table 9. Results achieved by the PSO-CNN-I in ASL alphabet database.

No.	No. Layers	No. Filters	Filter Size	Batch Size	Recognition Rate (%)
1	3	99	$[7 \times 7]$	107	98.85
2	3	104	$[9 \times 9]$	256	99.66
3	3	128	$[9 \times 9]$	256	99.70
4	3	128	$[7 \times 7]$	256	99.79
5	3	128	$[9 \times 9]$	256	99.72
6	3	128	$[7 \times 7]$	256	99.62
7	2	32	$[7 \times 7]$	256	98.18
8	3	109	$[7 \times 7]$	256	99.73
9	3	128	$[7 \times 7]$	197	99.75
10	3	128	$[7 \times 7]$	256	99.81
11	3	66	$[7 \times 7]$	181	99.31
12	3	118	$[7 \times 7]$	256	99.87
13	3	128	$[9 \times 9]$	256	99.67
14	3	128	$[7 \times 7]$	256	99.85
15	3	128	$[9 \times 9]$	256	99.61
16	3	128	$[9 \times 9]$	256	99.63
17	3	90	$[9 \times 9]$	256	99.66
18	3	128	$[7 \times 7]$	256	99.82
19	3	128	$[7 \times 7]$	256	99.79
20	3	128	$[7 \times 7]$	256	99.76
21	3	128	$[9 \times 9]$	256	99.68
22	3	128	$[9 \times 9]$	256	99.67
23	3	128	$[7 \times 7]$	256	99.75
24	3	123	$[7 \times 7]$	32	98.38
25	3	128	$[9 \times 9]$	256	99.64
26	3	128	$[7 \times 7]$	256	99.82
27	3	128	$[9 \times 9]$	215	99.56
28	3	128	$[7 \times 7]$	256	99.87
29	3	100	$[9 \times 9]$	256	99.64
30	3	128	$[7 \times 7]$	256	99.84
				Mean	99.58

Table 10. Results achieved by the PSO-CNN-I in ASL MNIST database.

No.	No. Layers	No. Filters	Filter Size	Batch Size	Recognition Rate (%)
1	3	128	$[9 \times 9]$	137	99.27
2	2	128	$[9 \times 9]$	218	99.54
3	2	128	$[7 \times 7]$	205	99.52
4	3	128	$[7 \times 7]$	136	99.33
5	2	128	$[9 \times 9]$	232	99.59
6	3	96	$[9 \times 9]$	107	98.82
7	2	118	$[7 \times 7]$	189	99.36
8	2	128	$[9 \times 9]$	256	99.59
9	2	112	$[9 \times 9]$	256	99.49
10	2	128	$[9 \times 9]$	256	99.60
11	2	128	$[7 \times 7]$	256	99.59
12	2	128	$[7 \times 7]$	256	99.61
13	2	128	$[9 \times 9]$	220	99.67
14	2	128	$[9 \times 9]$	256	99.57
15	2	128	$[9 \times 9]$	256	99.51
16	2	128	$[7 \times 7]$	237	99.55
17	2	128	$[7 \times 7]$	256	99.61

Table 10. *Cont.*

No.	No. Layers	No. Filters	Filter Size	Batch Size	Recognition Rate (%)
18	2	128	$[9 \times 9]$	256	99.58
19	2	128	$[9 \times 9]$	256	99.53
20	2	128	$[9 \times 9]$	256	99.65
21	2	128	$[7 \times 7]$	148	99.42
22	2	128	$[9 \times 9]$	256	99.51
23	2	128	$[9 \times 9]$	215	99.53
24	2	128	$[9 \times 9]$	255	99.56
25	2	128	$[9 \times 9]$	256	99.65
26	2	128	$[7 \times 7]$	256	99.57
27	2	128	$[9 \times 9]$	256	99.53
28	2	117	$[7 \times 7]$	129	99.98
29	3	128	$[5 \times 5]$	242	99.87
30	2	128	$[7 \times 7]$	256	99.55
				Mean	99.53

Table 11 presents the experimental results obtained when the approach is applied in the MSL alphabet database. As we can see in Table 11, the best accuracy reached by the CNN was 99.37% with a mean of 99.10%. In this case, the optimal architecture is as follows: one convolutional layer with 122 convolutional filters, filter size of 3×3, and batch size of 128.

Table 11. Results achieved by the PSO-CNN-I in MSL alphabet database.

No.	No. Layers	No. Filters	Filter Size	Batch Size	Recognition Rate (%)
1	2	101	$[7 \times 7]$	93	98.95
2	1	128	$[3 \times 3]$	56	98.95
3	1	110	$[3 \times 3]$	52	98.82
4	1	128	$[3 \times 3]$	121	99.20
5	1	128	$[3 \times 3]$	128	99.32
6	1	128	$[3 \times 3]$	128	99.07
7	1	128	$[3 \times 3]$	110	99.24
8	1	101	$[5 \times 5]$	114	98.82
9	1	128	$[3 \times 3]$	128	99.24
10	1	74	$[3 \times 3]$	88	98.95
11	1	128	$[3 \times 3]$	128	99.32
12	1	128	$[3 \times 3]$	32	98.48
13	1	128	$[3 \times 3]$	93	99.28
14	1	128	$[3 \times 3]$	97	99.11
15	1	128	$[3 \times 3]$	32	98.74
16	1	128	$[3 \times 3]$	72	99.32
17	1	128	$[3 \times 3]$	93	99.37
18	1	63	$[3 \times 3]$	47	98.44
19	1	128	$[3 \times 3]$	128	99.20
20	1	126	$[3 \times 3]$	128	99.28
21	1	128	$[3 \times 3]$	128	99.32
22	1	128	$[3 \times 3]$	83	99.20
23	1	128	$[3 \times 3]$	63	99.20
24	1	122	$[3 \times 3]$	128	99.37
25	1	128	$[3 \times 3]$	128	99.32
26	1	114	$[3 \times 3]$	84	99.32
27	1	128	$[3 \times 3]$	32	98.74
28	1	128	$[3 \times 3]$	89	99.28
29	1	43	$[3 \times 3]$	53	97.81
30	1	128	$[3 \times 3]$	72	98.99
				Mean	99.10

5.4. Optimization Results Obtained by the PSO-CNN-II Approach

The results presented in this section consist of 30 executions of the PSO-CNN-II approach applied in the ASL alphabet, ASL MNIST, and MSL alphabet databases; the objective is to maximize the recognition rate and minimize the value of AIC.

The experimental results obtained from the ASL alphabet database after applying the PSO-CNN-II optimization approach (Section 4.2) are presented in Table 12. In this test, the database was distributed so that 70% of the data were kept for the training phase and 30% of the data for testing. Table 12 shows the best recognition rate with a value of 99.23% and a mean of 98.69%. The best architecture found by the PSO for the CNN had the following structure: three convolutional layers where the first layer had 84 convolutional filters and 3×3 size filters; the second layer with 128 convolutional filters with the size of 9×9 and the third layer with 128 convolutional filters and 7×7 size filters. In this approach, the objective function is composed of the recognition rate and the AIC value; that is, the best recognition rate is evaluated first and then the AIC value, if it was the case that CNN achieved two or more architectures with the same recognition rate, the process takes the architecture with the minimum AIC, with the goal of achieving an optimal architecture with the fewest parameters and the highest recognition rate.

Table 12. Results achieved by the PSO-CNN-II in ASL alphabet database.

No.	No. Layers	Layer 1		Layer 2		Layer 3		Batch Size	AIC Value	(%) Recogn. Rate
		No. Filters	Filter Size	No. Filters	Filter Size	No. Filters	Filter Size			
1	3	128	[3 × 3]	128	[7 × 7]	128	[5 × 5]	256	764.78	98.99
2	3	128	[5 × 5]	121	[5 × 5]	128	[5 × 5]	213	750.78	98.73
3	3	84	[3 × 3]	128	[7 × 7]	128	[5 × 5]	84	676.78	99.23
4	2	45	[5 × 5]	128	[7 × 7]	0	0	0	340.78	98.15
5	3	32	[3 × 3]	128	[7 × 7]	128	[3 × 3]	256	572.78	98.86
6	3	128	[3 × 3]	128	[7 × 7]	128	[5 × 5]	256	764.78	98.96
7	3	84	[5 × 5]	128	[5 × 5]	128	[3 × 3]	256	676.78	98.85
8	3	128	[3 × 3]	128	[7 × 7]	128	[5 × 5]	256	764.78	99.02
9	3	32	[5 × 5]	128	[5 × 5]	128	[3 × 3]	256	572.78	98.9
10	3	124	[3 × 3]	128	[7 × 7]	128	[3 × 3]	256	756.78	98.64
11	3	32	[3 × 3]	128	[7 × 7]	128	[7 × 7]	256	572.78	98.93
12	3	32	[3 × 3]	128	[7 × 7]	128	[3 × 3]	256	572.78	98.53
13	3	128	[3 × 3]	128	[7 × 7]	128	[5 × 5]	256	764.78	99.01
14	3	73	[7 × 7]	128	[7 × 7]	108	[3 × 3]	256	614.78	97.91
15	3	128	[3 × 3]	128	[7 × 7]	128	[5 × 5]	256	764.78	99.06
16	2	128	[3 × 3]	128	[7 × 7]	0	0	0	506.78	98.23
17	2	88	[7 × 7]	128	[7 × 7]	0	0	0	426.78	97.4
18	3	32	[5 × 5]	128	[7 × 7]	128	[5 × 5]	256	572.78	99.06
19	2	128	[5 × 5]	119	[7 × 7]	0	0	0	488.78	98.1
20	3	116	[3 × 3]	128	[5 × 5]	128	[7 × 7]	252	740.78	98.96
21	3	49	[5 × 5]	128	[7 × 7]	128	[7 × 7]	256	606.78	98.93
22	2	128	[3 × 3]	128	[7 × 7]	0	0	0	506.78	98.19
23	3	32	[5 × 5]	128	[7 × 7]	128	[5 × 5]	256	572.78	98.96
24	2	32	[5 × 5]	128	[7 × 7]	0	0	0	314.78	98.04
25	2	128	[5 × 5]	81	[5 × 5]	0	0	0	412.78	98.92
26	3	32	[5 × 5]	128	[7 × 7]	128	[3 × 3]	256	572.78	98.58
27	3	32	[3 × 3]	128	[7 × 7]	128	[5 × 5]	256	572.78	98.88
28	3	128	[3 × 3]	128	[7 × 7]	128	[5 × 5]	256	764.78	99.02
29	3	128	[3 × 3]	128	[7 × 7]	128	[3 × 3]	256	764.78	99.08
30	3	128	[3 × 3]	128	[7 × 7]	128	[3 × 3]	256	764.78	98.71
									Mean	98.69

In Table 13, we present the results where the PSO-CNN-II was implemented in the ASL MNIST database. In this test, the best recognition rate was 99.80%, an AIC value of 506.79 and, a mean of 99.48%. The optimal parameters found by the PSO were the following: two-layer CNN architecture, the first layer had 128 filters of convolution and a filter size of 5 × 5; the second layer had 128 convolutional filters with a filter size of 9 × 9, and the batch size was 128.

Table 13. Results achieved by the PSO-CNN-II in ASL MNIST database.

No.	No. Layers	Layer 1		Layer 2		Layer 3		Batch Size	AIC Value	(%) Recogn. Rate
		No. Filters	Filter Size	No. Filters	Filter Size	No. Filters	Filter Size			
1	2	128	[5 × 5]	128	[9 × 9]	0	0	128	506.79	99.80
2	2	74	[9 × 9]	114	[9 × 9]	0	0	174	370.79	99.42
3	3	32	[5 × 5]	128	[9 × 9]	128	[5 × 5]	122	572.79	99.53
4	2	125	[5 × 5]	125	[9 × 9]	0	0	147	503.79	99.58
5	2	90	[5 × 5]	128	[9 × 9]	0	0	256	500.79	99.68
6	3	32	[3 × 3]	128	[9 × 9]	128	[9 × 9]	148	572.79	99.51
7	2	121	[7 × 7]	95	[9 × 9]	0	0	100	426.79	99.26
8	3	32	[7 × 7]	128	[9 × 9]	125	[9 × 9]	256	569.79	99.6
9	2	32	[9 × 9]	126	[9 × 9]	0	0	106	310.79	99.4
10	3	115	[7 × 7]	102	[9 × 9]	128	[7 × 7]	215	686.79	99.42
11	2	32	[9 × 9]	128	[9 × 9]	0	0	256	314.79	99.44
12	2	77	[7 × 7]	100	[9 × 9]	0	0	183	348.79	99.59
13	2	87	[7 × 7]	128	[9 × 9]	0	0	256	424.79	99.7
14	2	32	[9 × 9]	128	[9 × 9]	0	0	256	314.79	99.53
15	3	32	[5 × 5]	103	[9 × 9]	125	[9 × 9]	256	516.79	99.53
16	2	70	[9 × 9]	126	[9 × 9]	0	0	256	386.79	99.63
17	2	64	[7 × 7]	128	[9 × 9]	0	0	256	378.79	99.7
18	3	32	[7 × 7]	77	[9 × 9]	128	[9 × 9]	256	470.79	99.36
19	2	128	[7 × 7]	128	[9 × 9]	0	0	256	506.79	99.74
20	3	32	[3 × 3]	128	[9 × 9]	128	[5 × 5]	32	572.79	98.95
21	3	32	[7 × 7]	128	[9 × 9]	123	[7 × 7]	162	577.79	99.33
22	2	51	[9 × 9]	128	[9 × 9]	0	0	194	352.79	99.47
23	2	50	[7 × 7]	128	[9 × 9]	0	0	256	350.79	99.63
24	2	128	[7 × 7]	128	[9 × 9]	0	0	162	506.79	99.67
25	2	100	[5 × 5]	76	[5 × 5]	0	0	76	346.79	98.23
26	2	52	[9 × 9]	128	[7 × 7]	0	0	256	354.79	99.54
27	2	128	[5 × 5]	128	[9 × 9]	0	0	142	506.79	99.53
28	3	83	[3 × 3]	125	[9 × 9]	0	0	136	410.79	99.38
29	3	128	[5 × 5]	128	[9 × 9]	128	[9 × 9]	256	764.79	99.57
30	2	74	[7 × 7]	120	[9 × 9]	0	0	256	382.79	99.72
									Mean	99.48

In another experiment, the optimization approach was applied to the MSL alphabet database after 30 simulations. The results obtained are presented in Table 14, where the best recognition rate was 99.45% with an AIC of 248.79. The general mean for this study case was a value of 98.91%. In this optimization, one-layer CNN architecture was achieved, with 128 convolutional filters, 3 × 3 filter sizes, and 154 batch sizes.

5.5. Statistical Test between PSO-CNN-I and PSO-CNN-II Optimization Process

Table 15 presents a summary of the results obtained after the two approaches were applied to the three databases. We can see that good results were achieved in all the cases; we can analyze that for the ASL alphabet and the ASL MNIST, the PSO-CNN-I optimization approach was better with mean values of 99.58% and 99.53%, respectively. For the MSL alphabet database, the PSO-CNN-II optimization method achieved a better recognition rate with a mean value of 98.91%. Although, if the results were analyzed with respect to the AIC value, for the ASL MNIST and the MSL alphabet, the PSO-CNN-I reached the lowest values with AIC of 462.79 and 236.80, respectively, and for ASL alphabet, the PSO-CNN-II achieved a better AIC value. A low AIC value means that the CNN architecture required fewer parameters, so it is important to determine what is most relevant to any problem, the CNN accuracy, or to configure the CNN architectures with minimal parameters that can be implemented in real-time systems.

Table 14. Results achieved by the PSO-CNN-II in MSL alphabet.

No.	No. Layers	Layer 1		Layer 2		Layer 3		BATCH SIZE	AIC Value	(%) Recogn. Rate
		No. Filters	Filter Size	No. Filters	Filter Size	No. Filters	Filter Size			
1	1	128	[3 × 3]	0	0	0	0	32	248.79	98.74
2	1	128	[3 × 3]	0	0	0	0	163	248.79	99.28
3	1	116	[3 × 3]	0	0	0	0	105	224.79	98.99
4	1	81	[3 × 3]	0	0	0	0	32	154.79	98.44
5	1	128	[3 × 3]	0	0	0	0	149	248.79	98.90
6	1	128	[3 × 3]	0	0	0	0	221	248.79	98.57
7	1	128	[3 × 3]	0	0	0	0	57	248.79	99.24
8	1	67	[3 × 3]	0	0	0	0	246	126.73	97.77
9	1	118	[3 × 3]	0	0	0	0	113	228.79	99.16
10	1	128	[3 × 3]	0	0	0	0	154	248.79	99.45
11	1	103	[3 × 3]	0	0	0	0	92	198.79	99.03
12	1	65	[3 × 3]	0	0	0	0	32	122.79	98.32
13	1	128	[3 × 3]	0	0	0	0	94	248.79	99.07
14	1	128	[3 × 3]	0	0	0	0	90	248.79	99.11
15	1	112	[3 × 3]	0	0	0	0	97	216.79	99.24
16	1	128	[3 × 3]	0	0	0	0	32	248.79	98.74
17	1	128	[3 × 3]	0	0	0	0	46	248.79	98.65
18	1	128	[3 × 3]	0	0	0	0	199	248.79	98.32
19	1	128	[3 × 3]	0	0	0	0	244	248.79	99.03
20	1	120	[3 × 3]	0	0	0	0	32	232.79	99.07
21	1	128	[3 × 3]	0	0	0	0	105	248.79	99.16
22	1	128	[3 × 3]	0	0	0	0	77	248.79	99.03
23	1	108	[3 × 3]	0	0	0	0	84	208.79	99.07
24	1	54	[3 × 3]	0	0	0	0	32	100.79	98.44
25	1	102	[3 × 3]	0	0	0	0	102	196.79	99.03
26	1	128	[3 × 3]	0	0	0	0	114	248.79	99.20
27	1	119	[3 × 3]	0	0	0	0	256	230.79	98.61
28	1	98	[3 × 3]	0	0	0	0	122	188.79	99.20
29	1	128	[3 × 3]	0	0	0	0	83	248.79	99.07
30	1	128	[3 × 3]	0	0	0	0	135	248.79	99.37
									Mean	98.91

Table 15. Summary of the results obtained in the PSO-CNN-I and PSO-CNN-II approaches.

Database	PSO-CNN-I			PSO-CNN-II		
	Best	Mean	AIC	Best	Mean	AIC
ASL alphabet	99.87%	99.58%	764.79	99.23%	98.69%	676.78
ASL MNIST	99.98%	99.53%	462.79	99.80%	99.48%	506.79
MSL alphabet	99.37%	99.05%	236.80	99.45%	98.91%	248.79

To confirm if significant evidence exists between the architectures and to identify which is better, the Wilcoxon signed-rank test was applied [50]; this is a non-parametric test that is recommended to be applied when the numerical data are not normally distributed, as is the case with the experimental results of metaheuristic algorithms. The Wilcoxon test was performed to compare the PSO-CNN-I and PSO-CNN-II optimization processes, considering the results presented in Tables 9–14. The general description of the values used to execute the Wilcoxon test is presented in Table 16 and described below:

- A confidence level of 95% ($\alpha = 0.05$).
- The null hypothesis is given that (H_0): the PSO-CNN-I architecture (μ_1) is equal to PSO-CNN-II architecture (μ_2), expressed as $H_0 : \mu_1 = \mu_2$.
- The alternative hypothesis is (H_1): affirm that PSO-CNN-I architecture (μ_1) is greater than that PSO-CNN-II architecture (μ_2), expressed as $H_1 : \mu_1 > \mu_2$ (Affirmation).
- The objective is to reject the hypothesis null (H_0) and support the alternative hypothesis (H_1).

Table 16. General description of the Wilcoxon test.

	Description	Hypothesis
Null hypothesis	PSO-CNN-I architecture (μ_1) = PSO-CNN-II architecture (μ_2)	$H_0 : \mu_1 = \mu_2$
Alternative hypothesis	PSO-CNN-I architecture (μ_1) > PSO-CNN-II architecture (μ_2),	$H_1 : \mu_1 > \mu_2$ (Affirmation)

The first Wilcoxon test was applied for the ASL alphabet results (Tables 9 and 12). Table 17 shows the R^+, R^-, and the p-value (the p-values have been computed by using SPSS), where R^+ represents the sum of ranks for the problems in which the first algorithm outperformed the second, and $R-$ the sum of ranks for the opposite. The results obtained indicate an R^+ of 455, an R^- of 10 and the *p*-value of <0.001. Because the *p*-value is less than the alpha value of $\alpha = 0.05$, then we support the alternative hypothesis with a 95% level of evidence, and we can affirm that the PSO-CNN-I architecture is better than the PSO-CNN-II.

Table 17. Wilcoxon test results for the ASL alphabet.

Comparison PSO-CNN-I (μ_1)—PSO-CNN-II (μ_2)	R^+	R^-	*p*-Value
ASL alphabet	455	10	<0.001

Table 18 presents the results after the Wilcoxon test was applied for the ASL MNIST results (Tables 10 and 13). This test obtains the values $R^+ = 245.5$, $R^- = 189.5$, and the *p*-value = 0.545. Since the *p*-value is greater than the alpha value of $\alpha = 0.05$, the null hypothesis is accepted with a 95% level of evidence; therefore, we can affirm that evidence does not exist to determine that the PSO-CNN-I architecture is better than the PSO-CNN-II.

Table 18. Wilcoxon test results for the ASL MNIST.

Comparison PSO-CNN-I (μ_1)—PSO-CNN-II (μ_2)	R^+	R^-	*p*-Value
ASL MNIST	245.5	189.5	0.545

Finally, Table 19 presents the Wilcoxon test for the results of the MSL alphabet (Tables 11 and 14). The results obtained indicate the values $R^+ = 291$, $R^- = 115$, and the *p*-value = 0.045. We can see that the *p*-value is less than the alpha value of $\alpha = 0.05$; therefore, we support the alternative hypothesis with a level of evidence of 95%, and we can affirm that the PSO-CNN-I architecture is better than the PSO-CNN-II.

Table 19. Z-test results for the ASL MNIST.

Comparison PSO-CNN-I (μ_1)—PSO-CNN-II (μ_2)	R^+	R^-	*p*-Value
MSL alphabet	291	115	0.045

5.6. State-of-the-Art Analysis Comparison

To obtain more evidence about the performance of the optimization approaches presented in this paper, we make a comparative analysis (Table 20) against the state-of-art research, where CNN models are implemented in Alphabet Sign Language database recognition. The results presented in Table 20 represent the best recognition rate values reported by the authors, the detail of which is explained as follows: Zhao et al. [51] reports an accuracy of 89.32%, the CNN architecture has two convolutional layers, two pooling layers, the batch size is 150, and 80 iterations. The authors generated their own ASL database, this was captured in five people covering 24 letters of the alphabet, and each person's letters had about 528 photos.

Table 20. State-of-the-Art Comparison.

Reference	Recognition Rate (%)	Dataset
Y. Zhao and L.Wang [40]	89.32	ASL own
D.Rathi [41]	95.03	ASL MNIST
L.Y.Bin y Y.Huann [42]	95.00	ASL own
R. Dionisio [39]	97.64	ASL MNIST
PSO-CNN-I	99.98	ASL MNIST
PSO-CNN-II	99.80	ASL MNIST

Rathi [52] presents an optimization of the transfer learning model (based on CNN) and it was applied to the ASL MNIST database, using 27,455 images of 24 letters of the ASL alphabet. The data split was as follows, 80% of the data was for training, 10% for testing, and 10% of data for validation purposes with a training batch size of 100. The best recognition rate evidenced by the author was a value of 95.03%.

In Bin et al. [53], an architecture of four convolutional layers and two pooling layers was presented. The database was generated by the researchers themselves, taking characteristics of the ASL MINIST and consisting of 4800 images; the best accuracy reported by the authors was 95.00%.

Dionisio et al. [54], reported a recognition rate of 97.64% for the ASL MNIST with a six-layer convolutional architecture, three pooling layers, a filter size of 3×3, and a batch size of 128. The database was divided using 10% of data for phase testing, 10% for phase validation, and 80% for training.

Finally, we present the recognition rate achieved by our two approaches PSO-CNN-I and PSO-CNN-II with the best values of recognition rates of 99.98% and 99.80%, respectively. In the PSO-CNN-II, the best architecture obtained for the ASL MINIST was of two layers with 117 filters per layer with a size filter of 7×7 and batch size of 129. For the PSO-CNN-I, it was of two layers with 117 filters per layer with a size filter of 7×7 and batch size of 129.

As one can observe in Table 20, the highest performance was obtained by the proposed model (PSO-CNN-I) with a value of 99.98%, achieving an advantage over the rest of the approaches.

6. Conclusions and Future Work

In summary, in this paper, we present two approaches to optimize CNN architectures by implementing the PSO algorithm, these being applied to sign language recognition. The main contribution was to find some CNN hyper-parameters; in the proposals the number of convolutional layers, the size of the filter used in each convolutional layer, the number of convolutional filters, and the batch size were included. According to the experimentation and the results obtained in the two PSO-CNN optimization methodologies, we can conclude that the recognition rate increased in all case studies carried out, providing a robust performance with the minimum parameters. Overall, the recognition rates achieved by the three databases were as follows: for the ASL MNIST database, the best value was 99.98% and an average of 99.53% with the PSO-CNN-I approach. For the ASL alphabet database, the best accuracy was 99.87% and an average of 99.58% with PSO-CNN-I, and for the MSL alphabet, the best value was 99.45% and an average of 98.91% after applying the PSO-CNN-II approach. After a comparative analysis against other state-of-the-art works focused on sign language recognition (ASL and MSL), we can confirm that the optimization approaches of this work present competitive results.

This research focused on optimizing the number of convolutional layers, the filter size used in each convolutional layer, the number of convolutional filters, and the batch size. The results provide evidence of the importance of applying optimization algorithms to find the optimal parameters of convolutional neural network architectures.

As future work, the PSO algorithm could be applied to optimize other CNN hyperparameters, implement another version of the PSO algorithm or explore different evolutionary computational techniques, to produce more robust CNN architectures that will be implemented in different sign language datasets used in other countries. In the experimental test, the images were introduced as static images, but we are considering working with input images in real-time or capturing them through video. On the other hand, our idea is to be able to implement the use of this proposal in the development of assisted communication tools and to contribute to human—computer iteration applications that can be of support to the deaf community.

Author Contributions: Individual contributions by the authors are the following: formal analysis, C.I.G.; conceptualization, writing—review and editing G.E.M. and C.I.G.; methodology, G.E.M.; investigation, software, data curation, and writing—original draft preparation, J.F. All authors have read and agreed to the published version of the manuscript.

Funding: This research received no external funding.

Acknowledgments: We thank CONACYT for the financial support provided with the scholarship number: 954950 and our gratitude to the program of the Division of Graduate Studies and Research of the Tijuana Institute of Technology.

Conflicts of Interest: The authors declare no conflict of interest.

References

1. Hemanth, J.D.; Deperlioglu, O.; Kose, U. An enhanced diabetic retinopathy detection and classification approach using deep convolutional neural network. *Neural Comput. Appl.* **2020**, *32*, 707–721. [CrossRef]
2. Li, P.; Li, J.; Wang, G. Application of Convolutional Neural Network in Natural Language Processing. *IEEE Access* **2018**, 64–70. [CrossRef]
3. Simonyan, K.; Zisserman, A. Very deep convolutional networks for large-scale image recognition. *arXiv* **2015**, arXiv:1409.1556.
4. Liang, S.D. Optimization for Deep Convolutional Neural Networks: How Slim Can It Go? *IEEE Trans. Emerg. Top. Comput. Intell.* **2020**, *4*, 171–179. [CrossRef]
5. Goodfellow, I.; Bengio, Y.; Courville, A. *Deep Learning*; The MIT Press: Cambridge, MA, USA, 2016.
6. Sun, Y.; Xue, B.; Zhang, M.; Yen, G.G. Evolving Deep Convolutional Neural Networks for Image Classification. *IEEE Trans. Evol. Comput.* **2020**, *24*, 394–407. [CrossRef]
7. Sun, Y.; Yen, G.G.; Yi, Z. Evolving Unsupervised Deep Neural Networks for Learning Meaningful Representations. *IEEE Trans. Evol. Comput.* **2019**, *23*, 89–103. [CrossRef]
8. Ma, B.; Li, X.; Xia, Y.; Zhang, Y. Autonomous deep learning: A genetic DCNN designer for image classification. *Neurocomputing* **2020**, *379*, 152–161. [CrossRef]
9. Baldominos, A.; Saez, Y.; Isasi, P. Evolutionary convolutional neural networks: An application to handwriting recognition. *Neurocomputing* **2018**, *283*, 38–52. [CrossRef]
10. Poma, Y.; Melin, P.; Gonzalez, C.I.; Martinez, G.E. Optimization of Convolutional Neural Networks Using the Fuzzy Gravitational Search Algorithm. *J. Autom. Mob. Robot. Intell. Syst.* **2020**, *14*, 109–120. [CrossRef]
11. Poma, Y.; Melin, P.; Gonzalez, C.I.; Martinez, G.E. Filter Size Optimization on a Convolutional Neural Network Using FGSA. In *Intuitionistic and Type-2 Fuzzy Logic Enhancements in Neural and Optimization Algorithms*; Springer: Cham, Switzerland, 2020; Volume 862, pp. 391–403.
12. Poma, Y.; Melin, P.; Gonzalez, C.I.; Martinez, G.E. Optimal Recognition Model Based on Convolutional Neural Networks and Fuzzy Gravitational Search Algorithm Method. In *Hybrid Intelligent Systems in Control, Pattern Recognition and Medicine*; Springer: Cham, Switzerland, 2020; Volume 827, pp. 71–81.
13. Lee, W.-Y.; Park, S.-M.; Sim, K.-B. Optimal hyperparameter tuning of convolutional neural networks based on the parameter-setting-free harmony search algorithm. *Optik* **2018**, *172*, 359–367. [CrossRef]

14. Wang, B.; Sun, Y.; Xue, B.; Zhang, M. A hybrid differential evolution approach to designing deep convolutional neural networks for image classification. In Proceedings of the Australasian Joint Conference on Artificial Intelligence, Wellington, New Zealand, 11–14 December 2018; Springer: Cham, Switzerland, 2018; pp. 237–250.

15. Gülcü, A.; KUş, Z. Hyper-Parameter Selection in Convolutional Neural Networks Using Microcanonical Optimization Algorithm. *IEEE Access* **2020**, *8*, 52528–52540. [CrossRef]

16. Zhang, N.; Cai, Y.; Wang, Y.; Tian, Y.; Wang, X.; Badami, B. Skin cancer diagnosis based on optimized convolutional neural network. *Artif. Intell. Med.* **2020**, *102*, 101756. [CrossRef]

17. Tuba, E.; Bacanin, N.; Jovanovic, R.; Tuba, M. Convolutional Neural Network Architecture Design by the Tree Growth Algorithm Framework. In Proceedings of the 2019 International Joint Conference on Neural Networks (IJCNN), Budapest, Hungary, 17–19 July 2019; pp. 1–8.

18. Sun, Y.; Xue, B.; Zhang, M.; Yen, G.G. A particle swarm optimization based flexible convolutional autoencoder for image classification. *IEEE Trans. Neural Netw. Learn. Syst.* **2019**, *30*, 2295–2309. [CrossRef] [PubMed]

19. Singh, P.; Chaudhury, S.; Panigrahi, B.K. Hybrid MPSO-CNN: Multi-level Particle Swarm optimized hyperparameters of Convolutional Neural Network. *Swarm Evol. Comput.* **2021**, *63*, 100863. [CrossRef]

20. Wang, B.; Sun, Y.; Xue, B.; Zhang, M. Evolving deep convolutional neural networks by variable-length particle swarm optimization for image classification. In Proceedings of the 2018 IEEE Congress on Evolutionary Computation (CEC), Rio de Janeiro, Brazil, 8–13 July 2018; Volume 1–8.

21. Gonzalez, B.; Melin, P.; Valdez, F. Particle Swarm Algorithm for the Optimization of Modular Neural Networks in Pattern Recognition. *Hybrid Intell. Syst. Control Pattern Recognit. Med.* **2019**, *827*, 59–69.

22. Varela-Santos, S.; Melin, P. Classification of X-Ray Images for Pneumonia Detection Using Texture Features and Neural Networks. In *Intuitionistic and Type-2 Fuzzy Logic Enhancements in Neural and Optimization Algorithms: Theory and Applications*; Springer: Cham, Switzerland, 2020; Volume 862, pp. 237–253.

23. Miramontes, I.; Melin, P.; Prado-Arechiga, G. Particle Swarm Optimization of Modular Neural Networks for Obtaining the Trend of Blood Pressure. In *Intuitionistic and Type-2 Fuzzy Logic Enhancements in Neural and Optimization Algorithms: Theory and Applications*; Springer: Cham, Switzerland, 2020; Volume 862, pp. 225–236.

24. Peter, S.E.; Reglend, I.J. Sequential wavelet-ANN with embedded ANN-PSO hybrid electricity price forecasting model for Indian energy exchange. *Neural Comput. Appl.* **2017**, *28*, 2277–2292. [CrossRef]

25. Sánchez, D.; Melin, P.; Castillo, O. Comparison of particle swarm optimization variants with fuzzy dynamic parameter adaptation for modular granular neural networks for human recognition. *J. Intell. Fuzzy Syst.* **2020**, *38*, 3229–3252. [CrossRef]

26. Fernandes, F.E.; Yen, G.G. Particle swarm optimization of deep neural networks architectures for image classification. *Swarm Evol. Comput.* **2019**, *49*, 62–74. [CrossRef]

27. Santucci, V.; Milani, A.; Caraffini, F. An Optimisation-Driven Prediction Method for Automated Diagnosis and Prognosis. *Mathematics* **2019**, *7*, 1051. [CrossRef]

28. Zhou, G.; Moayedi, H.; Bahiraei, M.; Lyu, Z. Employing artificial bee colony and particle swarm techniques for optimizing a neural network in prediction of heating and cooling loads of residential buildings. *J. Clean. Prod.* **2020**, *254*, 120082. [CrossRef]

29. Karaboga, D.; Gorkemli, B.; Ozturk, C.; Karaboga, N. A comprehensive survey: Artificial bee colony (ABC) algorithm and applications. *Artif. Intell. Rev.* **2014**, *42*, 21–57. [CrossRef]

30. Xianwei, J.; Lu, M.; Wang, S.-H. An eight-layer convolutional neural network with stochastic pooling, batch normalization and dropout for fingerspelling recognition of Chinese sign language. *Spinger Multimed. Tools Appl.* **2019**, *79*, 15697–15715.

31. Hayami, S.; Benaddy, M.; El Meslouhi, O.; Kardouchi, M. Arab Sign language Recognition with Convolutional Neural Networks. In Proceedings of the 2019 International Conference of Computer Science and Renewable Energies (ICCSRE), Agadir, Morocco, 22–24 July 2019.

32. Huang, J.; Zhou, W.; Li, H.; Li, W. Attention-Based 3D-CNNs for Large-Vocabulary Sign Language Recognition. *IEEE Trans. Circ. Syst. Video Technol.* **2019**, *29*, 2822–2832. [CrossRef]

33. Kaggle. American Sign Language Dataset. 2018. Available online: https://www.kaggle.com/grassknoted/asl-alphabet (accessed on 10 February 2020).

34. Kaggle. Sign Language MNIST. 2017. Available online: https://www.kaggle.com/datamunge/sign-language-mnist (accessed on 8 February 2020).

35. Rastgoo, R.; Kiani, K.; Escalera, S. Sign Language Recognition: A Deep Survey. *Expert Syst. Appl.* **2021**, *164*, 113794. [CrossRef]

36. Hubel, D.H.; Wiesel, T.N. Receptive fields of single neurons in the cat's striate cortex. *J. Physiol.* **1959**, *148*, 574–591. [CrossRef]

37. Kim, P. *Matlab Deep Learning*; Apress: Seoul, Korea, 2017.

38. Cheng, J.; Wang, P.-s.; Li, G.; Hu, Q.-h.; Lu, H.-q. Recent advances in efficient computation of deep convolutional neural networks. *Front. Inf. Technol. Electron. Eng.* **2018**, *19*, 64–77. [CrossRef]

39. Zou, Z.; Shuai, B.; Wang, G. Learning Contextual Dependence with Convolutional Hierarchical Recurrent Neural Networks. *IEEE Trans. Image Process.* **2016**, *25*, 2983–2996.

40. Fukushima, K. A self-organizing neural network model for a mechanism of pattern recognition unaffected by shift in position. *Biol. Cybern.* **1980**, *36*, 193–202. [CrossRef]

41. Schmidhuber, J. Deep learning in neural networks: An overview. *Elsevier Neural Netw.* **2015**, *61*, 85–117. [CrossRef]

42. Aggarwal, C.C. *Neural Networks and Deep Learning*; Springer Nature: Cham, Switzerland, 2018.

43. Jang, J.; Sun, C.; Mizutani, E. *Neuro-Fuzzy and Soft Computing: A Computational Approach to Learning and Machine Intelligence;* Prentice-Hall: Upper Saddle River, NJ, USA, 1997.
44. Kennedy, J.; Eberhart, R.C. Particle swarm optimization. In Proceedings of the IEEE International Conference on Neural Networks IV, Washington, DC, USA, 27 November–1 December 1995; pp. 1942–1948.
45. Sandeep, R.; Sanjay, J.; Rajesh, K. A review on particle swarm optimization algorithms and their applications to data clustering. *J. Artif. Intell.* **2011**, *35*, 211–222.
46. Hasan, J.; Ramakrishnan, S. A survey: Hybrid evolutionary algorithms for cluster analysis. *Artif. Intell. Rev.* **2011**, *36*, 179–204. [CrossRef]
47. Fielding, B.; Zhang, L. Evolving Image Classification Architectures with Enhanced Particle Swarm Optimisation. *IEEE Access* **2018**, *6*, 68560–68575. [CrossRef]
48. Sedighizadeh, D.; Masehian, E. A particle swarm optimization method, taxonomy and applications. *Proc. Int. J. Comput. Theory Eng.* **2009**, *5*, 486–502. [CrossRef]
49. Gaxiola, F.; Melin, P.; Valdez, F.; Castro, J.R.; Manzo-Martínez, A. PSO with Dynamic Adaptation of Parameters for Optimization in Neural Networks with Interval Type-2 Fuzzy Numbers Weights. *Axioms* **2019**, *8*, 14. [CrossRef]
50. Derrac, J.; García, S.; Molina, D.; Herrera, F. A practical tutorial on the use of nonparametric statistical tests as a methodology for comparing evolutionary and swarm intelligence algorithms. *Swarm Evol. Comput.* **2011**, *1*, 3–18. [CrossRef]
51. Zhao, Y.; Wang, L. The Application of Convolution Neural Networks in Sign Language Recognition. In Proceedings of the 2018 Ninth International Conference on Intelligent Control and Information Processing (ICICIP), Wanzhou, China, 9–11 November 2018; pp. 269–272.
52. Rathi, D. Optimization of Transfer Learning for Sign Language Recognition Targeting. *Int. J. Recent Innov. Trends Comput. Commun.* **2018**, *6*, 198–203.
53. Bin, L.Y.; Huann, G.Y.; Yun, L.K. Study of Convolutional Neural Network in Recognizing Static American Sign Language. In Proceedings of the 2019 IEEE International Conference on Signal and Image Processing Applications (ICSIPA), Kuala Lumpur, Malaysia, 17–19 September 2019; pp. 41–45.
54. Rodriguez, R.; Gonzalez, C.I.; Martinez, G.E.; Melin, P. An improved Convolutional Neural Network based on a parameter modification of the convolution layer. In *Fuzzy Logic Hybrid Extensions of Neural and Optimization Algorithms: Theory and Applications;* Springer: Cham, Switzerland, 2021; pp. 125–147.

Article

Multicriteria Evaluation of Deep Neural Networks for Semantic Segmentation of Mammographies

Yoshio Rubio [†] and Oscar Montiel [*,†]

Instituto Politécnico Nacional-CITEDI, 1310 Instituto Politécnico Nacional Ave, Nueva Tijuana, Tijuana 22430, Baja California, Mexico; rrubio@citedi.mx
* Correspondence: oross@ipn.mx
† These authors contributed equally to this work.

Abstract: Breast segmentation plays a vital role in the automatic analysis of mammograms. Accurate segmentation of the breast region increments the probability of a correct diagnostic and minimizes computational cost. Traditionally, model-based approaches dominated the landscape for breast segmentation, but recent studies seem to benefit from using robust deep learning models for this task. In this work, we present an extensive evaluation of deep learning architectures for semantic segmentation of mammograms, including segmentation metrics, memory requirements, and average inference time. We used several combinations of two-stage segmentation architectures composed of a feature extraction net (VGG16 and ResNet50) and a segmentation net (FCN-8, U-Net, and PSPNet). The training examples were taken from the mini Mammographic Image Analysis Society (MIAS) database. Experimental results using the mini-MIAS database show that the best net scored a Dice similarity coefficient of 99.37% for breast boundary segmentation and 95.45% for pectoral muscle segmentation.

Keywords: breast segmentation; mammogram; deep learning; semantic segmentation

MSC: 68T20

Citation: Rubio, Y.; Montiel, O. Multicriteria Evaluation of Deep Neural Networks for Semantic Segmentation of Mammographies. *Axioms* **2021**, *10*, 180. https://doi.org/10.3390/axioms10030180

Academic Editor: Hsien-Chung Wu

Received: 5 July 2021
Accepted: 3 August 2021
Published: 5 August 2021

Publisher's Note: MDPI stays neutral with regard to jurisdictional claims in published maps and institutional affiliations.

1. Introduction

Computer-aided detection (CADe) systems are valuable tools to assist medical experts in detecting and diagnosing diseases. The aim of CADe for breast analysis is twofold: it reduces the chances of cancer being undetected in the earlier stages, and it can lower the number of unnecessary medical interventions (such as biopsies), mitigating the levels of anxiety and stress in the patients [1,2].

For mammogram CADe, accurate segmentation of the breast is a crucial step [3–5], as it can accelerate diagnosis and lower the number of false positives and false negatives [1,6,7]. Automatic mammogram segmentation identifies the different tissues in the breast and gives them a label such as pectoral muscle, fatty tissue, fibroglandular tissue, or nipple [6–8].

In breast segmentation, the most challenging task is to identify the pectoral muscle accurately. As a result, the brightness, position, and size of the pectoral muscle vary widely. The pectoral muscle may occupy most of the image or do not appear in it. Most of the time, the pectoral muscle appears in the upper part of the mammogram with a white triangular shape since the curvature and length of the lower edge fluctuates [7]. The pectoral muscle's brightness can appear similar to fibroglandular tissue in dense breasts or other structures such as parenchymal texture, artifacts, and labels in digitized mammograms [6,9].

CADe approaches for image segmentation began with traditional computer vision techniques such as edge detection and mathematical modeling. These systems evolved and started including machine learning approaches which, at present, constitute the core of a CADe system, becoming the primary option for medical image segmentation [10].

Although deep learning has shown excellent performance for medical image segmentation, the use of deep learning for mammogram segmentation is scarce [4,11].

This paper contributes with an extensive evaluation of mammogram segmentation using deep learning architectures. We combined several architectures to explore their performance under different scenarios. We obtained accuracy metrics, memory consumption, and inference time to compare the performance of the architectures. We present multicriteria evaluation results that can help obtain answers beyond simple accuracy metrics since other important characteristics can also be considered in selecting the best architecture for a given application. Additionally, we provide a comprehensive introduction to the topic of semantic segmentation using deep learning approaches and how different deep neural networks can be combined to achieve better results.

The paper is organized as follows: in Section 2, we summarized the different state-of-the-art approaches for breast segmentation. In Section 3 the theoretical background of deep neural network used in this work and key concepts about digitized mammograms are provided. In Section 4 we provide information about the designed experiments and how to reproduce our findings. In Section 5, we present our results using the mini-MIAS database and a comparative against other methods. Finally, in Section 6 we discuss our results and provide our opinion about future work.

2. Related Work

Traditional methods for breast segmentation combine geometric models and iterative calculation of a threshold to distinguish between different breast tissues. Mustra and Grgic [5] used polar representation to identify the round part of the breast. They found the breast line using a combination of morphological thresholding and contrast limited adaptive histogram equalization. To extract the pectoral muscle, they used a mixture of thresholding and cubic polynomial fitting.

Liu et al. [8] developed a method for pectoral muscle segmentation based on statistical features using the Anderson–Darling test to identify the pectoral muscle's boundary pixels. They eliminated the regions outside the pectoral muscle's probable area by assuming the position of the pectoral muscle (upper-left corner of the mammogram) and found the final boundary of the muscle by using an iterative process that searches for edge continuity and orientation.

More recently, Taghanaki et al. [2] used a mixture of geometric rules and intensity thresholding to identify the breast region and the pectoral muscle. Vikhe and Thool [7] developed an intensity-based algorithm to segment the pectoral region. They estimated the pectoral region through a binary mask of the breast obtained by a thresholding technique followed by an intensity filter. Then, with a thresholding technique for rows separated by a pixel interval, several estimated pectoral muscle outline points were found.

Rampun et al. [3] developed a mammogram model to estimate the pectoral region's position and the breast's orientation. First, they found the breast region using Otsu's thresholding, then removed the noise using an anisotropic diffusion filtering; after that, the initial breast boundary was found using the image's median and standard deviation. The Chan–Vese model was used to obtain a more precise boundary. Finally, using an extended breast and Canny edge detection model, the pectoral muscle was found.

The use of deep learning in medical imaging has incremented since 2015 and is now an essential topic for research. In the medical field, deep learning has been used to detect, classify, enhance, and segmentation [12]. Despite this trend, the number of works published for mammograms with deep learning is low.

In this regard, Dalmış et al. [13] segmented breast MRI's using several sets of U-nets. In the first experiment, they trained two U-nets, the first one for the breast area's segmentation and the second one to segment the fibroglandular tissue. In the second experiment, a single U-net with three classes to identify the background, the fatty tissue, and the fibroglandular tissue was trained. They used the SØrensen–Dice similarity coefficient to evaluate the experiments, obtaining 0.811 for the first experiment and 0.85 for the second.

Dubrovina et al. [4], used a convolutional neural network (CNN) for pixel-wise classification of mammograms. In their work, they classify mammograms patches into five classes: pectoral muscle, fibroglandular tissue, nipple, breast tissue, and background.

In a recent work, de Oliveira et al. [11] used several semantic segmentation nets for mammogram segmentation. They used three different net architectures (FCNs, U-net, and SegNet) for the segmentation of the pectoral muscle, breast region, and background.

Rampun et al. [14] used a convolutional neural network for pectoral segmentation in mediolateral oblique mammograms. The network used by Rampun et al. was a modified holistically nested edge detection network. In a more recent work, Ahmed et al. [15] used two different architectures, Mask-RCNN and the DeepLab V3, for the semantic segmentation of cancerous regions in mammograms.

3. Theoretical Background

This section provides the mathematical foundation of the convolutional neural networks that we used to perform semantic segmentation of the breast; we explain key concepts about mammograms to facilitate the understanding of the methods, evaluation, and results. All the tested architectures of DNN are also explained.

3.1. Supervised Learning Foundations

The general problem in machine learning can be summarized as the following: given a collection of input values X^i and a set of adjustable weights W, calculate an approximation function $F(X^i, W)$ that estimates output values Y^i, see (1). The output Y^i can be seen as the recognized class label of the given pattern X^i, as scores, or as probabilities associated with each class [16].

$$Y^i = F(X^i, W) \tag{1}$$

The loss function is calculated by Equation (2), where D measures the discrepancy between the desired valued $\hat{Y}^i$ and the output given by our approximation function F.

$$\mathcal{L} = D(Y^i, \hat{Y}^i) \tag{2}$$

The average loss function, $\mathcal{L}_t(W)$ is the average of the errors $\mathcal{L}$ over a set of labeled examples called the training set $\{X^i, Y^i\}$. A simple learning problem consists in finding the value W that minimizes $\mathcal{L}_t(W)$. Commonly, the system's performance is estimated by using a disjoint set of samples called the test set, Z^i [16].

A method to minimize the loss function, is by estimating the impact of small variations in the parameters W on the loss value. This is measured by the gradient of the loss function $\mathcal{L}$ with respect W. Generally, W is a real-valued vector, in which $\mathcal{L}(W)$ is continues and differentiable [16]. The most simple minimization procedure is by using gradient descent, where W is adjusted in each step by:

$$W_t = W_{t-1} - \eta \frac{\partial \mathcal{L}(W)}{\partial W} \tag{3}$$

where η is called the learning rate, which indicates how much the algorithm would change by the calculated error.

Another popular minimization procedure is the stochastic gradient descent. This algorithm consists in updating the W using a noisy or approximated version of the average gradient. This can be represented as randomly selecting a subset of the input data X^i and calculate an approximate gradient for each batch. Since over a set of iterations the addition of the loss functions would be calculated randomly for most of the data, the average of the different trials would be very similar to the real gradient. The stochastic gradient descent is represented as:

$$W_t = W_{t_1} - \eta \sum \frac{\partial \mathcal{L}_x(W)}{\partial W} \tag{4}$$

The main difference, is that in this case, $\mathcal{L}(W)$ is calculated over a subset of X^i called x. In the most simple case, W is updated using only a single example. With this procedure, the calculated gradient fluctuates around an average trajectory that usually converges faster than the original gradient descent [16].

Feedforward networks use backpropagation to efficiently calculate gradients of differentiable layers. The basic feedforward networks is built as a group of cascade elements (neurons) each one implementing a function $Y_n = F(Y_{n-1}, W_n)$, where Y_n is the output of the module, W_n is the vector of tunable parameters, and Y_{n-1} is the input of the module. The input Y_0 to the first module is the input pattern X, and the subsequent layers that calculate Y_i are called hidden layers. If the partial derivative of F^i with respect to Y_n is known, then the partial derivatives of F^i with respect to W_n and Y_{n-1} can be calculated using rule chain as:

$$\frac{\partial F^i}{\partial W_n} = \frac{\partial F}{\partial W}(W_n, Y_{n-1})\frac{\partial F^i}{\partial Y_n} \tag{5}$$

$$\frac{\partial F^i}{\partial Y_{n-1}} = \frac{\partial F}{\partial Y}(W_n, Y_{n-1})\frac{\partial F^i}{\partial Y_n} \tag{6}$$

where $(\partial F/\partial W)(W_n, Y_{n-1})$ is the Jacobian of F with respect to W evaluated at the point (W_n, Y_{n-1}), and $(\partial F/\partial Y)(W_n, Y_{n-1})$ is the Jacobian of F with respect to X. The full gradient is obtained by getting the average of the gradients over all the training patterns.

3.2. Convolutional Neural Networks

Multilayer networks can learn complex, high-dimensional, nonlinear mappings from a large set of training examples. This ability makes them a clear candidate for image recognition tasks. A typical pipeline of an image recognition system uses a feature extractor method to obtain the most relevant characteristics of the image and then use them as feature vectors to train the network. Most of these feature extractors are hand-crafted, which makes appealing the possibility of creating features extractors that learn the most relevant features. Using traditional neural networks (also called fully connected networks) for feature extraction and classification has some evident limitations.

The first one is the size of the network. Traditional images are multi-channel matrices with millions of pixels. To represent this information on a fully connected network, the number of weights for each fully connected layer would be in the millions even if the image is resized to a certain degree. The increment in the number of trainable parameters increases the memory requirements to represent the weights of the network.

Additionally, there is no built-in invariance of translation or local distortion of the inputs in these types of networks. In theory, a fully connected network could learn to generate outputs that are invariants to distortion or some degree of translation. The downside is that it would result that several learning units have to learn similar weight patterns that are positioned at different locations in the input. Covering the space of all the possible variations would need many training instances; hence it would demand considerable memory requirements.

Another deficiency of fully connected networks is that the topology of the input vector is ignored. Images have a solid two-dimensional local structure where pixels have a high correlation with their neighbors. These local correlations in images are the main reason for using hand-crafted feature extractors before a classification stage.

Convolutional neural networks (CNN) address the issues of fully connected networks and achieve shift, scale, and distortion invariance by using three elements: local receptive fields, shared weights, and spatial subsampling. Local receptive field neurons are used to extract elementary visual features such as edges, endpoints, corners, and textures. Subsequent layers combine these low-order features to detect higher-order features. On a CNN, learning units in a layer are organized in planes in which all units share the same set of weights, this allows the network to be invariant to shifts or distortions of the input.

The set of outputs of the learning units is called a feature map [16]. The shared weights are adjusted during the learning phase to detect a set of relevant features.

A convolutional layer comprises several feature maps, this allows multiple features to be extracted at each location. The operation of the feature maps is equivalent to a convolution, followed by an additive bias and a squashing function, giving its name of convolutional network [17]. A convolutional layer can be represented as:

$$Conv(Y_n, K) = \sum_i \sum_j \sum_k K_{\{i,j,k\}} Y_{n\{i,j,k\}} \tag{7}$$

where K is the kernel that acts on the input map Y_n using a window of size k. The kernel of this convolutional layer is the set of connection weights used by the units in the feature map. In Figure 1, an arbitrary input map is feed into a convolutional filter with size 3×3. To obtain the output, each element in their respective position $K_{\{i,j\}}$ would be multiplied with their pairs in the input map at position $Y_{n\{i,n\}}$ and then they would be summed up. In this specific example, the operation is as follows: $(1 \times 1) + (2 \times -1) + (1 \times -1) + (1 \times 0) + (3 \times 1) + (2 \times -1) + (1 \times 0) + (1 \times 0) + (4 \times -1) = -5$. The convolutional filter would move over the input map given a step size. The output map would be the encoded features from the input map given the convolutional filter.

The shape, size, and composition of the kernel helps the network to learn specific features that the network detects as important. Convolutional layers are robust to shift and distortions of the inputs since the feature maps only shifts by the same amount as the shift in the input image.

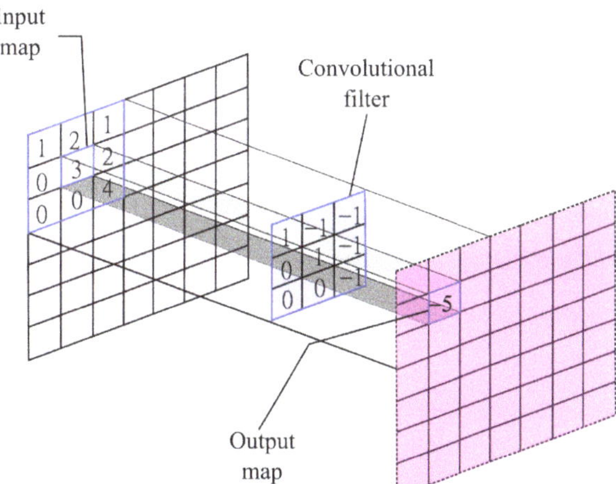

Figure 1. Example of a convolutional operation.

After a feature is detected, its exact position becomes irrelevant, and only the approximate position relative to other features is important. Knowing its precise position can be harmful to achieving robustness to slight variations of the input. In order to reduce precision, spatial resolution reduction of the feature maps is used in convolutional networks.

Pooling layers can perform a subsampling using a wide array of mathematical operations over a feature map. These operations can be as simple as the maximum value over a given window (max pooling), the average value (average pooling), or more complex, such as taking the average of the whole feature map (global average pooling). Similar to the convolutional layers, the mathematical operation of pooling layers acts upon a fixed window of size p. In Figure 2, two different pooling layers of size $p = 2$ are applied over an given feature map of size 4×4. Given $p = 2$, the feature map is downsized to 2×2. The colors indicate the area of effect of the pooling layer. As an example, on the orange layer, the output of the max-pooling is 23 given that the maximum value is 23, on the contrary, for the average pooling (rounded to the next integer value) the output is 15 given that $(23 + 23 + 0 + 12)/4 = 15$. Selecting over different pooling layers would affect the learning capacity of the network.

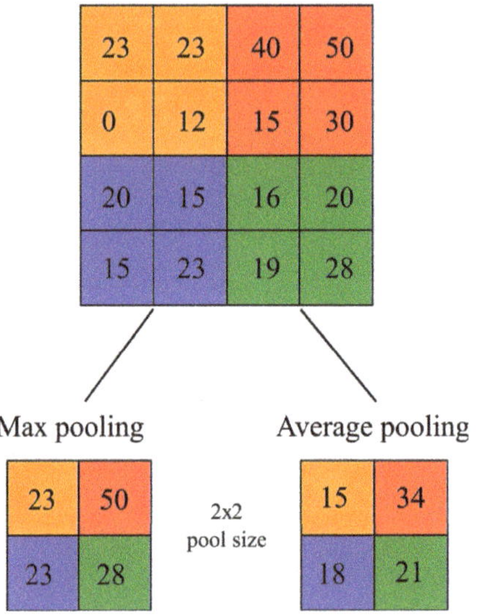

Figure 2. Example of max pooling and average pooling.

Using a combination of convolutional and pooling layers over a progressing reduction of spatial reduction compensated by an increasing number of feature maps helps achieve invariance to the geometric transformation of the input.

At the end of the convolutional and pooling layers, a fully connected network is used. Since all the weights in the system are learned using backpropagation, and at the end there is a fully connected network, in essence, a CNN is learning to extract its own features and classify them.

CNNs are an example of deep neural networks (DNNs), which are neural networks with many layers between the input layer and the output layer (hidden layers), hence the name.

3.3. Fully Convolutional Networks

Fully convolutional networks (FCN) are a particular type of CNNs, where all the layers compute a nonlinear filter (a convolutional filter). This allows FCNs to naturally operate on any input size and produce an output of the corresponding spatial dimensions [18].

Commonly, detection CNNs take fixed-sized inputs and produce nonspatial outputs. Fully connected layers can also be viewed as convolutions with kernels that cover the entire input. This can transform traditional CNN into FCNs that take input of any size and make spatial output maps.

These outputs maps are typically reduced due to the subsampling provided by the pooling layers, reducing the resolution of the FCNs by a factor equal to the pixel stride of the receptive fields of the output units; this is to prevent the coarse outputs from being connected with upsampling layers that act as convolutional layers with a fractional input stride of $1/s$. These fractional stride convolutions are typically called *transpose convolutions*.

A network with these characteristics can perform classification at a pixel level or semantic segmentation. In semantic segmentation, every pixel of the image is associated with a class. This association helps to understand *what* is happening in the image and *where* it is happening [18].

3.4. Deep Learning Architectures for Semantic Segmentation

Many deep neural networks (DNNs) have been proposed for semantic segmentation, most of these approaches are based on a derivation of the following: convolutional neural networks, fully convolutional networks [18], U-Net [19] variations, convolutional residual networks, recurrent neural networks [10], densely connected convolutional networks [20], and DeepLab variations [21–23].

In this work, we used several DNNs to test their performance in mammogram segmentation. Similar to Siam et al. [24], we combined pairs of feature extraction networks and segmentation networks for semantic segmentation. For the feature extraction section, we used the VGG16 and the ResNet50; for the segmentation part, we used FCN8, U-Net, and PSPNnet. These composite structures summarize the spectrum of DNNs for semantic segmentation.

In the following lines, we describe the base networks used in this work in detail, followed by the description of the specific encoder–decoder pairs used for the breast segmentation.

3.4.1. VGG16

Simonyan and Zisserman [25] proposed VGG16, which is a well-known architecture that won the 2014 ImageNet Challenge. The VGG16 has 16 weight layers (convolutional and fully connected layers), with a total of 134 million trainable parameters, see Figure 3.

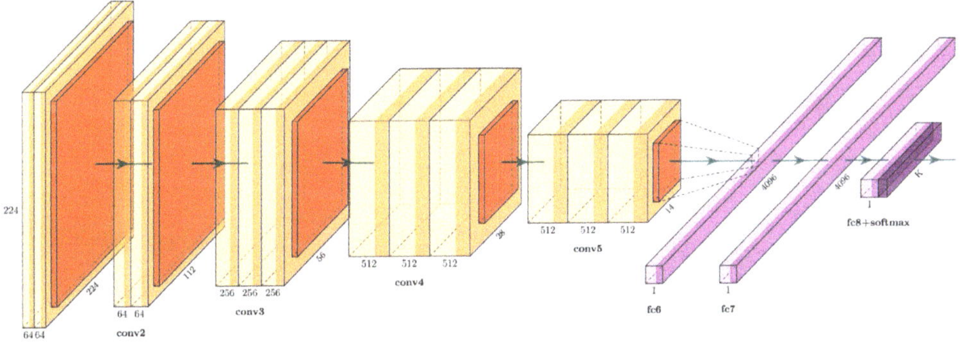

Figure 3. Architecture of the VGG-16.

The input layer has a fixed size of 224×224, followed by five convolutional sections. Each convolutional section have 3×3 filters with a stride of 1, followed by a 2×2 max-pooling filter with stride 2. The activation function used in each block are rectified linear units (ReLU) layers, defined as:

$$ReLU(Y_n) = max(0, Y_{n-1}). \tag{8}$$

The end layers are three fully connected layers: the first two with 4096 channels, and the third one is a softmax layer with the number of channels adjusted to the number of classes. Table 1 describes each layer in detail.

Table 1. Layer description of the VGG16.

Type	Filters	Size, Stride
Conv_1	64×2	$3 \times 3, 1$
Pool_1	–	2
Conv_2	128×2	$3 \times 3, 1$
Pool_2	–	2
Conv_3	256×3	$3 \times 3, 1$
Pool_3	–	2
Conv_4	512×3	$3 \times 3, 1$
Pool_4	–	2
Conv_5	512×3	$3 \times 3, 1$
Pool_5	–	2
FConv	4096	7×7
FConv	4096	1×1
Softmax	–	–

3.4.2. ResNet50

A big obstacle for earlier networks with depth above thirty layers was accuracy saturation, followed by a fast accuracy degradation. ResNet was proposed by He et al. [26], winning the ILSVRC classification task in 2015. ResNet addresses the degradation problem with residual blocks, see Figure 4, which are *shortcut connections* between layers using identity mapping, and adding their outputs of the stack layers.

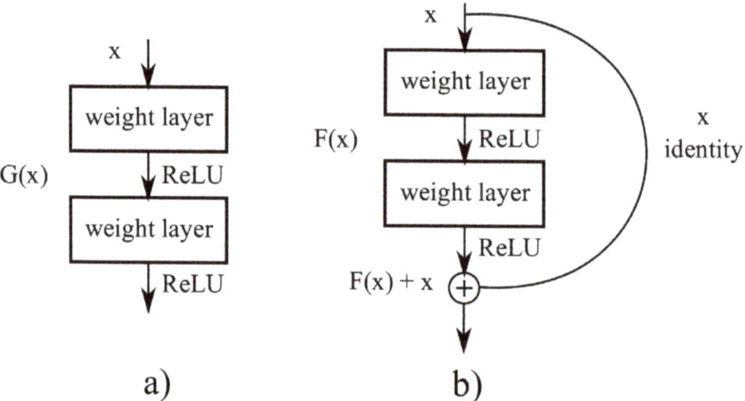

Figure 4. ResNets address the accuracy saturation and accuracy degradation issue with residual blocks. (**a**) Traditional DNN stacking where each layer feeds into the next layer. (**b**) In residual blocks the output of a layers is added to a layer deeper in the block.

ResNets have five sections: the first section is a convolutional layer followed by a max-pooling layer; the other four sections are residual blocks that repeat different times. ResNet50 is the fifty layer variant of ResNet, see Figure 5, their four residual blocks are called *bottlenecks*. Each *bottleneck* block has three convolutional layers, the first and the third one have 1×1 kernels, and the second layers have a 3×3 kernel. A description of the layers in ResNet50 can be found in Table 2.

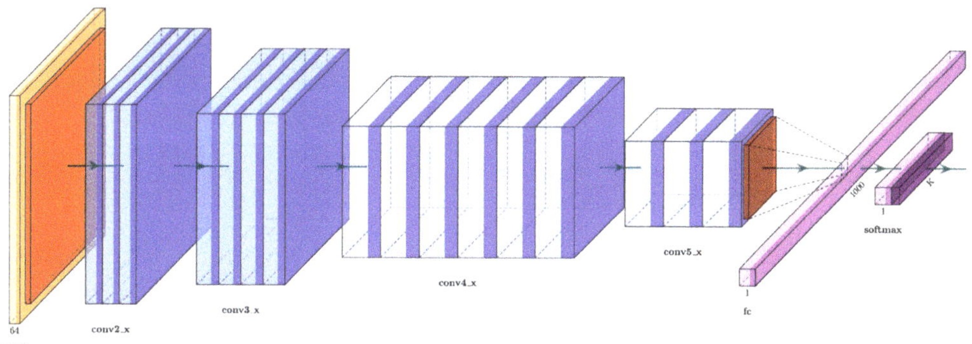

Figure 5. Architecture of the ResNet50.

Table 2. Layer description of ResNet50.

Type	Filters	Size, Stride
Conv_1	64	$7 \times 7, 2$
Pool_1	–	$3 \times 3, 2$
	64	$1 \times 1, 1$
Conv_2 $\times$ 3	64	$3 \times 3, 1$
	256	$1 \times 1, 1$
	128	$1 \times 1, 1$
Conv_3 $\times$ 4	128	$3 \times 3, 1$
	512	$1 \times 1, 1$
	256	$1 \times 1, 1$
Conv_4 $\times$ 6	256	$3 \times 3, 1$
	1024	$1 \times 1, 1$
	512	$1 \times 1, 1$
Conv_5 $\times$ 3	512	$3 \times 3, 1$
	2048	$1 \times 1, 1$
GAP	–	1×1
Softmax	–	–

3.4.3. FCN-8

Fully convolutional networks were one of the first networks designed for semantic segmentation [18]. Their main advantage was their capacity to take an input of arbitrary size, generating a correspondingly sized output.

The fully convolutional versions of classification networks (such as Alexnet, VGG16, GoogLeNet) add skip connections at the end of convolutional blocks, and they add a convolutional filter in the output and fuse it with an upsampled region at the end of the network. The upsampling layers are *transposed convolutional* layers.

In FCNs, *fully convolutional layers replace the fully connected layers at the end of the classifiers*: instead of having p neurons interconnected, the layer will have p convolutional layers. At the end of the network, a softmax layer classifies every pixel into a class.

The FCN8 has two skip connections and three upsampling layers. The first upsampling layer feeds directly from the fully convolutional layers and has a stride of 32. The second and third upsampling layers are fed from the skip connections and have 16 and 8 strides, respectively. The two skip connections come from upper pooling layers, and each one passes through a convolutional layer. An example of an FCN-8 for an arbitrary network is shown in Figure 6.

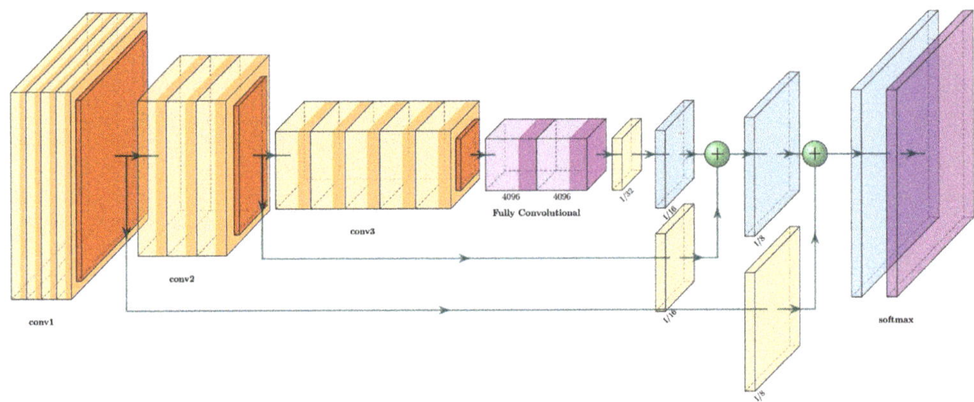

Figure 6. Architecture of the FCN-8 of a generic DNN.

3.4.4. U-Net

The U-Net architecture has two parts: a contracting path to capture context and an expanding path for localization [19]. The contracting path consists of several blocks of two 3×3 convolutional layers, followed by a ReLU layer and a 2×2 max pooling with stride 2. At each downsampling, the number of feature channels is doubled. The expansive paths have several 2×2 transpose convolution layers for upsampling, a concatenation with the corresponding feature map in the contracting path, and two 3×3 convolutional layers followed by a ReLU layer. The final layer is a 1×1 convolutional layer followed by a softmax layer.

The architecture of the classic U-Net can be seen in Figure 7.

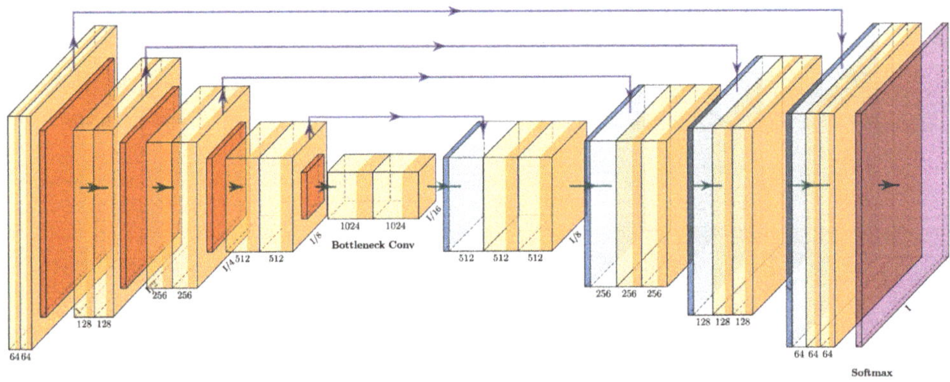

Figure 7. Architecture of the U-Net.

3.4.5. PSPNet

The pyramid scene parsing network (PSPNet) was proposed by Zhao et al. [27] to incorporate suitable global features for semantic segmentation and scene parsing tasks. To obtain global information, PSPNet relies on a pyramid pooling module. This module uses a hierarchical prior, which contains information at different pyramid scales and varying among different sub-regions.

In Figure 8, we can see the pyramid pooling module inside PSPNet. Once the image features are extracted, the multiple pooling layers at several sizes extract global information, and then a 1×1 convolutional layer is used to flatten the information, followed by an upsampling layer to obtain the original size of the feature map. The final step is to concatenate each feature obtained by the pyramid pooling module, and a convolutional layer generates the final prediction.

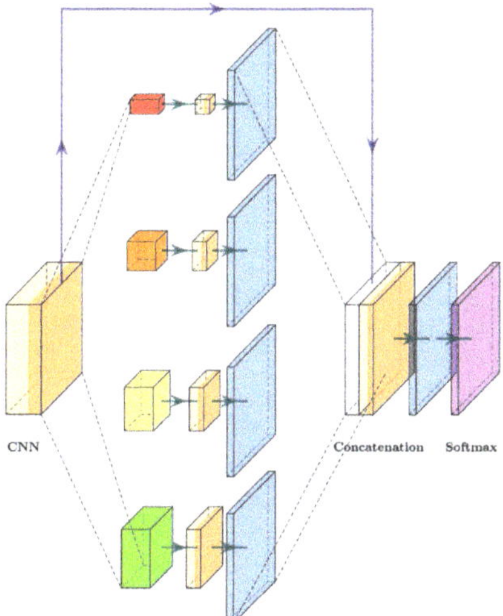

Figure 8. Pyramid pooling module.

3.4.6. VGG16+FCN8

The VGG16+FCN8 is the fully convolutional version of the VGG16. It uses the same architecture as VGG16, but it replaces the two fully connected layers with fully convolutional layers and uses skip connections. This net has three streams before the final upsampling (transposed convolution layer): the output of the fully convolutional layers, a skip connection from pool_4, and a skip connection from pool_3. Every stream passes through a convolutional layer before adding them to the output. The architecture of VGG16+FCN8 can be seen in Figure 9 and is described in Table 3.

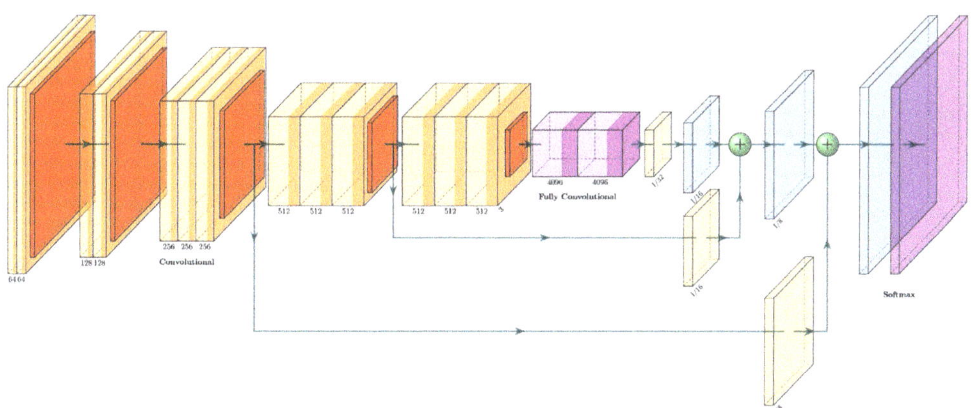

Figure 9. Architecture of the VGG16+FCN8.

Table 3. Layer description of the VGG16+FCN8.

Type	Filters	Conected to
VGG16	–	–
Conv_6 × 2	4096	Conv_5
Conv_7	2	Conv_6
TransConv_1	1	Conv_7
Conv_8	1	Pool_4
Add_1	–	–
TransConv_2	1	Add_1
Conv_9	1	Pool_4
Add_2	–	–
TransConv_3	1	Add_2
Softmax	–	TransConv_3

3.4.7. VGG16+U-Net

The VGG16+U-Net has the same shape as the U-Net, with a contracting path and an expanding path. The contracting path is integrated by four convolutional blocks of the VGG16 (the last block was removed), with an added single convolutional layer before the expanding path. The expanding path has three upsampling blocks connected with the convolutional blocks. The concatenations between the two paths are performed between Conv_1-Upsampling_3, Conv_2-Upsampling_2, and Conv_3-Upsampling_1. The architecture of the VGG16+U-Net can be seen in Figure 10 and is described in Table 4.

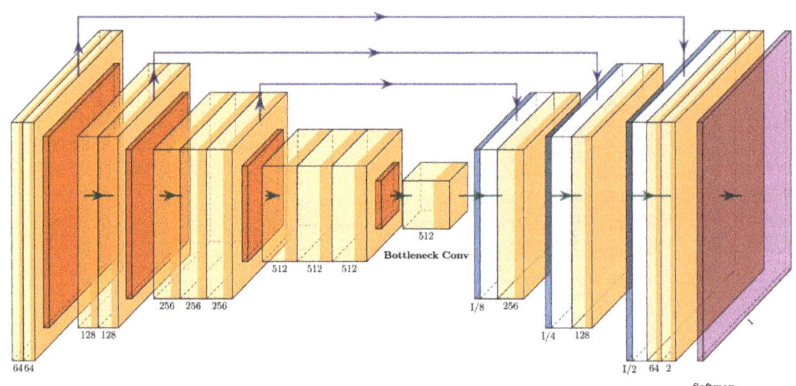

Figure 10. Architecture of the VGG16+U-Net.

Table 4. Layer description of the VGG16+U-Net.

Type	Filters	Conected to
VGG16	–	–
Conv_5	512	Conv_4
Upsampling_1	512	Conv_5
Concat_1	768	Conv_3
Conv_6	256	Concat_1
Upsampling_2	256	Conv_6
Concat_2	384	Conv_2
Conv_7	128	Concat_2
Upsampling_3	128	Conv_7
Concat_3	192	Upsampling_3
Conv_8	64	Concat_3
Conv_9	2	Conv_8
Softmax	–	Conv_9

3.4.8. ResNet+PSPNet

The architecture of the ResNet+PSPNet is simple: it eliminates the fully connected layers of the ResNet50 and adds a pyramid pooling module with two convolutional layers before the final upsampling layers. The last upsampling layer has the same dimension as the first residual block (conv2_x), while the upsampling layers in the pyramid pooling module are half the dimensions of conv5_x. The architecture of ResNet+PSPNet can be seen in Figure 11 and is described in Table 5.

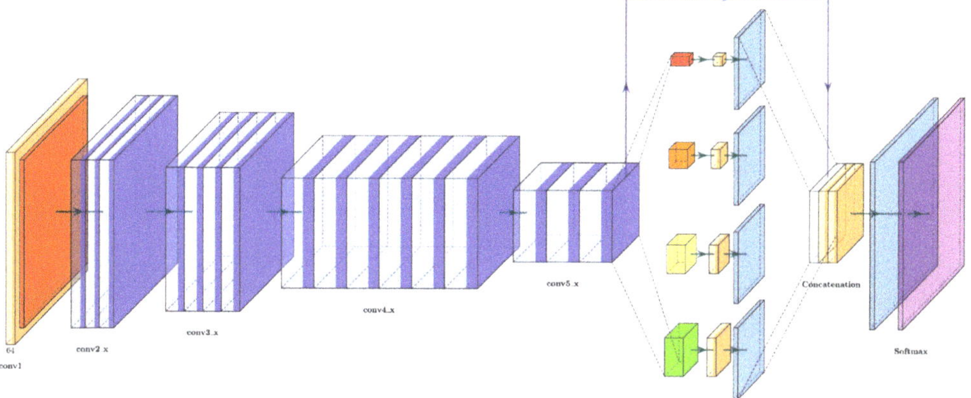

Figure 11. Architecture of the ResNet50+PSPNet.

Table 5. Layer description of the ResNet50+PSPNet.

Type	Filters	Conected to
ResNet50	–	–
AVGPool_1	2048	Conv5_x
Conv_6	512	AVGPool_1
Upsampling_1	512	Conv_6
AVGPool_2	2048	Conv5_x
Conv_7	512	AVGPool_2
Upsampling_2	512	Conv_7
AVGPool_3	2048	Conv5_x
Conv_8	512	AVGPool_3
Upsampling_3	512	Conv_8
AVGPool_4	2048	Conv5_x
Conv_9	512	AVGPool_4
Upsampling_4	512	Conv_9
Concat_1	4096	Conv_6
		Conv_7
		Conv_8
		Conv_9
Conv_10	512	Concat_1
Conv_11	2	Conv_10
Softmax	–	Conv_11

3.4.9. ResNet+U-Net

Since VGG16+Unet uses three upsampling blocks in the expanding path, the same number was used for ResNet+U-Net, and only three of the four residual blocks of the original Resnet50 were used for the contracting path. The concatenations between the two paths are carried out between conv2_x-up_3, conv3_x-up_2, and conv4_x-up_1. The architecture of the ResNet50+U-Net can be seen in Figure 12 and is described in Table 6.

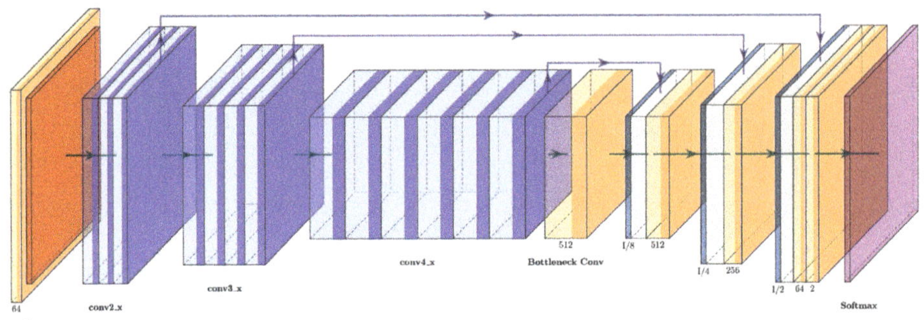

Figure 12. Architecture of the ResNet50+U-Net.

Table 6. Layer description of the ResNet50+U-Net.

Type	Filters	Conected to
ResNet50	–	–
Conv_5	512	Conv4_x
Upsampling_1	512	Conv_5
Concat_1	1024	Conv4_x
Conv_6	256	Concat_1
Upsampling_2	256	Conv_6
Concat_2	512	Conv3_x
Conv_7	128	Concat_2
Upsampling_3	128	Conv_7
Concat_3	192	Conv2_x
Conv_8	64	Concat_3
Conv_9	2	Conv_8
Softmax	–	Conv_9

3.5. Digitized Mammograms

Mammograms have two standard views: craniocaudal (CC) and mediolateral oblique (MLO). In the CC view, the radiologist has a higher appreciation of the anterior, central, and middle area of the breast [28]. The MLO view complements the CC view; it is taken with an oblique angle, providing a lateral image of the pectoral muscle and the breast. Most of the current research in the automatic segmentation of mammograms focuses on the MLO view. In this view, there are different components, such as the breast region (sometimes divided in fibroglandular tissue and fat), the pectoral muscle, and the nipple.

The breast region is generally more extensive than the pectoral muscle, with a round appearance and most of its pixels near the center or in the mammogram's lower region. The breast's gray level intensity depends on its density (the proportion of glandular tissue and fat); the higher the breast's density, the brighter it appears.

The pectoral muscle is in the upper region of the mammogram. In most cases, the pectoral muscle appears a bright right triangle, with its origin depending on the mammogram's orientation. The size, brightness, and curvature of the muscle vary on a case-by-case basis [7].

When the mammogram comes from digitized films, it may contain medical labels, noise, and other bright elements that are not part of the breast's anatomy, e.g., adhesive tape or abnormal spots [3,4,7]. Elements with such characteristics are cataloged as artifacts. Figure 13 shows a typical digitized mammogram and its components.

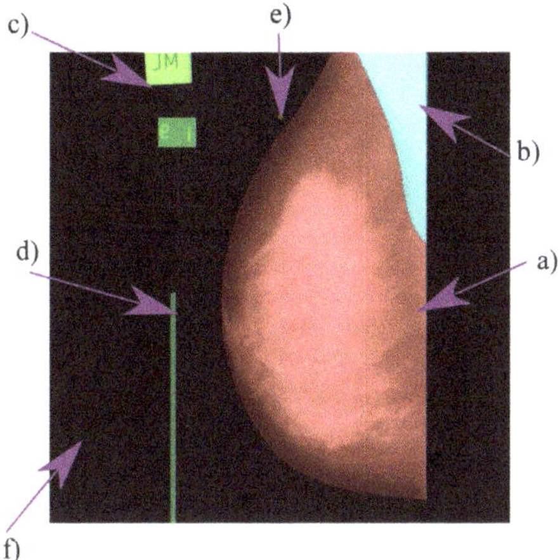

Figure 13. Typical components of a digitized mammogram: (**a**) breast region, (**b**) pectoral muscle, (**c**) medical label, (**d**) digitization artifact, (**e**) salt noise, and (**f**) background.

4. Method

As detailed in Section 2, although deep learning is the dominant technique for segmentation in medical imaging, the number of works that study the performance of architectures for breast segmentation is limited; therefore, the main motivation for this work is to compare the performance of several DNNs in breast segmentation and shed some light on selecting the appropriate network for breast segmentation given the specific purpose.

Modern nets have a better overall performance for multi-label segmentation, but this does not necessarily reflect higher performance on classes with lower pixel-count or classes that resemble geometric shapes (such as the pectoral muscle and breast).

To address all these questions, we performed experiments with four different DNNs: VGG16+FCN8, VGG16+U-Net, ResNet+PSPNet, and ResNet+UNet. The experiments are designed to study the segmentation quality in "One" vs. "All scenario" and "multi-class scenario", then measure how well the smaller class (the pectoral muscle) is identified.

The selected dataset was the mini-MIAS database [29]. The database has a total of 322 images of digitized mammograms, with a resolution of 200 microns and a size of 1024×1024 pixels. To train the networks, we used the labels provided by Oliver et al. [30]. The images are labeled in three classes: breast region, pectoral muscle, and background.

Since the pectoral muscle is much smaller than the breast region (and sometimes does not appear), and the amount of artifacts varies widely between the images, a set of four different training batches was developed.

For the first two experiments, the "One" vs. "All approach" was used. In the first experiment, the pectoral muscle was segmented, and all the other pixels were seen as the background. For the second experiment, the breast region was segmented, with the other classes set as background. In the third experiment, the breast area was identified as the union of the pectoral muscle and the breast region, segmented from the background. The three classes (breast region, pectoral muscle, and background) were segmented individually for the final experiments.

All the experiments used identical hyper-parameters. Each net was trained for 100 epochs using the Adam optimizer with an initial learning rate of 10^{-4} and an exponential decay adjusted to achieve a 10^{-7} in the last epoch.

Since the amount of training examples is small (only 322 image), dividing the set into training, validation, and testing sets could be prone to overfitting, or it will not assess the performance of the models correctly. A common way to solve this issue is to use K-fold cross-validation in which a set is divided randomly into K batches of N/K examples and trained with $K-1$ batches and leave one for testing. The process is repeated K times, always leaving a different fold as testing set. This means that each training example is used $K-1$ times as part of the training set, and 1 time as part of the test set. Accuracy metrics are calculated for each individual test set, and the average is reported. Since K-fold cross validation averages the performance of the model over all the data, there is a higher confidence in the evaluation of the model.

We used 5-fold cross-validation to evaluate each architecture's performance; this means 20% of the dataset (64 images) is selected randomly as a test set and the other 80% (258 images) for training. Thus, the model is trained five times, in which each time, the training set and test set are different. Additionally, we used data augmentation in the training process that consisted of a random crop, rotation, translation, padding, and shear. All the networks were trained on a Titan X GPU with 12 GB of RAM.

5. Results

To measure the performance of each architecture, we used two similarity metrics. The first metric is the Jaccard index, also known as the Intersection over Union [31], which is defined as the cardinality of the intersection of two sets divided by the cardinality of the union of the same sets, see Equation (9). The second metric is the SØrensen–Dice (shortened to Dice) coefficient, which is equal to twice the cardinality of the intersection of both sets divided by the number of elements of both sets, see Equation (10).

$$J(M,N) = \frac{|M \cap N|}{|M \cup N|} \tag{9}$$

$$D(M,N) = \frac{2|M \cap N|}{|M| + |N|} \tag{10}$$

The experiments were labeled as A (pectoral muscle vs. all), B (breast region vs. all), C (breast area vs. background), and D (three class segmentation). Tables 7 and 8 indicate the mean value of the 5-fold cross-validation for both segmentation metrics. The columns are labeled PM for pectoral muscle, BC for background, BR for breast region interest, and BR+PM refers to the breast area, including the breast region of interest and the pectoral muscle.

Table 7. Segmentation metrics for experiments A and B. Best results in bold.

	A		B	
Net	**PM**	**BC**	**BR**	**BC**
VGG+FCN8	J: 0.8985	J: 0.9888	J: 0.9681	J: 0.9563
	D: 0.9465	D: 0.9943	D: 0.9838	D: 0.9777
VGG+U-Net	J: 0.8644	J: 0.9852	J: 0.9641	J: 0.9503
	D: 0.9272	D: 0.9925	D: 0.9817	D: 0.9745
ResNet+PSPNet	J: 0.9130	J: 0.9903	J: 0.9627	J: 0.9488
	D: **0.9545**	D: **0.9951**	D: 0.9810	D: 0.9737
ResNet+U-Net	J: 0.8954	J: 0.9885	J: 0.9705	J: 0.9595
	D: 0.9448	D: 0.9942	D: **0.9850**	D: **0.9793**

Table 8. Segmentation metrics for experiments C and D. Best results in bold.

Net	C		D		
	BR+PM	BC	BR	PM	BC
VGG+FCN8	J: 0.9863	J: 0.9715	J: 0.9674	J: 0.8934	J: 0.9701
	D: 0.9931	D: 0.9855	D: 0.9834	D: 0.9437	D: 0.9848
VGG+U-Net	J: 0.9856	J: 0.9697	J: 0.9630	J: 0.8483	J: 0.9656
	D: 0.9927	D: 0.9846	D: 0.9811	D: 0.9179	D: 0.9825
ResNet+PSPNet	J: 0.9798	J: 0.9575	J: 0.9617	J: 0.9024	J: 0.9551
	D: 0.9898	D: 0.9782	D: 0.9805	D: **0.9487**	D: 0.9770
ResNet+U-Net	J: 0.9876	J: 0.9741	J: 0.9704	J: 0.8995	J: 0.9734
	D: **0.9937**	D: **0.9869**	D: **0.9850**	D: 0.9471	D: **0.9865**

In experiment A, ResNet+PSPNet had the highest Dice coefficient for pectoral muscle segmentation, followed by VGG+FCN8, Resnet+U-Net, and VGG+U-Net. For experiment B, ResNet+U-Net had the highest Dice coefficient, followed by VGG+FCN8, VGG+U-Net, and Resnet+PSPNet. In both experiments, the VGG+FCN8 had the second-best performance differing 0.2% from the best reported in the experiment in A, and 0.12% from the best DNN in experiment B.

As the results for experiment B indicate, all the nets have very similar performance (0.98 Dice) for breast region segmentation. The main difference in the performance of the architecture comes when segmenting the pectoral muscle of dense mammograms.

With the density parameter in the mini-MIAS database, we obtained the performance for each density class in experiment A. The three density classes in the mini-MIAS database are F (fatty), G (fatty-glandular), and D (dense-glandular), and the results for each class can be seen in Figure 14.

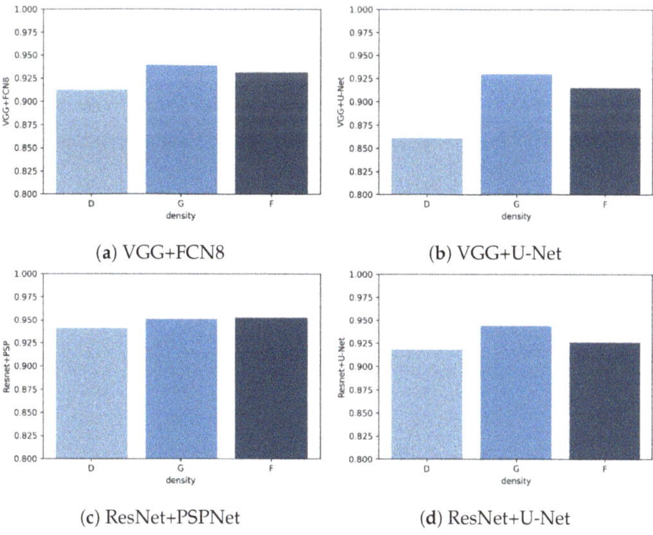

<table><tr><td style="text-align:center">(a) VGG+FCN8</td><td style="text-align:center">(b) VGG+U-Net</td></tr><tr><td style="text-align:center">(c) ResNet+PSPNet</td><td style="text-align:center">(d) ResNet+U-Net</td></tr></table>

Figure 14. Jaccard index for each architecture by density class.

Class D had the lowest Jaccard metric in every net, while G class had the highest Jaccard index. The best net all-around was ResNet+PSPNet, which has very similar results for every class and achieves the highest value for each class. The most significant difference between classes is observed in VGG+U-Net in which the difference between class D and class G is almost 0.06.

Dense mammograms have the highest amount of error for all the nets, but it is higher for the two U-Net based nets. Figure 15 shows the outline of the pectoral muscle on three dense breasts. As can be seen, both VGG+FCN8 (in pink) and ResNet+PSPNet (in blue) have an outline very similar to the ground truth. Although ResNet+PSPNet (in green) seems to overshoot the pectoral line in Figure 15b,c, the shape of the pectoral line is more natural than the line attained by VGG+FCN8. In the case of Resnet+U-Net, it tends to under-segment the pectoral muscle with an unnatural shape in the lower part of the pectoral region. Finally, VGG+U-Net (in purple) does not perform well for very dense mammograms and generates the most unnatural pectoral line of all the architectures.

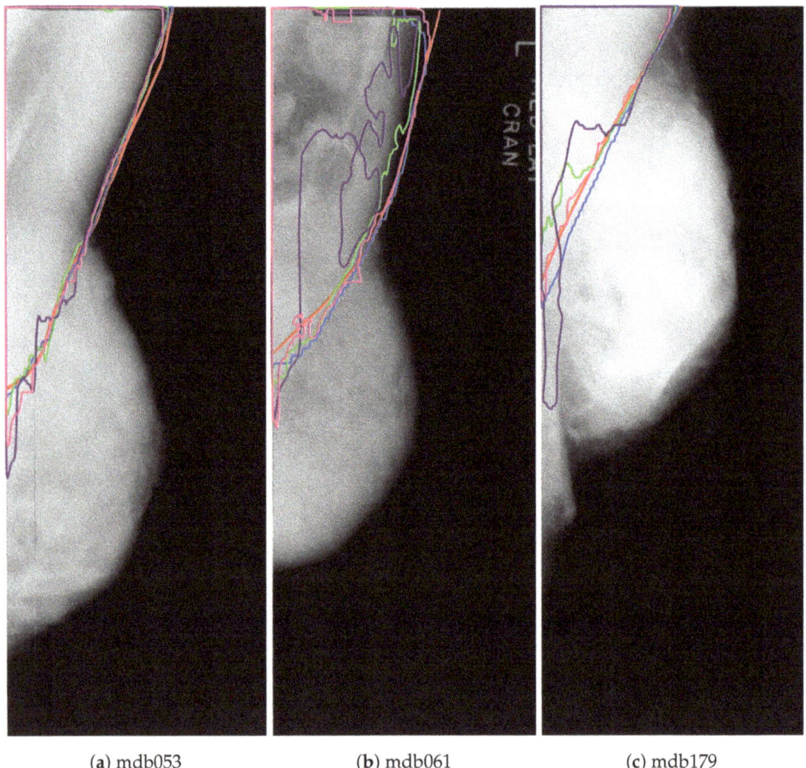

(**a**) mdb053　　　　(**b**) mdb061　　　　(**c**) mdb179

Figure 15. Pectoral muscle segmentation for the different DNNs: ground truth (red), ResNet50+PSPNet (blue), ResNet50+U-Net (green), VGG+FCN8 (pink), and VGG+U-Net (purple).

The Jaccard index and Dice coefficient are higher in experiments C than those on B, near 0.99 Dice in all the architectures, but with similar ranking in between networks. In experiment D, the breast region segmentation is almost identical to experiment B, with lower performance for the muscle segmentation than in experiment A and better background segmentation.

The reduction in the Jaccard and Dice metrics for the pectoral muscle class in experiment D is more evident in dense mammograms, as Figure 16 shows.

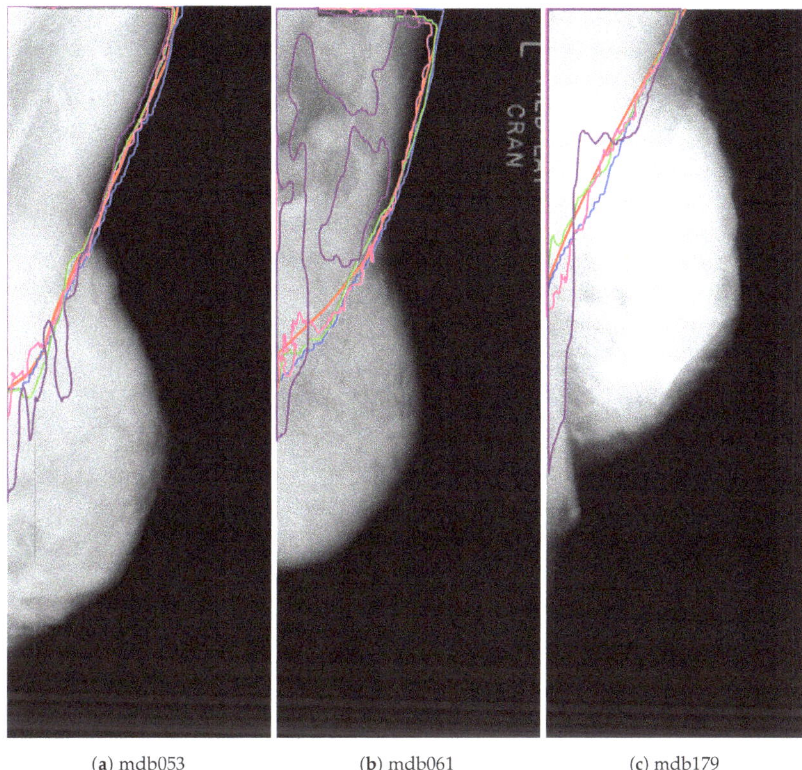

(**a**) mdb053   (**b**) mdb061   (**c**) mdb179

Figure 16. Pectoral muscle segmentation for the dense class using the different DNNs: ground truth (red), ResNet50+PSPNet (blue), ResNet50+U-Net (green), VGG+FCN8 (pink), and VGG+U-Net (purple).

From all of the experiments, there are interesting observations to be made. As the results show, U-Net based architectures do not perform well in classes with low pixel-count and have a lower performance with shallower nets. U-Net's performance was maximized in bigger classes and deeper nets, as indicated by experiments B and C where it achieved the highest results of all the tested architectures. Although, ResNet+PSPNet is a more modern architecture, VGG+FCN8 outperformed it in the breast class in every experiment. On the other hand, the best performance for the low pixel-count class (pectoral muscle) was obtained by ResNet+PSPNet, as indicated in experiments A and D.

We compared the results of the best architectures with different state-of-the-art segmentation approaches, and they are summarized in Table 9. For the breast area segmentation (BR+PM), the best DNN performs better than any other method in the state-of-the-art, which is not surprising since the breast area has a high pixel count and can be easily segmented by a DNN.

For the breast area (BR), the results are also higher than the other two methods compared. In the pectoral muscle, the best DNN was 2.6% below the best method, achieving 0.9545 compared to the 0.98 of the best result in the state-of-the-art.

Table 9. Comparison of the best net with other proposals. Best results in bold.

Method	BR+PM	PM	BR
Best DNN	J: 0.9876 D: **0.9937**	J: 0.9130 D: 0.9545	J: 0.9705 D: **0.9850**
Nagi et al. [1]	J: 0.9236 D: 0.9598	J: 0.6170 D: 0.7361	J: —— D: ——
Mustra and Grgic [5]	J: 0.9525 D: 0.9751	J: —— D: ——	J: —— D: ——
Olsen [32]	J: 0.9436 D: 0.9704	J: —— D: ——	J: —— D: ——
Shen et al. [33]	J: —— D: ——	J: 0.9125 D: 0.9496	J: —— D: ——
Oliver et al. [30]	J: —— D: 0.9600	J: —— D: 0.8300	J: —— D: 0.9700
Taghanaki et al. [2]	J: —— D: ——	J: 0.9700 D: **0.9800**	J: —— D: ——
Rampun et al. [3]	J: 0.9760 D: 0.9880	J: 0.9210 D: 0.9580	J: 0.9510 D: 0.9730
Rampun et al. [14]	J: —— D: ——	J: 0.9460 D: 0.9750	J: —— D: ——

Multiple Objective Evaluation

The performance of the implemented architectures does not provide a definitive answer on which network performs the best under the tested conditions. A way to balance these results is to test other important features on deep architectures, such as temporal and spatial requirements. The aforementioned can be achieved by trying to maximize the accuracy metrics and minimize the inference time and memory requirements for each network. This approach is an example of a multiobjective optimization problem (MOP) since we are trying to find a compromise between the computational resources and the segmentation metrics of each network.

Multiobjective optimization relies on concepts such as Pareto optimal set and Pareto front [34]. A vector of decision variables $\vec{x}^* \in \mathcal{F}$ is a Pareto optimal if there is no other $\vec{x} \in \mathcal{F}$ such that $f_i(\vec{x}) \leq f_i(\vec{x}^*)$ for all $i = 1, \ldots, k$ and $f_j(\vec{x}) \leq f_j(\vec{x}^*)$ for at least one j. In this definition, the feasible region is represented by $\mathcal{F}$. In MOP, it is common to have a set of solutions called the Pareto optimal set [34]. The vectors $\vec{x}^*$ of this set are called nondominated solutions. A vector of solution $\vec{u}$ is said to dominate $\vec{v}$, $\vec{u} \preceq \vec{v}$, if and only if $u_i \leq v_i \wedge \exists i \in \{i, \ldots n\} : u_i \leq v_i$. Using this nomenclature, the Pareto optimal set, P^*, can be defined as:

$$\mathcal{P}^* := \{x \in \mathcal{F} | \neg \exists x' \in \mathcal{F}, \vec{f}(x') \preceq \vec{f}(x)\}. \tag{11}$$

The representation of the nondominated vectors included in the Pareto optimal set is called the Pareto front, which can be represented as:

$$\mathcal{PF}^* := \{\vec{u} = \vec{f} = (f_1(x), \ldots, f_k(x)) | x \in \mathcal{P}*\}. \tag{12}$$

For calculating the trade-off between different experiments and the temporal and spatial requirements, we use a similar approach as in [35]. The idea is to rank each network in a Pareto front, taking into account the performance in each parameter. We calculated the inference time in seconds and the memory in megabytes for a standard inference on an image; the results can be seen in Table 10. At first glance, VGG16+U-Net requires less time to do an inference and less memory than the others networks, while VGG16+FCN8 took the most time for an inference (0.1445 s), and ResNet+U-Net needed the most memory (1057 MB).

Table 10. Temporal and spatial requirements of each architecture.

Network	Time (s)	Memory (MB)
VGG16+FCN8	0.1445	826
VGG16+U-Net	0.0699	613
ResNet+PSPNet	0.0717	727
ResNet+U-Net	0.0797	1057

To rank the solutions, we base on the distance between the solutions. Since in some experiments, the distance between metrics is minimal, we established a difference of at least 1% between each parameter was enough to indicate that one network dominated over another and they had a lower rank. The results can be seen in Table 11.

Table 11. Ranking for every network at a given metric. The experiment is indicated in parenthesis.

Network	Time	Memory	PM(A)	BR(B)	PM+BR(C)	PM(D)
VGG16+FCN8	III	III	II	I	I	I
VGG16+U-Net	I	I	III	II	I	II
ResNet+PSPNet	I	II	I	II	II	I
ResNet+U-Net	II	IV	II	I	I	I

The ranking in Table 11 corroborates the previous findings: the best-performing networks taking only into consideration the accuracy metrics are VGG+FCN8 and ResNet+U-Net; however, if we consider the results of inference time and memory needed for each network, the best performing architecture is Resnet+PSPNet followed by VGG+U-Net. The results of the ranking systems are important on limited since the expert can decide to use a network that does not yield the best results but overall has good performance metrics and a lower computational power to implement.

6. Discussion and Future Work

This paper proposes an extensive analysis of mammogram semantic segmentation using DNNs. Four different encoder–decoder pairs were trained and evaluated in four different experimental setups.

Our results show that all the architectures perform well (near 0.99 in Dice) for breast border segmentation, with higher results than state-of-the-art approaches. For breast region segmentation, all the architectures had a Dice metric superior to 0.98.

The highest variation was attained in the experiments that included the isolation of the pectoral muscle. The two architectures that use the U-Net segmentation approach have the smallest Dice value, surpassed by a very shallow net such as VGG16+FCN8. This issue is transcendental since U-Net is the go-to architecture for semantic segmentation for many medical imaging articles. Our research work reveals that other architectures may obtain higher performance for classes with low pixel-count in medical images.

The highest variation was due to breast density for the pectoral muscle segmentation since the lowest Dice coefficient was obtained in dense mammograms in all four architectures; however, ResNet+PSPNet had very similar results for the three classes proving to be the best architecture for low pixel-count classes in mammograms. The main drawback was its performance on the breast region class, where it had the worst Dice metric of all the architectures.

Our results indicate that the trained DNNs have good performance for mammogram segmentation. The best combination for mammogram segmentation seems to be a union of the output of the breast line given by ResNet+U-Net and the pectoral segmentation given by ResNet+PSPNet.

The multiobjective evaluation of memory and time requirements showed the importance of balancing the best performance against these two parameters. This evaluation

axioms

Article

RainPredRNN: A New Approach for Precipitation Nowcasting with Weather Radar Echo Images Based on Deep Learning

Do Ngoc Tuyen [1], Tran Manh Tuan [2,*], Xuan-Hien Le [3], Nguyen Thanh Tung [2], Tran Kim Chau [3], Pham Van Hai [1], Vassilis C. Gerogiannis [4,*] and Le Hoang Son [5,*]

[1] School of Information and Communication Technology, Hanoi University of Science and Technology, Hanoi 01000, Vietnam; dongoctuyenn@gmail.com (D.N.T.); haipv@soict.hust.edu.vn (P.V.H.)
[2] Faculty of Computer Science and Engineering, Thuyloi University, Hanoi 10000, Vietnam; tungnt@tlu.edu.vn
[3] Faculty of Water Resources Engineering, Thuyloi University, Hanoi 10000, Vietnam; hienlx@tlu.edu.vn (X.-H.L.); kimchau_hwru@tlu.edu.vn (T.K.C.)
[4] Department of Digital Systems, Faculty of Technology, University of Thessaly, Geopolis, 41500 Larissa, Greece
[5] VNU Information Technology Institute, Vietnam National University, Hanoi 01000, Vietnam
* Correspondence: tmtuan@tlu.edu.vn (T.M.T.); vgerogian@uth.gr (V.C.G.); sonlh@vnu.edu.vn (L.H.S.)

Abstract: Precipitation nowcasting is one of the main tasks of weather forecasting that aims to predict rainfall events accurately, even in low-rainfall regions. It has been observed that few studies have been devoted to predicting future radar echo images in a reasonable time using the deep learning approach. In this paper, we propose a novel approach, RainPredRNN, which is the combination of the UNet segmentation model and the PredRNN_v2 deep learning model for precipitation nowcasting with weather radar echo images. By leveraging the abilities of the contracting-expansive path of the UNet model, the number of calculated operations of the RainPredRNN model is significantly reduced. This result consequently offers the benefit of reducing the processing time of the overall model while maintaining reasonable errors in the predicted images. In order to validate the proposed model, we performed experiments on real reflectivity fields collected from the Phadin weather radar station, located at Dien Bien province in Vietnam. Some credible quality metrics, such as the mean absolute error (MAE), the structural similarity index measure (SSIM), and the critical success index (CSI), were used for analyzing the performance of the model. It has been certified that the proposed model has produced improved performance, about 0.43, 0.95, and 0.94 of MAE, SSIM, and CSI, respectively, with only 30% of training time compared to the other methods.

Keywords: radar image prediction; rain radar; deep learning; precipitation nowcasting; UNet; PredRNN_v2

MSC: 62M45

Citation: Tuyen, D.N.; Tuan, T.M.; Le, X.-H.; Tung, N.T.; Chau, T.K.; Van Hai, P.; Gerogiannis, V.C.; Son, L.H. RainPredRNN: A New Approach for Precipitation Nowcasting with Weather Radar Echo Images Based on Deep Learning. *Axioms* **2022**, *11*, 107. https://doi.org/10.3390/axioms11030107

Academic Editor: Oscar Humberto Montiel Ross

Received: 3 January 2022
Accepted: 22 February 2022
Published: 28 February 2022

Publisher's Note: MDPI stays neutral with regard to jurisdictional claims in published maps and institutional affiliations.

1. Introduction

Precipitation nowcasting from high-resolution radar data is essential in many branches such as water management, agriculture, aviation, emergency planning, and so on. It aims to make detailed and plausible predictions of future radar images based on past radar images with information about the amount, timing, and location of rainfall. This problem is significant to nowcasting the rainfall events in the next few hours with tropical depression in a given direction, entering from one area to another [1]. In such a case with heavy rainfall in the past few days, which is expected to continue to increase in the coming days, the prediction would help localities to ensure the safety of dams and essential dikes, to avoid unexpected flood discharges, causing flooding and inundation for the downstream area. According to the report on the assessment of disaster events in the 21st century implemented by the Centre for Research on the Epidemiology of Disasters [2], floods cause more negative impacts on people than any other natural catastrophe. Additionally, rain has a detrimental impact on travel demand and travel time, as well as on road traffic accidents, in metropolitan areas worldwide [3–5].

In recent years, deep learning has been applied to many areas [6–8]. Various variants of convolutional neural network (CNN) and recurrent neural network (RNN) architectures have been modified and applied in a variety of domains to produce suitable versions and solve specific problems [9–12]. Several studies on applications of deep learning for time-series data problems are briefly reviewed as follows.

Khiali et al. [13] presented a new approach that is a combination of graph-based techniques in order to design a new clustering framework for satellite time-series images. Spatiotemporal features are firstly extracted, which are then represented in their movements in the graph. Based on their similar characteristics, spatiotemporal clusters are produced. Since transmission lines often undergo various faults and errors, which caused terrible economic damage, Fahim et al. [14] proposed a robust self-attention CNN (SAT-CNN) that uses time-series image extracted features for detection and classification of faults. By adding the discrete wavelet transform (DWT) preprocessing method, the proposed model shows the superiority of the performance compared to others. Since in the traditional approaches of exploiting time-series, there may be human intervention in extracting features, Li et al. [15] introduced a novel approach that uses various computer vision algorithms in order to automatically extract features from time-series imagery after the images are transformed into recurrence plots. The method showed significant performance in two datasets: the largest forecasting competition dataset (M4) and the tourism forecasting competition dataset.

Precipitation nowcasting has attracted many researchers' attention [16–18]. In recent years, computer vision with deep learning has shown dramatic promise. Agrawal et al. [19] applied one of the most popular models, UNet, in order to forecast precipitation and produced favorably comparable results. In 2021, Fernández and Mehrkanoon [20] used deep learning for weather nowcasting by presenting a novel UNet-based architecture model, Broad-UNet. To learn more complex abstract features of input images, this model alters convolution layers and pooling layers by asymmetric parallel convolutions and atrous spatial pyramid pooling (ASPP), respectively. Thus, the Broad-UNet model exhibits great performance compared to others. In order to support meteorologists nowcasting short-term weather with a large volume of satellite and radar images, Ionescu et al. [21] introduced the family of the CNN architecture, DeePS. By using five satellite products to collect satellite image data, the model was analyzed and compared with other CNN-based models and was found to reach a 3.84% MAE score for an entire dataset.

By applying deep learning methods in supporting meteorologists to predict future disastrous weather, Zhang et al. [22] proposed a high-performance model for predicting changes in weather radar echo shape, which is based on the combination of the conventional CNN and the long short-term memory (LSTM) network. In practice, their model produces significant results in various evaluation methods, such as the critical success index (CSI) and the Heidke skill score, compared to ConvLSTM and TrajGRU models. Trebing et al. [23] noticed that numerical weather prediction methods lack the ability for short-term forecasts using the latest available information. They introduced the application of deep learning in a novel comparable performance neural network, small attention UNet (SmaAt-UNet), which uses only 25% of the trainable network parameters. Additionally, Le et al. [24] firstly applied the LSTM neural network to perform flood forecasting on Da River in Vietnam. The suggested model was evaluated by the Nash–Sutcliffe efficiency (NSE) score in different prediction cases and produced considerably high performances (around 90% NSE). In 2021, Le et al. [25] also compared different deep learning models for forecasting river streamflow. Various state-of-the-art models were reviewed and evaluated, such as StackedLSTM and BiLSTM.

Although the above-mentioned articles have contributed considerably to the fields of forecasting and nowcasting by applying various state-of-the-art deep learning algorithms, few can manage and apply spatiotemporal and temporal features in both long- and short-term time-series imagery. However, in precipitation nowcasting, not enough studies are available that have applied time-series imagery to predict future scenes [26,27]. Recently, Wang et al. [28] released a deep learning model that has proven powerful in processing time-series image datasets. Despite the original model, PredRNN_v2 working well in most cases, we noticed that this model took a tremendous amount of time for training and testing (in terms of radar dataset), in particular if we want to retrain the model with a larger dataset later down the road. This describes the motivation for this paper, that is, to design a new deep learning method to improve the overall process and work well with multistep prediction.

Therefore, in this paper, we aim to introduce a novel deep learning approach in precipitation nowcasting with valuable collected radar datasets to overcome the above limitations. The proposed model is a combination of the power of UNet [29] and PredRNN_v2 [28] with the purpose of reducing training and testing time while preserving the complex spatial features of radar data. Our model benefits from the robustness of PredRNN_v2 in managing both spatiotemporal and temporal information of time-series images. Additionally, the contracting and expanding paths of UNet have a vital role in reducing the size of inputs, while it still captures the high-level features of original images.

In the implementation, we set up our case study to allow our model to have the ability to predict images following one hour (6 timesteps with a 10 min gap) so that the model still produces comparable performances. By such a design, the computation time of the training and testing phases of the proposed model is reduced remarkably, approximately 30% smaller than that of the original PredRNN_v2. In addition, our model produces impressive performances compared to others by evaluating it with various quality assessments.

The organization of the rest of the paper is as follows. The data preparation and the background underlying the proposed model are described in Section 2. In Section 3, by leveraging the advantages of the encoder–decoder architecture, we introduce the most suitable model for solving the above-mentioned problems. In Section 4, we present the environment setup and the implementation served for the evaluation and the comparison of the suggested model with others, as well as discuss the comparison results. Finally, conclusions and future development directions are described in Section 5.

2. Data and Background

2.1. Background

2.1.1. Convolutional LSTM (ConvLSTM)

Since traditional standard LSTMs, which are special RNN architectures [30], have a significant drawback in simultaneously modeling the spatiotemporal information of inputs, the hidden states, and the output memory cells, the ConvLSTM [31] with various improvements can tackle the problem of the former version (FC-LSTM). First, in order to encode the spatial structure information, all the inputs $\mathcal{X}_1$, …, $\mathcal{X}_t$, the output cells $\mathcal{C}_1, …, \mathcal{C}_t$, and the hidden states $\mathcal{H}_1$, …, $\mathcal{H}_t$ are 3D tensors ($\mathbb{R}^{P \times M \times N}$), in which M and N are rows and columns representing, respectively, the spatial dimensions. Second, since '$*$' and '$\odot$' denote the convolution operator and the Hadamard product, all the gates i_t, f_t, o_t are also 3D tensors, which are responsible for transferring the information in different conditions. The equations of ConvLSTM are described as follows:

$$
\begin{aligned}
g_t &= \tanh\left(\mathcal{W}_{xg} * \mathcal{X}_t + \mathcal{W}_{hg} * \mathcal{H}_{t-1} + b_g\right) \\
i_t &= \sigma(\mathcal{W}_{xi} * \mathcal{X}_t + \mathcal{W}_{hi} * \mathcal{H}_{t-1} + \mathcal{W}_{ci} \odot \mathcal{C}_{t-1} + b_i) \\
f_t &= \sigma\left(\mathcal{W}_{xf} * \mathcal{X}_t + \mathcal{W}_{hf} * \mathcal{H}_{t-1} + \mathcal{W}_{cf} \odot \mathcal{C}_{t-1} + b_f\right) \\
\mathcal{C}_t &= f_t \odot \mathcal{C}_{t-1} + i_t \odot g_t \\
o_t &= \sigma(\mathcal{W}_{xo} * \mathcal{X}_t + \mathcal{W}_{ho} * \mathcal{H}_{t-1} + \mathcal{W}_{co} \odot \mathcal{C}_t + b_o) \\
\mathcal{H}_t &= o_t \odot \tanh(\mathcal{C}_t)
\end{aligned}
\tag{1}
$$

Since the last two dimensions of the standard FC-LSTM are equal to 1, FC-LSTMs can be considered as the particular case of ConvLSTM. Although the ConvLSTM has a crucial role in paving the way for processing time-series image datasets for many real-life problems, some points can be further improved in this architecture. First, the memory states C_t are merely dependent on the hierarchical representation of the features of other layers due to the states being updated horizontally of the corresponding layers. This means the operator in the first layer of the current timestamp t will not know what features are memorized in the previous top layer of the timestamp $t-1$. Second, since the hidden states $\mathcal{H}_t$ are the output of the operation on the two gates o_t and C_t, which means $\mathcal{H}_t$ will contain both long-term and short-term information, the performance of the model will be considerably limited by these spatiotemporal variations.

2.1.2. Spatiotemporal LSTM with Spatiotemporal Memory Flow (ST-LSTM)

By combining the novel spatiotemporal long short-term memory (ST-LSTM) as the basic building block with the architecture of the spatiotemporal memory flow design, Wang et al. [32] introduced the predictive recurrent neural network (PredRNN), which overcomes the limitations of the former version ConvLSTM. The equations of ST-LSTM are presented as follows:

$$
\begin{aligned}
g_t &= \tanh\left(W_{xg} * \mathcal{X}_t + W_{hg} * \mathcal{H}^l_{t-1} + b_g\right) \\
i_t &= \sigma\left(W_{xi} * \mathcal{X}_t + W_{hi} * \mathcal{H}^l_{t-1} + b_i\right) \\
f_t &= \sigma\left(W_{xf} * \mathcal{X}_t + W_{hf} * \mathcal{H}^l_{t-1} + b_f\right) \\
C^l_t &= f_t \odot C^l_{t-1} + i_t \odot g_t \\
g'_t &= \tanh\left(W'_{xg} * \mathcal{X}_t + W_{mg} * \mathcal{M}^{l-1}_t + b'_g\right) \\
i'_t &= \sigma\left(W'_{xi} * \mathcal{X}_t + W_{mi} * \mathcal{M}^{l-1}_t + b'_i\right) \\
f'_t &= \sigma\left(W'_{xf} * \mathcal{X}_t + W_{mf} * \mathcal{M}^{l-1}_t + b'_f\right) \\
\mathcal{M}^l_t &= f'_t \odot \mathcal{M}^{l-1}_t + i'_t \odot g'_t \\
o_t &= \sigma\left(W_{xo} * \mathcal{X}_t + W_{ho} * \mathcal{H}^l_{t-1} + W_{co} \odot C^l_t + W_{mo} * \mathcal{M}^l_t + b_o\right) \\
\mathcal{H}^l_t &= o_t \odot \tanh\left(W_{1\times1} * \left[C^l_t, \mathcal{M}^l_t\right]\right)
\end{aligned}
\tag{2}
$$

Two significant improvements are introduced by the PredRNN model: the spatiotemporal memory cell $\mathcal{M}^l_t$ and how the novel cells are updated in the zigzag direction. Two memory cells contain the temporal and spatiotemporal information: the conventional cell C^l_t is propagated horizontally from the previous corresponding layer at $t-1$ timestamp to the current time step, and the novel cell $\mathcal{M}^l_t$ is delivered vertically from the lower layer $l-1$ in the meantime. In the first improvement, by presenting gate structures for $\mathcal{M}^l_t$, the final hidden output $\mathcal{H}^l_t$ benefits from containing information of both gates C^l_t and $\mathcal{M}^l_t$. Secondly, the spatiotemporal memory cell is delivered in the zigzag style (i.e., information is conveyed upward first and then forward overtime between layers), which means that from the first layer where $l=1$, $\mathcal{M}^0_t = \mathcal{M}^L_{t-1}$ (L stack ST-LSTM layers). The mechanism makes the long-term and short-term dynamics resulting in the hidden output by twisting the pairs of memory states (horizontally and vertically).

2.1.3. Spatiotemporal LSTM with Memory Decoupling

In practice, by using t-SNE [33] for visualizing memory data at every timestamp, the authors in [32] noticed that the memory states are not distinguished between each other automatically and decoupled spontaneously. Based on the PredRNN, the authors established a new loss function, which is the combination of the standard mean square error loss and the new decoupling loss:

$$
\mathcal{L} = \mathcal{L}_{MSE} + \mathcal{L}_{decouple}
\tag{3}
$$

in which $\mathcal{L}_{MSE}$ is the conventional loss function for the former version PredRNN, and $\mathcal{L}_{decouple}$ is the novel memory decoupling regularization loss function, which is described as follows:

$$\Delta \mathcal{C}_t^l = W_{decouple} * (i_t \odot g_t)$$
$$\Delta \mathcal{M}_t^l = W_{decouple} * (i_t' \odot g_t')$$
$$\mathcal{L}_{decouple} = \sum_t \sum_l \sum_c \frac{|\Delta \mathcal{C}_t^l, \Delta \mathcal{M}_{tc}^l|}{||\Delta \mathcal{C}_t^l||_c \cdot ||\Delta \mathcal{M}_t^l||_c} \tag{4}$$

where $W_{decouple}$ is the parameter of a convolution layer added after the memory cells $\mathcal{C}_t^l$ and $\mathcal{M}_t^l$ at each timestep. By this means, two memory states are separated to train on different aspects of spatiotemporal and temporal information. Further, the new convolution layer is removed in the predicting phase, making the size of the entire model unchanged. That makes a novel version of the PredRNN, PredRNN_v2 [28].

2.2. Study Area

In this study, we utilized a radar echo dataset, which was retrieved from the Phadin weather radar station, located in Dien Bien province, Vietnam. The Phadin station, located at 21.58° N and 103.52° S, is under the direct management of the Northwest Aero-Meteorological Observatory and has the primary task of providing short-term forecast information on meteorology and climate for the provinces in this region. Officially launched and operating since March 2019, this is a doppler weather radar station and operates in dual-polarization mode. This means that it is capable of transmitting and receiving pulses of radio waves in both vertical and horizontal directions (Figure 1).

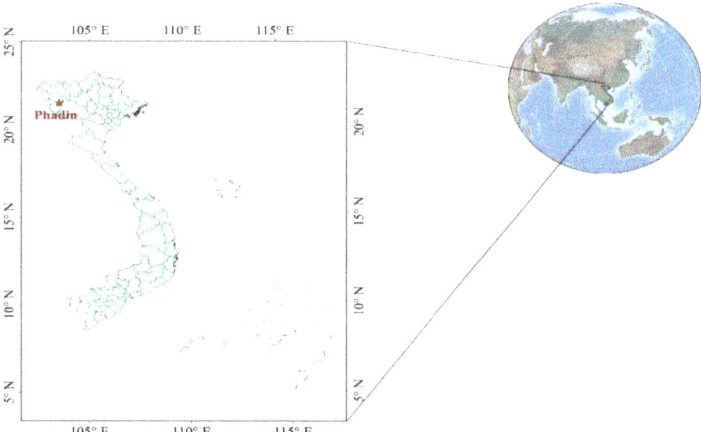

Figure 1. Geographical area of the region of study.

As a result, it can provide super-high-resolution weather observations and cover a large area with an effective scanning radius of up to 300 km. For the issue of precipitation nowcasting based on weather radar, the collected data here is understood as the composite reflectivity images of radio pulses, in which, these images are grayscale images, and each image represents a transmission and reception of a weather radar signal. With an area coverage of 300 km $\times$ 300 km (equivalent to the effective range of radar), these reflectivity images have a spatial resolution of 150 $\times$ 150 pixels and a corresponding temporal resolution of 10 min. A total of 2429 weather radar composite reflectivity images were collected during rainfall events that took place between June and July 2020 (in the rainy season of Vietnam). Several weather radar images are illustrated in Figure 2.

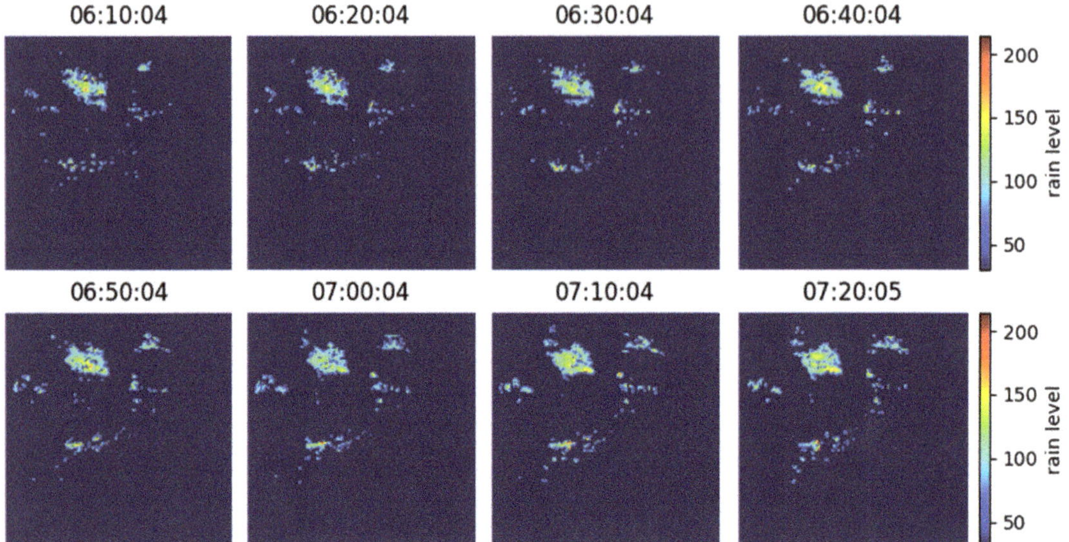

Figure 2. Samples of radar reflectivity images were recorded in the period between 6:10 a.m. and 7:20 a.m. on 23 June 2020. The pixels with high value (white) denote raining areas. In contrast, the low-value pixels are not raining areas.

2.3. Data Preparation

In neural networks, the dataset split ratio depends mainly on the data characteristics, the total number of collected samples, and the actual model being trained. The single hold-out method [34] is one of the simplest data resampling strategies that will be applied to our training strategies. In order to train our model effectively and to produce excellent model performance, we randomly divide our gathered dataset into separated three parts: training, validation, and testing set with a ratio of 80:10:10, respectively. This means that in 2429 images of the dataset, the training, validation, and testing set will contain precisely 1947, 242, and 242 images, respectively, to the above ratio. In particular:

- Training set: The weights and biases of the model will be trained and updated on the samples of the set until reaching convergence.
- Validation set: An unbiased evaluation will be calculated to see how fit the model is on the training set. This set helps to improve the model performance by fine-tuning the model,
- Testing set: This set informs us about the final accuracy of the model after completing the training phase.

The details of how the dataset was divided are presented in Table 1, as follows:

Table 1. Quantity and size of each dataset.

Dataset	Quantity	Size
Training Set	1947	150×150
Validation Set	242	150×150
Testing Set	242	150×150

In order to make the images as inputs for our model, we stack all images into one array and take the continuous sliding window of the stack sequentially until the index reaches the end. For each consecutive part, we define a number of frames of the head as input and the remainder as output in which a number of timesteps exist in the past. For example, we slide the consecutive image in the array with 10 frames wider per window: 5 for the input and 5 for the output.

2.4. Evaluation Criteria

In the context of computer vision, the mean absolute error (MAE) [35], which is referred to as L_1 loss function in some particular problems, is interpreted as a measure of the difference between every pixel of the predicted image and the ground truth (true value) of that image. Mathematically, the MAE score takes total absolute errors in the entire testing dataset that will be divided by the number of observations. The MAE measure is described in Formula (5) below, where $\hat{y}_i$ and y_i are the ith predicted images and the ith ground truth in the testing set, respectively, and the subtraction operation is an element-wise operation:

$$MAE = \frac{1}{n} \sum_{i=1}^{n} |\hat{y}_i - y_i| \tag{5}$$

Another measure that also has a significant impact on assessing the performance of the model is the structural similarity index measure (SSIM) [36]. The SSIM index calculates the image quality degradation after some processing phase, especially propagating through the deep learning model. Formula (6) below explains the SSIM measure mathematically:

$$SSIM(y, \hat{y}) = \frac{\left(2\mu_y\mu_{\hat{y}} + c_1\right)\left(2\sigma_{y\hat{y}} + c_2\right)}{\left(\mu_y^2 + \mu_{\hat{y}}^2 + c_1\right)\left(\sigma_y^2 + \sigma_{\hat{y}}^2 + c_2\right)} \tag{6}$$

In Formula (6), μ and σ are respectively the average and variance of the label y and the prediction $\hat{y}$, and $\sigma_{y\hat{y}}$ is the covariance of two images (y and $\hat{y}$). Furthermore, c_1 and c_2 are two variables responsible for stabilizing the division and are presented as follows:

$$\begin{aligned} c_1 &= (k_1 L)^2 \\ c_2 &= (k_2 L)^2 \end{aligned} \tag{7}$$

where $k_1 = 0.01$ and $k_2 = 0.03$ are set by default, and $L = 2^{\#bits\ per\ pixel} - 1$ is the dynamic range of the pixel value of the image.

Third, we also use the critical success index (CSI) [37], which is considered as the threat score to evaluate how well our model performs compared to former models. Suppose that we use the four quantities of the confusion matrix [38], which is described in Table 2.

Table 2. Confusion matrix.

		Ground Truth	
		Rain	**No Rain**
Predicted	Rain	TP	FP
	No Rain	FN	TN

In Table 2, true positive (TP) is the number of ground-truth-positive pixels (Rain) that were correctly predicted. False positive (FP) corresponds to the number of ground-truth-negative pixels (No Rain) that were predicted incorrectly. False negative (FN) is the number of ground-truth-positive pixels that were not predicted. True negative (TN) corresponds to the number of ground-truth-negative pixels that were correctly predicted as negative. The CSI score is shown as follows in Equation (8):

$$CSI = \frac{TP}{TP + FP + FN} \tag{8}$$

In the current paper, since our modification focuses on reducing the processing time of the deep learning models, the training and testing time is also evaluated as a crucial factor for assessing the performance of models. The last criterion that we include in the evaluation phase is the multiply-accumulate operations (MACs), i.e., a MAC has one multiply operation and one add operation. We clearly detail the model implementation and evaluation results in Section 4.

3. Proposed RainPredRNN

In this article, by utilizing the strength of the PredRNN_v2 model, we propose the new modified model, RainPredRNN, which can be fitted into the problems of processing time-series radar images for predicting images in the following time step. Our model uses the contracting-expansive path of the UNet model as the encoder and decoder paths before and after forwarding input to the ST-LSTM layers, which will reduce the huge number of operations required to be calculated.

3.1. Benefit of the Encoder–Decoder Path

Since the robustness of the UNet model has been verified in various domains over the years from its first publication [29], we borrow the UNet-based architecture in order to make our modifications. Thus, the proposed model will benefit from the abilities of the contracting-expansive path and concatenation technique.

Encoder path: Two 3×3 convolution layers are repeatedly included to capture the context of original images, and each layer is followed by a rectified linear unit (ReLU) for making the model nonlinear and batch normalization (regularization). In order to reduce the spatial dimensions, a pooling layer, which is a max-pooling layer, is applied right after these convolution layers. After each of the above operations, the original inputs are cut in half the spatial dimensions and double the number of feature channels to produce high-level feature maps.

Decoder path: First, the model must upsample the feature map produced by the encoder path to return to its original shape gradually. Secondly, after each upsampling operator, the number of feature channels will be cut in half by a 2×2 transpose convolution layer. In addition, a concatenation technique will be used from the corresponding feature map in the encoder path in order to avoid vanishing the gradients when training. Third, two 3×3 convolution layers with ReLU and batch normalization operations are applied. At the final layer, a 1×1 convolution layer is used to map every pixel to the desired number of classes.

3.2. Unified RainPredRNN

In order to take advantage of the UNet model, we will borrow the key characteristics of its architecture: contracting-expansive path and concatenation technique. Firstly, every original image is propagated through the encoder path with one max-pooling layer coming between four 3×3 convolution layers. By doing this, the high-level valuable contexts of the inputs will be captured and stored in the feature maps before processing by the spatiotemporal LSTM layers (ST-LSTM). Since various image resizing algorithms cause the loss of image information considerably and transform images improperly, the encoder path keeps as much context as possible and still reduces the spatial dimension of original images.

Since ST-LSTM is designed with many gates and a huge number of floating-point operations, the larger the inputs come in, the more calculations are operated. After the encoding path, the original inputs are halved by the width and height and have more spatial information. Thus, the computation time of ST-LSTM will be reduced significantly in both the forward and backward propagation strategies. The visualization of our modification is shown in Figure 3 as follows:

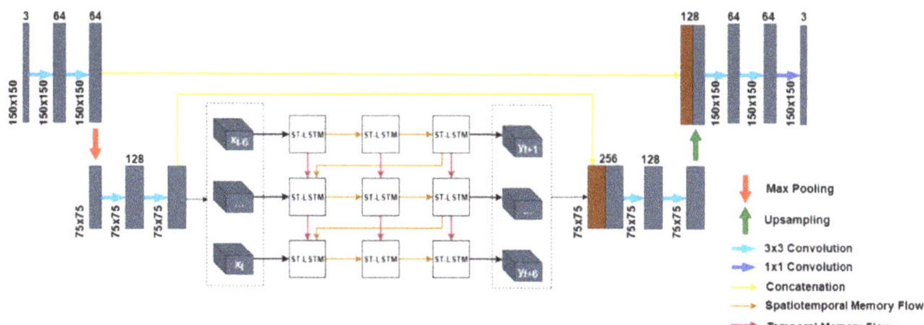

Figure 3. Unified RainPredRNN model. The boxes with the text "ST-LSTM" denote the conventional spatiotemporal LSTM, the gray boxes represent images in different processing levels of the model, and the brown boxes are the copy of cropped feature maps. While the encoding path has a role in reducing the spatial dimension of inputs for lightweight computation of the stacked ST-LSTM, the decoding path processes the output of the stacked ST-LSTM to recover back to the original size of the output.

The expansive path was added right after ST-LSTM layers, the outputs of the layers were taken as the inputs. At this time, the skip-connection technique was applied to take the cropped of the corresponding feature map in the encoding path to obtain more information and avoid the gradient vanishing problem. By using one upsampling layer, we obtained the original spatial dimension of the original images. We noticed that our modification remarkably reduced training and testing time, while the model still produced the same evaluated scores as the former version. We detail our experimental results in Section 4.

3.3. Implementation

In this subsection, to be able to predict six next frames (1 h in advance), all models are set up in a proper manner. In practice, after conducting various experiments, we empirically chose the best-fit hyperparameters for our model and resources.

To clarify our implementation in detail, we describe our hyperparameters, which were practically most suitable with our dataset. Our modification model RainPredRNN comprises the critical characteristics of the UNet architecture presented in Section 3.1, in which the kernel size of convolution layers is set to 3×3, and both stride and padding were equal to 1. With this choice, our model captures the objects (rain) of our dataset because the pixels move slowly.

In the main body of the model, we put two consecutive ST-LSTM layers together, which were set up with 64 hidden states each and a 3×3 filter of the inside convolution layers. Because the size of our input image is quite small, the number of stacked ST-LSTM layers with the hidden states was kept at a moderate size. In addition, the total input length was fixed to 12 frames, with the first six consecutive images for input and the last six ones for ground truth. Thus, to compare the performance of all models, we trained each one in 100 epochs, in which the batch size was set equal to 4, and the learning rate was set equal to 0.001 during the training phase. All the models were evaluated with the above-mentioned criteria. The results are shown in the following section, where, in particular, the three baseline models (PredRNN, PredRNN_v2, and RainPredRNN) are implemented and compared. Finally, all hyperparameters were chosen based on the knowledge about the dataset and by different scenarios practically.

For implementing conveniently, we used the state-of-the-art machine learning PyTorch library written in the Python programming language. These software libraries are free open-source software for communities who want to develop and build machine learning models in research and production. In addition, to visualize the model's results, we also imported and implemented the Matplotlib library. All algorithms and models used in the paper are listed in the Appendix A.

4. Results and Discussion

To conveniently implement and debug our source code properly, we prepared a single powerful workstation running on the Windows 10 64-bit operating system. The machine was equipped with one 12 Gb GPU card, GeForce RTX 2080 Ti. In order to run the proposed deep learning model RainPredRNN on the GPU, we also installed the compatible version 10.1 of the CUDA driver, which can be integrated with the NVIDIA card.

In Figure 4, we observed that the models all converged to the same point as training and validation progressed, approximately $2.5e - 4$ and $4e - 4$ of the loss value, respectively. The detail of evaluation scores is listed in Table 3, in which MAE, CSI, and SSIM are estimated in the test set. From the table, it can be seen that the SSIM measure of all models is not significantly different (around 0.94), which means that the quality of the images is not degraded after propagating.

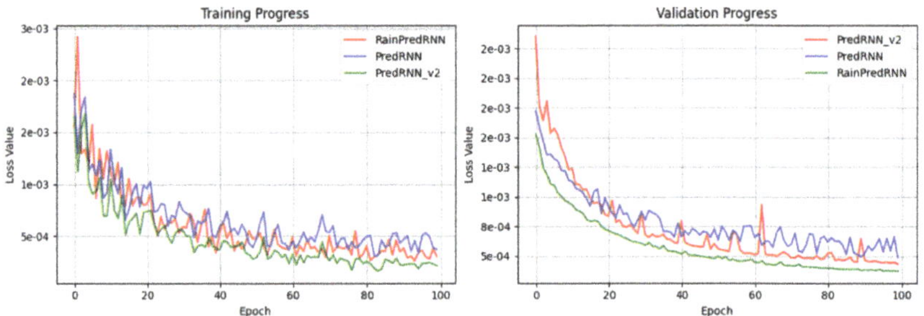

Figure 4. Training and validation loss of models.

Table 3. Evaluation scores of our modification model with others.

Model	MAE	CSI	SSIM	Training Time (hour)	MACs(G)
PredRNN	0.4535	0.9455	0.9397	15.1	101.469
PredRNN_v2	0.4157	0.9420	0.9430	15.53	103.885
RainPredRNN	0.4301	0.9590	0.9412	**4.46**	**54.705**

It is considered that the model RainPredRNN takes only below 30% of the training time of PredRNN and PredRNN_V2. We can benefit from this point. It will have significant meaning in the future if new training data arrive, and we want to produce a new version. The MAC value of RainPredRNN is one second compared to the others, at about 54 billion, so we can conclude that our modification remarkably reduced the number of operations that need to be processed when training and testing. In addition, our model still has great performance compared to former models. From the result, the models certainly produce the predicted image with high quality and resolution.

Figure 5 shows the input and the ground truth that we used to test our model with the former versions. A prediction example of three models is depicted in Figures 6 and 7. The quality of our model tends to have a higher resolution and be more precise than that of the models PredRNN and PredRNN_v2.

From these results, we can infer that the family of the PredRNN model is suitable for the problem of precipitation nowcasting, and our proposed model can help reduce training and testing significantly and also produces high-quality future images in a short time.

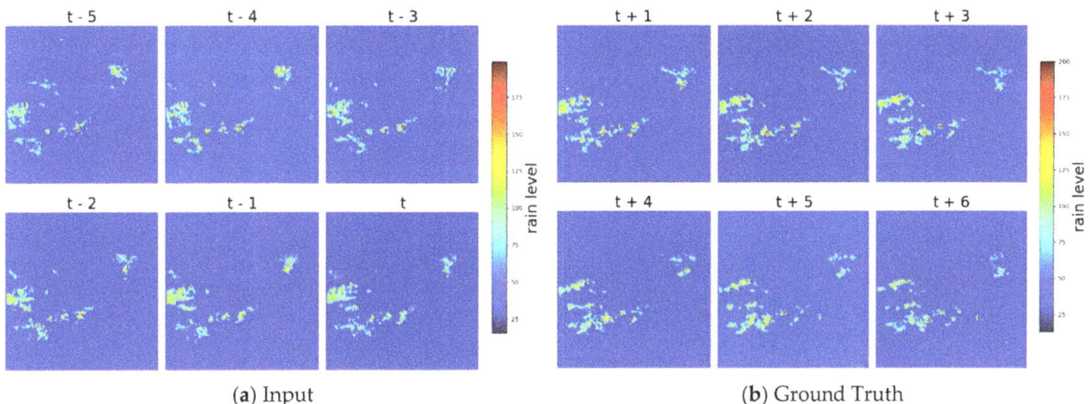

(a) Input   (b) Ground Truth

Figure 5. Consecutive image input and ground truth for comparison: (**a**) input; (**b**) ground truth.

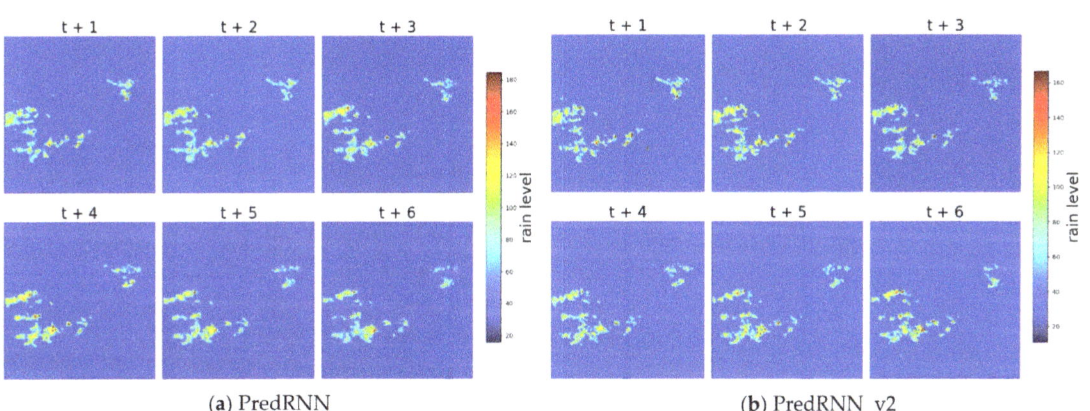

(a) PredRNN   (b) PredRNN_v2

Figure 6. Predicted image of compared models: (**a**) six next predicted frames of model PredRNN; (**b**) next six predicted frames of PredRNN_v2.

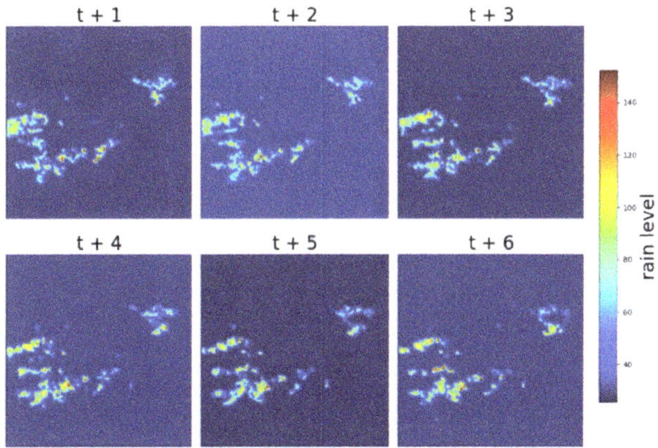

Figure 7. Six consecutive frames output of RainPredRNN model.

5. Conclusions

In this paper, we proposed a new deep learning model, RainPredRNN, for precipitation nowcasting with weather radar echo images. This model is a combination of UNet and PredRNN_v2 with the purpose of reducing training and testing time while preserving the complex spatial features of radar data. RainPredRNN manages both spatiotemporal and temporal information of time-series images. Additionally, the contracting and expanding paths of UNet have a vital role in reducing the size of inputs, while it still captures the high-level features of original images. The experiments on real data from the Phadin weather radar station, located in Dien Bien province, Vietnam, have clearly affirmed that RainPredRNN reduces training and testing significantly and also produces high-quality future images in a short time. The proposed approach has produced comparable results in which the training time is equal to less than 30% training time and 50% MAC value of the former versions.

However, some limitations of the proposed model remain, such as the validation measures are not outstanding compared to the former measures. We retained the core ST-LSTM layer as a building block, so the layer needs to be modified down the road. In the future, we hope that we will also make further improvements to the accuracy of the model for precipitation nowcasting.

Author Contributions: Conceptualization, methodology, software: D.N.T., T.M.T., L.H.S.; data curation, writing—original draft preparation: D.N.T., T.M.T., X.-H.L.; visualization, investigation: T.K.C., P.V.H.; software, validation: D.N.T., T.M.T.; supervision: L.H.S., V.C.G.; writing—reviewing and editing: N.T.T., V.C.G., L.H.S. All authors have read and agreed to the published version of the manuscript.

Funding: This research was funded by the Thuyloi University Foundation for Science and Technology.

Acknowledgments: The authors would like to acknowledge the editors and reviewers who provided valuable comments and suggestions that improved the quality of the manuscript.

Conflicts of Interest: The authors declare that they do not have any conflict of interests. This research does not involve any human or animal participation. All authors have checked and agreed with the submission.

Appendix A

List of all algorithms and models used in the paper:

1. Adam is the learning algorithm that is used in the training phase to seek the convergence point of the model. The parameters of the model are modified until the model converges.
2. Convolutional LSTM is the general version of LSTM that is designed to tackle the problem of processing image inputs. By replacing the multiply operator with the convolution operator of spatial structure information, the model successfully encodes the spatial structure information of input.
3. PredRNN combines the spatiotemporal LSTM (ST-LSTM) as the building block with the memory flow technique. The ST-LSTM introduces improvements to the memory cells, which contain the information of the flow (the memory flow technique) in both horizontal and vertical directions.
4. PredRNN_v2 introduces the new component of the final loss function. This improvement trains the model more effectively and successfully, but the overall size of the model remains.
5. RainPredRNN is the proposed model, which is a combination of the strength of the PredRNN_v2 model and the UNet model. The model borrows the contracting and expansive path of the UNet model for processing input to reduce the computational operators of the overall model. From that, the proposed model produces satisfactory results in a short time.

References

1. Wapler, K.; de Coning, E.; Buzzi, M. Nowcasting. In *Reference Module in Earth Systems and Environmental Sciences*; Elsevier: Amsterdam, The Netherlands, 2019. [CrossRef]
2. CRED. *Natural Disasters*; Centre for Research on the Epidemiology of Disasters (CRED): Brussels, Belgium, 2019. Available online: https://www.cred.be/sites/default/files/CREDNaturalDisaster2018.pdf (accessed on 10 October 2021).
3. Keay, K.; Simmonds, I. Road accidents and rainfall in a large Australian city. *Accid. Anal. Prev.* **2006**, *38*, 445–454. [CrossRef]
4. Chung, E.; Ohtani, O.; Warita, H.; Kuwahara, M.; Morita, H. Effect of rain on travel demand and traffic accidents. In Proceedings of the IEEE Intelligent Transportation Systems, Vienna, Austria, 16 September 2005; pp. 1080–1083.
5. Sun, Q.; Miao, C.; Duan, Q.; Ashouri, H.; Sorooshian, S.; Hsu, K.-L. A Review of Global Precipitation Data Sets: Data Sources, Estimation, and Intercomparisons. *Rev. Geophys.* **2018**, *56*, 79–107. [CrossRef]
6. LeCun, Y.; Bengio, Y.; Hinton, G. Deep learning. *Nature* **2015**, *521*, 436–444. [CrossRef] [PubMed]
7. Sit, M.A.; Demiray, B.Z.; Xiang, Z.; Ewing, G.; Sermet, Y.; Demir, I. A Comprehensive Review of Deep Learning Applications in Hydrology and Water Resources. *Water Sci. Technol.* **2020**, *82*, 2635–2670. [CrossRef] [PubMed]
8. Gao, Z.; Shi, X.; Wang, H.; Yeung, D.; Woo, W.; Wong, W. Deep Learning and the Weather Forecasting Problem: Precipitation Nowcasting. In *Deep Learning for the Earth Sciences: A Comprehensive Approach to Remote Sensing, Climate Science, and Geosciences*; John Wiley & Sons, Ltd.: Hoboken, NJ, USA, 2021; pp. 218–239. [CrossRef]
9. Simonyan, K.; Zisserman, A. Very Deep Convolutional Networks for Large-Scale Image Recognition. *arXiv* **2014**, arXiv:1409.1556.
10. Zhao, Z.-Q.; Zheng, P.; Xu, S.-T.; Wu, X. Object Detection With Deep Learning: A Review. *IEEE Trans. Neural Netw. Learn. Syst.* **2019**, *30*, 3212–3232. [CrossRef]
11. Le, X.H.; Lee, G.; Jung, K.; An, H.-U.; Lee, S.; Jung, Y. Application of Convolutional Neural Network for Spatiotemporal Bias Correction of Daily Satellite-Based Precipitation. *Remote Sens.* **2020**, *12*, 2731. [CrossRef]
12. Ayzel, G.; Scheffer, T.; Heistermann, M. RainNet v1.0: A convolutional neural network for radar-based precipitation nowcasting. *Geosci. Model Dev.* **2020**, *13*, 2631–2644. [CrossRef]
13. Khiali, L.; Ienco, D.; Teisseire, M. Object-oriented satellite image time series analysis using a graph-based representation. *Ecol. Inform.* **2018**, *43*, 52–64. [CrossRef]
14. Fahim, S.R.; Sarker, Y.; Sarker, S.K.; Sheikh, R.I.; Das, S.K. Self attention convolutional neural network with time series imaging based feature extraction for transmission line fault detection and classification. *Electr. Power Syst. Res.* **2020**, *187*, 106437. [CrossRef]
15. Li, X.; Kang, Y.; Li, F. Forecasting with time series imaging. *Expert Syst. Appl.* **2020**, *160*, 113680. [CrossRef]
16. Ravuri, S.; Lenc, K.; Willson, M.; Kangin, D.; Lam, R.; Mirowski, P.; Fitzsimons, M.; Athanassiadou, M.; Kashem, S.; Madge, S.; et al. Skilful precipitation nowcasting using deep generative models of radar. *Nature* **2021**, *597*, 672–677. [CrossRef] [PubMed]
17. Li, D.; Liu, Y.; Chen, C. MSDM v1.0: A machine learning model for precipitation nowcasting over eastern China using multisource data. *Geosci. Model Dev.* **2021**, *14*, 4019–4034. [CrossRef]
18. Chen, L.; Cao, Y.; Ma, L.; Zhang, J. A Deep Learning-Based Methodology for Precipitation Nowcasting with Radar. *Earth Space Sci.* **2020**, *7*, e2019EA000812. [CrossRef]
19. Agrawal, S.; Barrington, L.; Bromberg, C.; Burge, J.; Gazen, C.; Hickey, J. Machine Learning for Precipitation Nowcasting from Radar Images. *arXiv* **2019**, arXiv:1912.12132.
20. Fernández, J.G.; Mehrkanoon, S. Broad-UNet: Multi-scale feature learning for nowcasting tasks. *Neural Netw.* **2021**, *144*, 419–427. [CrossRef]
21. Ionescu, V.-S.; Czibula, G.; Mihuleţ, E. DeePS at: A deep learning model for prediction of satellite images for nowcasting purposes. *Procedia Comput. Sci.* **2021**, *192*, 622–631. [CrossRef]
22. Zhang, L.; Huang, Z.; Liu, W.; Guo, Z.; Zhang, Z. Weather radar echo prediction method based on convolution neural network and Long Short-Term memory networks for sustainable e-agriculture. *J. Clean. Prod.* **2021**, *298*, 126776. [CrossRef]
23. Trebing, K.; Stańczyk, T.; Mehrkanoon, S. SmaAt-UNet: Precipitation nowcasting using a small attention-UNet architecture. *Pattern Recognit. Lett.* **2021**, *145*, 178–186. [CrossRef]
24. Le, X.H.; Ho, H.V.; Lee, G.; Jung, S. Application of Long Short-Term Memory (LSTM) Neural Network for Flood Forecasting. *Water* **2019**, *11*, 1387. [CrossRef]
25. Le, X.H.; Nguyen, D.H.; Jung, S.; Yeon, M.; Lee, G. Comparison of Deep Learning Techniques for River Streamflow Forecasting. *IEEE Access* **2021**, *9*, 71805–71820. [CrossRef]
26. Kreklow, J.; Tetzlaff, B.; Burkhard, B.; Kuhnt, G. Radar-Based Precipitation Climatology in Germany—Developments, Uncertainties and Potentials. *Atmosphere* **2020**, *11*, 217. [CrossRef]
27. Otsuka, S.; Kotsuki, S.; Ohhigashi, M.; Miyoshi, T. GSMaP RIKEN Nowcast: Global Precipitation Nowcasting with Data Assimilation. *J. Meteorol. Soc. Jpn.* **2019**, *97*, 1099–1117. [CrossRef]
28. Wang, Y.; Wu, H.; Zhang, J.; Gao, Z.; Wang, J.; Yu, P.; Long, M. PredRNN: A Recurrent Neural Network for Spatiotemporal Predictive Learning. *arXiv* **2021**, arXiv:2103.09504.
29. Ronneberger, O.; Fischer, P.; Brox, T. U-Net: Convolutional Networks for Biomedical Image Segmentation. *arXiv* **2015**, arXiv:1505.04597.
30. Cho, K.; van Merriënboer, B.; Gulcehre, C.; Bahdanau, D.; Bougares, F.; Schwenk, H.; Bengio, Y. Learning phrase representations using RNN encoder-decoder for statistical machine translation. *arXiv* **2014**, arXiv:1406.1078.

31. Shi, X.; Chen, Z.; Wang, H.; Yeung, D.-Y.; Wong, W.-K.; Woo, W.-C. Convolutional LSTM Network: A Machine Learning Approach for Precipitation Nowcasting. *arXiv* **2015**, arXiv:1506.04214.
32. Wang, Y.; Long, M.; Wang, J.; Gao, Z.; Yu, P. PredRNN: Recurrent Neural Networks for Predictive Learning using Spatio-temporal LSTMs. In *Advances in Neural Information Processing Systems*; Guyon, I., Luxburg, U.V., Bengio, S., Wallach, H., Fergus, R., Vishwanathan, S., Garnett, R., Eds.; Curran Associates, Inc.: Red Hook, NY, USA, 2017; Volume 30.
33. Maaten, L.; Hinton, G. Visualizing Data using t-SNE. *J. Mach. Learn. Res.* **2008**, *9*, 2579–2605.
34. Berrar, D. Cross-Validation. In *Encyclopedia of Bioinformatics and Computational Biology*; Elsevier: Amsterdam, The Netherlands, 2019.
35. Chai, T.; Draxler, R.R. Root mean square error (RMSE) or mean absolute error (MAE)?—Arguments against avoiding RMSE in the literature. *Geosci. Model Dev.* **2014**, *7*, 1247–1250. [CrossRef]
36. Wang, Z.; Bovik, A.; Sheikh, H.; Simoncelli, E. Image quality assessment: From error visibility to structural similarity. *IEEE Trans. Image Process* **2004**, *13*, 600–612. [CrossRef]
37. Schaefer, J. The Critical Success Index as an Indicator of Warning Skill. *Weather. Forecast.* **1990**, *5*, 570–575. [CrossRef]
38. Townsend, J. Theoretical analysis of an alphabetic confusion matrix. *Percept. Psychophys.* **1971**, *9*, 40–50. [CrossRef]

Article

Cubical Homology-Based Machine Learning: An Application in Image Classification

Seungho Choe [†] and Sheela Ramanna *

Department of Applied Computer Science, University of Winnipeg, Winnipeg, MB R3B 2E9, Canada; choe-s@webmail.uwinnipeg.ca
* Correspondence: s.ramanna@uwinnipeg.ca
† This study is part of Seungho Choe's MSc Thesis of Cubical homology-based Image Classification—A Comparative Study, defended at the University of Winnipeg in 2021.

Abstract: Persistent homology is a powerful tool in topological data analysis (TDA) to compute, study, and encode efficiently multi-scale topological features and is being increasingly used in digital image classification. The topological features represent a number of connected components, cycles, and voids that describe the shape of data. Persistent homology extracts the birth and death of these topological features through a filtration process. The lifespan of these features can be represented using persistent diagrams (topological signatures). Cubical homology is a more efficient method for extracting topological features from a 2D image and uses a collection of cubes to compute the homology, which fits the digital image structure of grids. In this research, we propose a cubical homology-based algorithm for extracting topological features from 2D images to generate their topological signatures. Additionally, we propose a novel score measure, which measures the significance of each of the sub-simplices in terms of persistence. In addition, gray-level co-occurrence matrix (GLCM) and contrast limited adapting histogram equalization (CLAHE) are used as supplementary methods for extracting features. Supervised machine learning models are trained on selected image datasets to study the efficacy of the extracted topological features. Among the eight tested models with six published image datasets of varying pixel sizes, classes, and distributions, our experiments demonstrate that cubical homology-based machine learning with the deep residual network (ResNet 1D) and Light Gradient Boosting Machine (lightGBM) shows promise with the extracted topological features.

Keywords: cubical complex; cubical homology; image classification; deep learning; persistent homology

Citation: Choe, S.; Ramanna, S. Cubical Homology-Based Machine Learning: An Application in Image Classification. *Axioms* **2022**, *11*, 112. https://doi.org/10.3390/axioms11030112

Academic Editor: Oscar Humberto Montiel Ross

Received: 10 January 2022
Accepted: 24 February 2022
Published: 3 March 2022

Publisher's Note: MDPI stays neutral with regard to jurisdictional claims in published maps and institutional affiliations.

1. Introduction

The origin of topological data analysis (TDA) and persistent homology can be traced back to H. Edelsbrunner, D. Letscher, and A. Zomorodian [1]. More recently, TDA has emerged as a growing field in applied algebraic topology to infer relevant features for complex data [2]. One of the fundamental methods in computational topology is persistent homology [3,4], which is a powerful tool to compute, study, and encode efficiently multi-scale topological features of nested families of simplicial complexes and topological spaces [5]. Simplices are building blocks used to study the shape of data and a simplicial complex is its higher-level counterpart. The process of shape construction is commonly referred to as filtration [6]. There are many forms of filtrations and a good survey is presented in [7]. Persistent homology extracts the birth and death of topological features throughout a filtration built from a dataset [8]. In other words, persistent homology is a concise summary representation of topological features in data and is represented in a persistent diagram or barcode. This is important since it tracks changes and makes it possible to analyze data at multiple scales since the data structure associated with topological features is a multi-set, which makes learning harder. Persistent diagrams are then mapped into metric spaces with additional structures useful for machine learning tasks [9]. The application of TDA in machine learning (also known as TDA pipeline) in several fields is well documented [2]. The TDA pipeline consists of using data (e.g., images, signals) as input and then filtration operations are applied to obtain persistence diagrams. Subsequently, ML classifiers such as support vector machines, tree classifiers, and neural networks are applied to the persistent diagrams.

The TDA pipeline is an emerging research area to discover descriptors useful for image and graph classification learning and in science—for example, quantifying differences between force networks [10] and analyzing polyatomic structures [11]. In [12], microvascular patterns in endoscopy images can be categorized as *regular* and *irregular*. Furthermore, there are three types of regular surface of the microvasculature: *oval*, *tubular*, and *villous*. In this paper, topological features were derived with persistence diagrams and the *q-th norm* of the *p*-th diagram is computed as

$$N_q = \left[\sum_{A \in \mathrm{Dgm}_p(f)} \mathrm{pers}(A)^q \right]^{\frac{1}{q}},$$

where $\mathrm{Dgm}_p(f)$ denotes the *p*-th diagram of f and $\mathrm{pers}(A)$ is the persistence of a point A in $\mathrm{Dgm}_p(f)$. Since N_q is a norm of *p*-th Betti number with restriction (or threshold) s, it will obtain the *p*-th Betti number of $\mathbb{M}_s$, where $\mathbb{M}$ is the rectangle covered by pixels. Then, $\mathbb{M}$ is mapped to $\mathbb{R}$ by a signed distance function. A naive Bayesian learning method that combines the results of several Adaboost classifiers is then used to classify the images. The authors in [13] introduce a multi-scale kernel for persistence diagrams that is based on scale space theory [14]. The focus is on the stability of persistent homology since any occurrence of small changes in the input affects both the 1-Wasserstein distance and persistent diagrams. Experiments on two benchmark datasets for 3D shape classification/retrieval and texture recognition are discussed.

Vector summaries of persistence diagrams is a technique that transforms a persistence diagram into vectors and summarizes a function by its minimum through a pooling technique. The authors in [15] present a novel pooling within the bag-of-words approach that shows a significant improvement in shape classification and recognition problems with the Non-Rigid 3D Human Models SHREC 2014 dataset.

The topological and geometric structures underlying data are often represented as point clouds. In [16], the RGB intensity values of each pixel of an image are mapped to the point cloud $P \in \mathbb{R}^5$ and then a feature vector is derived. Computing and arranging the persistence of point cloud data by descending order makes it possible to understand the persistence of features. The extracted topological features and the traditional image processing features are used in both vector-based supervised classification and deep network-based classification experiments on the CIFAR-10 image dataset. More recently, multi-class classification of point cloud datasets was discussed in [17]. In [8], a random forest classifier was used to classify the well-known MNIST image dataset using the voxel structure to obtain topological features.

In [18], persistent diagrams were used with neural network classifiers in graph classification problems. In TDA, Betti numbers represent counts of the number of homology groups, such as points, cycles, and so on. In [19], the similarity of the brain networks of twins is measured using Betti numbers. In [20,21], persistent barcodes were used to visualize brain activation patterns in resting-state functional magnetic resonance imaging (rs-fMRI) video frames. The authors used a geometric Betti number that counts the total number of connected cycles forming a vortex (nested, usually non-concentric, connected cycles) derived from the triangulation of brain activation regions.

The success of deep learning [22] in computer vision problems has led to its use in deep networks that can handle barcodes [23]. Hofer et al. used a persistence diagram as a topological signature and computed a parametrized projection from the persistence diagram, and then leveraged it during the training of the network. The output of this process is stable when using the 1-Wasserstein distance. Classification of 2D object shapes and social network graphs was successfully demonstrated by the authors. In [24], the authors apply topological data analysis to the classification of time series data. A 1D convolutional neural network is used, where the input data are a *Betti sequence*. Since machine learning models rely on accurate feature representations, multi-scale representations of features are becoming increasingly important in applications involving computer vision and image analysis. Persistence homology is able to bridge the gap between geometry and topology, and persistent homology-based machine learning models have been used in various areas, including image classification and analysis [25].

However, it has been shown that the implementations of persistent homology (of simplicial complexes) are inefficient for computer vision since it requires excessive computational resources [26] due to the formulations based on triangulations. To mitigate the problem of complexity, cubical homology was introduced, which allows the direct application of its structure [27,28]. Simply, cubical homology uses a collection of cubes to compute the homology, which fits the digital image structure of grids. Since there is neither skeletonization nor triangulation in the computation of cubical homology, it has advantages in the fast segmentation of images for extracting features. This feature of cubical homology is the motivation for this work in exploring the extraction of topological features from 2D images using this method.

The focus in this work is twofold: (i) on the extraction of topological features from 2D images with varying pixel sizes, classes, and distributions using cubical homology; (ii) to study the effect of extracted 1D topological features in a supervised learning context using well-known machine learning models trained on selected image datasets. The work presented in this paper is based on the thesis by Choe [29]. Figure 1 illustrates our proposed approach. Steps 2.1 and 2.2 form the core of this paper, namely the generation of 1D topological signatures using a novel *score* that is proposed by this study. This score allows us to filter out low persistence features (or noise). Our contribution is as follows: (i) we propose a cubical homology-based algorithm for extracting topological features from 2D images to generate their topological signatures; (ii) we propose a score, which is used as a measure of the significance of the subcomplex calculated from the persistence diagram. Additionally, we use gray-level co-occurrence matrix (GLCM) and contrast limited adapting histogram equalization (CLAHE) for obtaining additional image features, in an effort to improve the classification performance, and (iii) we discuss the results of our supervised learning experiments of eight well-known machine learning models trained on six different published image datasets using the extracted topological features.

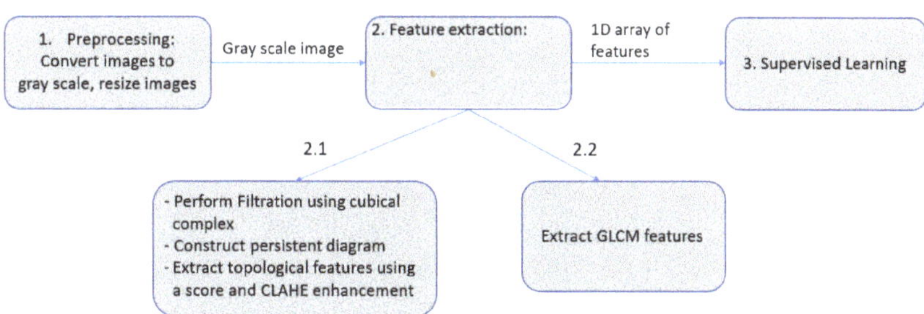

Figure 1. Classification pipeline.

The paper is organized as follows. Section 2 gives basic definitions for simplicial, cubical, and persistent homology used in this work. Section 3.2 illustrates the feature engineering process, including the extraction of topological and other subsidiary features. Section 3.1 introduces the benchmark image datasets used in this work. Section 4 gives the results and analysis of using eight machine learning models trained on each dataset. Finally, Section 5 gives the conclusions of this study.

2. Basic Definitions

We recall some basic definitions of the concepts used in this work. A simplicial complex is a space or an object that is built from a union of points, edges, triangles, tetrahedra, and higher-dimensional polytopes. Homology theory is in the domain of algebraic topology related to the connectivity in multi-dimensional shapes [26].

2.1. Simplicial Homology

Graphs are mathematical structures used to study pairwise relationships between objects and entities.

Definition 1. *A graph is a pair of sets, $G = (V, E)$, where V is the set of vertices (or nodes) and E is a set of edges.*

Let S be a subset of a group G. Then, the subgroup generated by S, denoted $\langle S \rangle$, is the subgroup of all elements of G that can be expressed as the finite operation of elements in S and their inverses. For example, the set of all integers, $\mathbb{Z}$, can be expressed by the operation of elements $\{1\}$ so $\mathbb{Z}$ is the subgroup generated by $\{1\}$.

Definition 2. *A rank of a group G is the size of the smallest subset that generates G.*

For instance, since $\mathbb{Z}$ is the subgroup generated by $\{1\}$, rank($\mathbb{Z}$)=1.

Definition 3. *A simplex complex on a set V is a family of arbitrary cardinality subsets of V closed under the subset operation, which means that, if a set S is in the family, all subsets of S are also in the family. An element of the family is called a simplex or face.*

Definition 4. *Moreover, $p - simplex$ can be defined to the convex hull of $p + 1$ affinely independent points $x_0, x_1, \cdots, x_p \in \mathbb{R}^d$.*

For example, in a graph, 0-simplex is a point, 1-simplex is an edge, 2-simplex is a triangle, 3-simplex is a tetrahedron, and so on (see Figure 2).

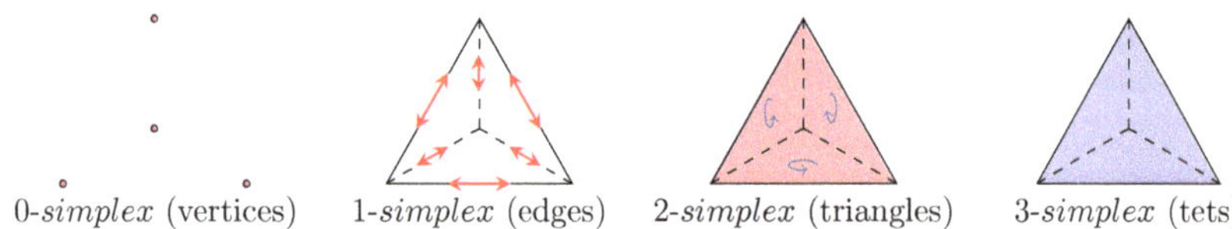

Figure 2. Examples of p-simplex for $p = 0, 1, 2, 3$ in tetrahedron. A 0-simplex is a point, a 1-simplex is an edge with a convex hull of two points, a 2-simplex is a triangle with a convex hull of three distinct points, and a 3-simplex is a tetrahedron with a convex hull of four points [30].

Chain, Boundary, and Cycle

To extend simplicial homology to persistent homology, the notion of *chain, boundary,* and *cycle* is necessary [31].

Definition 5. *A p-chain is a subset of p-simplices in a simplicial complex K. Assume that K is a triangle. Then, a 1-chain is a subset of 1-simplices—in other words, a subset of the three edges.*

Definition 6. *A boundary, generally denoted ∂, of a p-simplex is the set of $(p - 1)$-simplices' faces.*

For example, a triangle is a 2-simplex, so the boundary of a triangle is a set of 1-simplices which are the edges. Therefore, the boundary of the triangle is the three edges.

Definition 7. *A cycle can be defined using the definitions of chain and boundary. A p-cycle c is a p-chain with an empty boundary. More simply, it is a path where the starting point and destination point are the same.*

2.2. Cubical Homology

Cubical homology [27] is efficient since it allows the direct use of the cubical structure of the image, whereas simplicial theory requires increasing the complexity of data. While the simplicial homology is built with the triangle and its higher-dimensional structure, such as a tetrahedron, cubical homology consists of *cubes*. In cubical homology, each cube has a unit size and the n-cube represents its dimension. For example, 0-cubes are points, 1-cubes are lines with unit length, 2-cubes are unit squares, and so on [27,32,33].

Definition 8. *Here, 0-cubes can be defined as an interval,*

$$[m] = [m, \ m], \ m \in \mathbb{Z}, \tag{1}$$

which generate subsets $I \in \mathbb{R}$, such that

$$I = [m, \ m + 1], \ m \in \mathbb{Z}. \tag{2}$$

Therefore, I is called a 1-cube, or *elementary interval*.

Definition 9. *An n-cube can be expressed as a product of elementary intervals as*

$$Q = I_1 \times I_2 \times \cdots \times I_n \subseteq \mathbb{R}^n, \tag{3}$$

where Q indicates that n-cube $I_i (i = 1, \ 2, \ \cdots, \ n)$ is an elementary interval.

A d-dimensional image is a map $\mathcal{I} : I \subseteq \mathbb{Z}^d \to \mathbb{R}$.

Definition 10. *A pixel can be defined as an element $v \in I$, where $d = 2$. If $d > 2$, v is called a voxel.*

Definition 11. *Let $\mathcal{I}(v)$ be the intensity or grayscale value. Moreover, in the case of binary images, we consider a map $\mathcal{B} : I \subseteq \mathbb{Z}^d \to \{0, \ 1\}$.*

A voxel is represented by a d-cube and, with all of its faces added, we have

$$\mathcal{I}'(\sigma) := \min_{\sigma \text{ face of } \tau} \mathcal{I}(\tau). \tag{4}$$

Let K be the cubical complex built from the image I, and let

$$K_i := \{\sigma \in K | \mathcal{I}'(\sigma) \le i\}, \tag{5}$$

be the i-th *sublevel* set of K. Then, the set $\{K_i\}_{i \in \mathrm{Im}(I)}$ defines a filtration of the cubical complexes. Thus, the pipeline to filtration from an image with a cubical complex is as follows:

$$\text{Image} \to \text{Cubical complex} \to \text{Sublevel sets} \to \text{Filtration}$$

Moreover, *chain, boundary,* and *cycle* in cubical homology can be defined in the same manner as in Section 2.1.

2.3. Persistent Homology

In topology, there are subcomplices of complex K and *cubes* are created (*birth*) and destroyed (*death*) by filtration. Assume that K^i ($0 \le i \le$, $i \in \mathbb{Z}$) is a subcomplex of filtered complex K such that

$$\varnothing \subseteq K^0 \subseteq K^1 \subseteq \cdots \subseteq K^n = K,$$

and $\mathcal{Z}_k^i$, $\mathcal{B}_k^i$ are its corresponding cycle group and boundary group.

Definition 12. *The kth homology group [1] can be defined as*

$$\mathcal{H}_k = \mathcal{Z}_k / \mathcal{B}_k. \tag{6}$$

Definition 13. *The p-persistent kth homology group of K^i [1] can be defined as*

$$\mathcal{H}_k^{i,p} = \mathcal{Z}_k^i / \mathcal{B}_k^{i+p} \cap \mathcal{Z}_k^i. \tag{7}$$

Definition 14. *A persistence is a lifetime of these attributes based on the filtration method used [1].*

One can plot the birth and death times of the topological features as a barcode, also known as a *persistence barcode*, shown in Figure 3. This diagram graphically represents the topological signature of the data. Illustration of persistence is useful when detecting a change in terms of topology and geometry, which plays a crucial role in supervised machine learning [34].

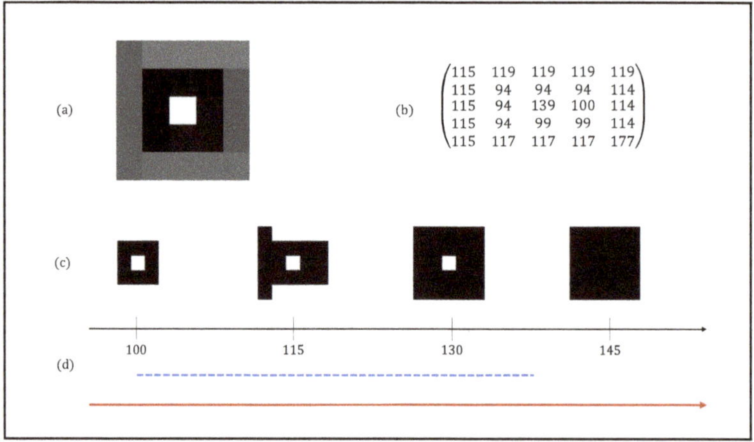

Figure 3. An example of persistent homology for grayscale image. (**a**) A given image, (**b**) a matrix of gray level of given image, (**c**) the filtered cubical complex of the image, (**d**) the persistence barcode according to (**c**).

3. Materials and Methods

3.1. Image Datasets

In this section, we give a brief description of the six published image datasets used in this work. Datasets used for benchmarking were collected from various sources that include *Mendeley Data* (https://data.mendeley.com/ accessed on 7 January 2022), *Tensorflow dataset* (https://www.tensorflow.org/datasets accessed on 7 January 2022), and *Kaggle competition* (https://www.kaggle.com/competitions accessed on 7 January 2022). The *concrete crack images* dataset [35] contains a total of 40,000 images, where each image consists of 227×227 pixels. These images were collected from the METU campus building and consist of two classes: 20,000 images where there are no cracks in the concrete (positive) and 20,000 images of concrete that is cracked (negative). A crack on an outer wall occurs with the passage of time or due to natural aging. It is important to detect these cracks in terms of evaluating and predicting the structural deterioration and reliability of buildings. Samples of the two types of images are shown in Figure 4.

The *APTOS blindness detection* dataset [36] is a set of retina images taken by fundus photography for detecting and preventing diabetic retinopathy from causing blindness (https://www.kaggle.com/c/aptos2019-blindness-detection/overview accessed on 7 January 2022). This dataset has 3662 images and consists of 1805 images diagnosed as non-diabetic (labeled as 0) retinopathy and 1857 images diagnosed as diabetic retinopathy, as shown in Figure 5. Figure 6 shows the distribution of examples in the four classes using a severity range from 1 to 4 with the following interpretation: 1: Mild, 2: Moderate, 3: Severe, 4: Proliferative DR.

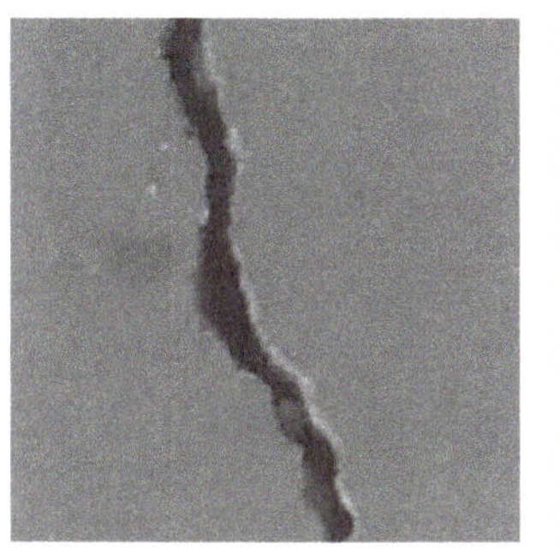
(**a**) Negative crack image

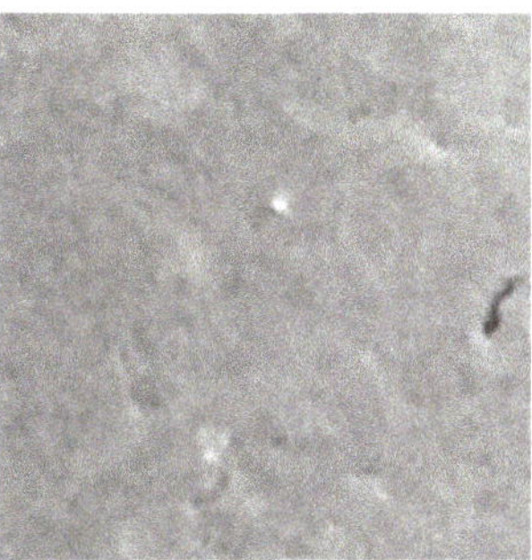
(**b**) Positive crack image

Figure 4. Sample images of the concrete crack dataset.

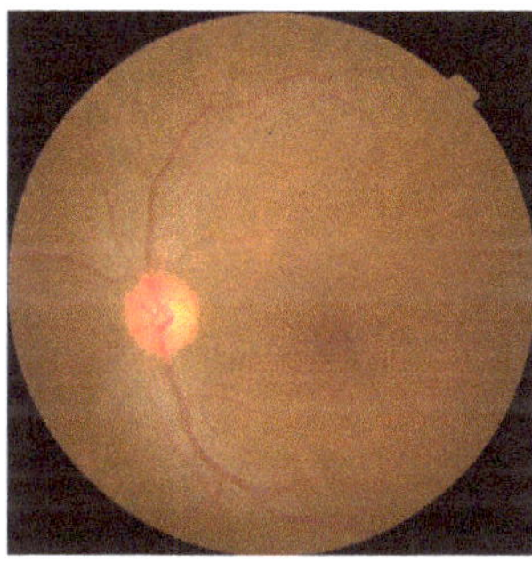
(**a**) Non-diabetic retinopathy

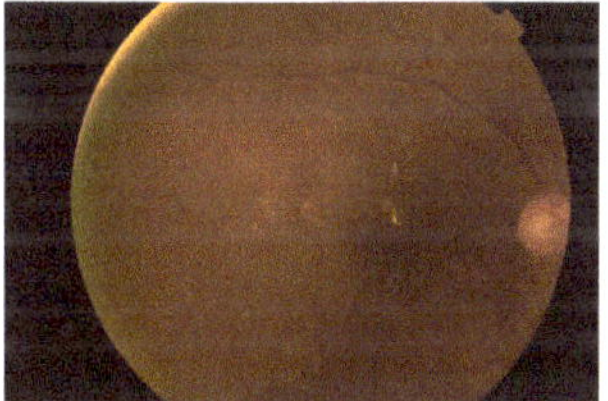
(**b**) Diabetic retinopathy

Figure 5. Sample images of APTOS dataset. (**a**) is a picture of Non-diabetic retionpathy which is ordinary case and (**b**) is a picture of diabetic retinopathy which cas cause blindness.

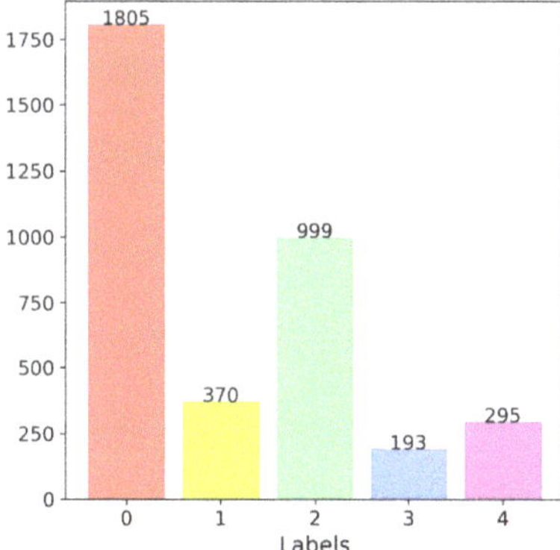

Figure 6. Data distribution for APTOS 2019 blindness detection dataset. This dataset can be classified as non-diabetic and diabetic. Around 50% of images are categorized as non-diabetic retinopathy (label 0) and diabetic retinopathy is subdivided according to the severity range from 1 to 4.

The *pest classification in mango farms* dataset [37] is a collection of 46,500 images of mango leaves affected by 15 different types of pests and one normal (unaffected) mango leaf, as shown in Figure 7. Some of these pests can be detected visually. Figure 8 shows the data distribution of examples in the 15 classes of pests and one normal class.

Figure 7. Sample images of pest classification in mango farms.

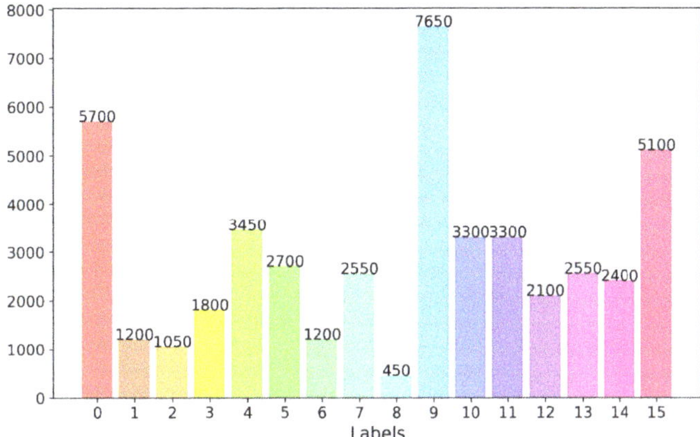

Figure 8. Data distribution for pest classification in mango farms dataset. Labels are assigned in alphabetical order.

The *Indian fruits* dataset [38] contains 23,848 images that cover five popular fruits in India: apple, orange, mango, pomegranate, and tomato. This dataset includes variations of each fruit, resulting in 40 classes. This dataset was already separated into training and testing sets by the original publishers of the dataset, as shown in Figure 9. Note that this dataset has an imbalanced class distribution.

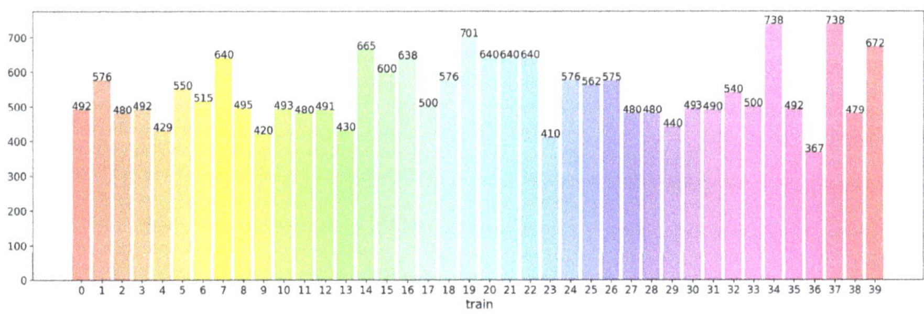

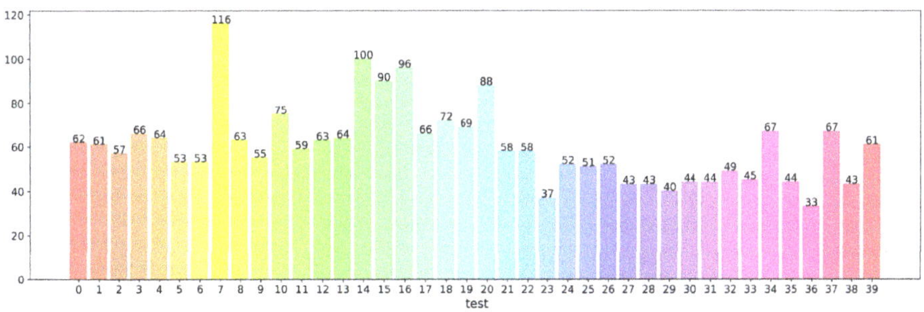

Figure 9. Data distribution for the Indian fruits dataset. These data are already split by the original publishers of the dataset by the ratio of 9 to 1. Labels are assigned in alphabetical order.

The *colorectal histology* dataset [39] contains 5000 histological images of different tissue types of colorectal cancer. It consists of 8 classes of tissue types with 625 images for each class, as shown in Figure 10.

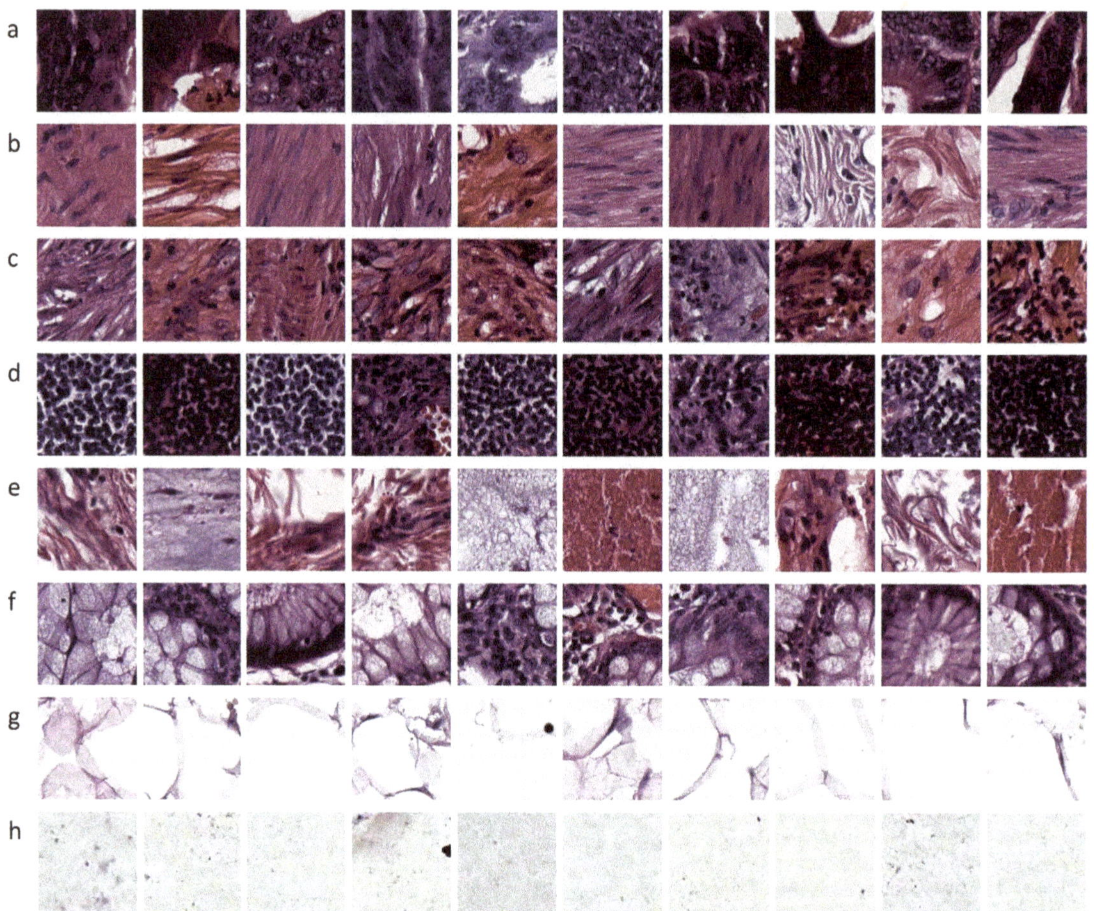

Figure 10. Example of colorectal cancer histology. (**a**) Tumor epithelium, (**b**) simple stroma, (**c**) complex stroma, (**d**) immune cell conglomerates, (**e**) debris and mucus, (**f**) mucosal glands, (**g**) adipose tissue, (**h**) background.

The *Fashion MNIST* dataset [40] is a collection of 60,000 training images of fashion products, as shown in Figure 11. It consists of 28×28 grayscale images of products from 10 classes. Since the dataset contains an equal number of images for each class, there are 6000 test images in each class, resulting in a balanced dataset.

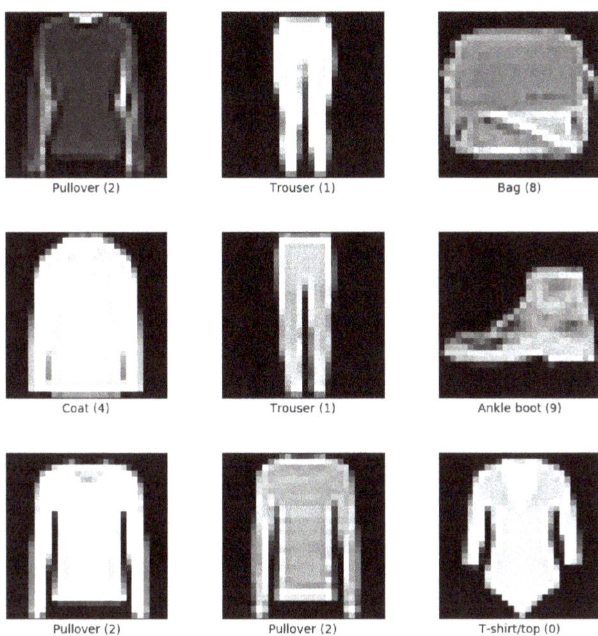

Figure 11. Example of the Fashion MNIST dataset.

Table 1 gives the dataset characteristics in terms of the various image datasets used in this work. Moreover, we provide the preprocessing time per image. For example, the feature extraction time for the concrete dataset was 5 h 12 min.

Table 1. Dataset details with preprocessing times.

Dataset	Size	Num of Classes	Pixel Dimension	Balanced	Time in Sec/Image
Concrete [1]	40,000	2	227×227	Yes	0.4713
Mangopest [2]	46,000	16	from 500×333 to 1280×853	No	0.5394
Indian fruits [3]	23,848	40	100×100	No	0.4422
Fashion MNIST [4]	60,000	10	28×28	Yes	0.4297
APTOS [5]	3662	5	227×227	No	0.5393
Colorectal histology [6]	5000	8	150×150	Yes	0.3218

[1] Çağlar Fırat Özgenel, 23 July 2019, https://data.mendeley.com/datasets/5y9wdsg2zt/2; [2] Kusrini Kusrini et al., accessed on 27 February 2020, https://data.mendeley.com/datasets/94jf97jzc8/1; [3] prabira Kumar sethy, accessed on 12 June 2020, https://data.mendeley.com/datasets/bg3js4z2xt/1; [4] Han Xiao et al., accessed on 28 August 2017, https://github.com/zalandoresearch/fashion-mnist; [5] Asia Pacific Tele-Ophthalmology Society, accessed on 27 June 2019, https://www.kaggle.com/c/aptos2019-blindness-detection/overview; [6] Kather, Jakob Nikolas et al., 26 May 2016, https://zenodo.org/record/53169#.XGZemKwzbmG.

3.2. Methods—Feature Engineering

In this section, we describe the feature engineering process. The main purpose of this process is to obtain a 1-dimensional array from each image in the dataset. Each point from the persistence diagram plays a significant role in the extraction of the topological features. Moreover, the *gray-level co-occurrence matrix* (GLCM) supports these topological features as additional signatures. Because every image dataset is not identical in size and some images have very high resolution, resizing every image to 200×200 and converting them to grayscale guarantees a relatively constant duration of extraction (approximately 4 s) regardless of its original size.

Algorithm 1 gives the method for extracting topological features from a dataset. In this algorithm, β_0 and β_1 are Betti numbers derived from Equation (6), where the dimension of ith homology is called the ith Betti number of K. β_0 gives the number of connected components and β_1 gives the number of holes. Betti numbers represent the count of the number of topological features.

Algorithm 1 Extraction of Topological Features.

Input: $N \leftarrow$ number of dataset
    for $i = 1, 2, \cdots, N$ **do**
        $img \leftarrow$ load ith image from dataset
        $img \leftarrow$ resize img to (200, 200) and convert to grayscale
        $PD_0 \leftarrow$ set of points of β_0 in persistence diagram of img with cubical complex
        $PD_1 \leftarrow$ set of points of β_1 in persistence diagram of img with cubical complex
        $PD_0 \leftarrow$ sort PD_0 in descending order of *persistence*
        $PD_1 \leftarrow$ sort PD_1 in descending order of *persistence*
        $d_i \leftarrow$ project each point in PD_0 to $[0, 1]$
        $d_i \leftarrow d_i +$ project each point in PD_1 to $[1, 2]$
        $fimg \leftarrow$ adapt CLAHE filter to img
        $fPD_0 \leftarrow$ set of points of β_0 in persistence diagram of $fimg$ with cubical complex
        $fPD_1 \leftarrow$ set of points of β_1 in persistence diagram of $fimg$ with cubical complex
        $fPD_0 \leftarrow$ sort fPD_0 in descending order of *persistence*
        $fPD_1 \leftarrow$ sort fPD_1 in descending order of *persistence*
        $d_i \leftarrow d_i +$ project each point in fPD_0 to $[0, 1]$
        $d_i \leftarrow d_i +$ project each point in fPD_1 to $[1, 2]$
        $d_i \leftarrow d_i +$ convert img to GLCM with distances (1, 2, 3), directions ($0°, 45°, 90°, 135°$),
    and properties (*energy, homogeneity*)
Output: $D(d_1, d_2, \cdots, d_N)$

3.3. Projection of Persistence Diagrams

The construction of a persistence diagram is possible once the filtration (using cubical complex) is completed. The dth persistence diagram, $\mathcal{D}_d$, contains all of the d-dimensional topological information. These are a series of points with a pair of (*birth, death*), where *birth* indicates the time at which the topological features were created and the *death* gives the time at which these features are destroyed. From here, *persistence* is defined using the definition of *birth* and *death* as

$$pers(birth, death) := death - birth, \ where \ (birth, death) \in \mathcal{D}_d. \tag{8}$$

Low-persistence features are treated as having low importance, or 'noise', whereas high-persistence features are regarded as *true* features [1]. However, using *persistence* as a result of a projection of a topological feature to a 1-dimensional value is not helpful, because it is impossible to distinguish the features which have the same *persistence* but different values for *birth*. Therefore, we propose a measure (*score*) to compensate for this limitation of *persistence*, shown in Equation (9).

$$score_d(birth, death) := \begin{cases} 0 & if\ persistence < threshold \\ d + \left(\dfrac{e^{\sin\left(\frac{death}{255} \cdot \frac{\pi}{2}\right)} - 1}{e - 1} \right)^3 - \left(\dfrac{e^{\sin\left(\frac{birth}{255} \cdot \frac{\pi}{2}\right)} - 1}{e - 1} \right)^3 & if\ persistence \geq threshold \end{cases} \tag{9}$$

Since the sinusoidal term is increasing and has the value $[0, 1]$ when the input is $[0, \frac{\pi}{2}]$, the $score_d$ has a value range from d to $d + 1$. Hence, it is easy to distinguish the dimension and persistence of each feature. Moreover, a higher exponent emphasizes a feature that has longer persistence and, significantly, ignores a feature that has shorter persistence. This in is keeping with ideas underlying homology groups, where longer the persistence, the higher the significance of the homology. Conversely, the homology group that has short persistence is considered noise, which degrades the quality of the digital image and, as such, is less significant as a topological feature [10]. By ignoring such noise, using a threshold (as a parameter) allows us to separate useful features from noise. The optimal threshold (value = 10) was determined experimentally by comparing the performance of machine learning models. In summary, the *score* takes into account not only the *persistence*, but also other aspects such as the *dimension, birth, and death* of topological features.

3.4. Contrast Limited Adapting Histogram Equalization (CLAHE)

When pixel values are concentrated in a narrow range, it is hard to perceive features visually. Histogram equalization makes the distribution of pixel values in the image balanced, thereby enhancing the image. However, this method often results in degrading the content of the image and also amplifying the noise. Therefore, it produces undesirable results. Contrast limited adapting histogram equalization (CLAHE) is a well-known method for compensating for the weakness of histogram equalization by dividing an image into small-sized blocks and performing histogram equalization for each block [41]. After completing histogram equalization in all blocks, bilinear interpolation makes the boundary of the tiles (blocks) smooth. In this paper, we used the following hyperparameters: `clipLimit=7` and `tileGridSize=((8, 8))`. An illustration of the CLAHE method on the APTOS data is given in Figure 12.

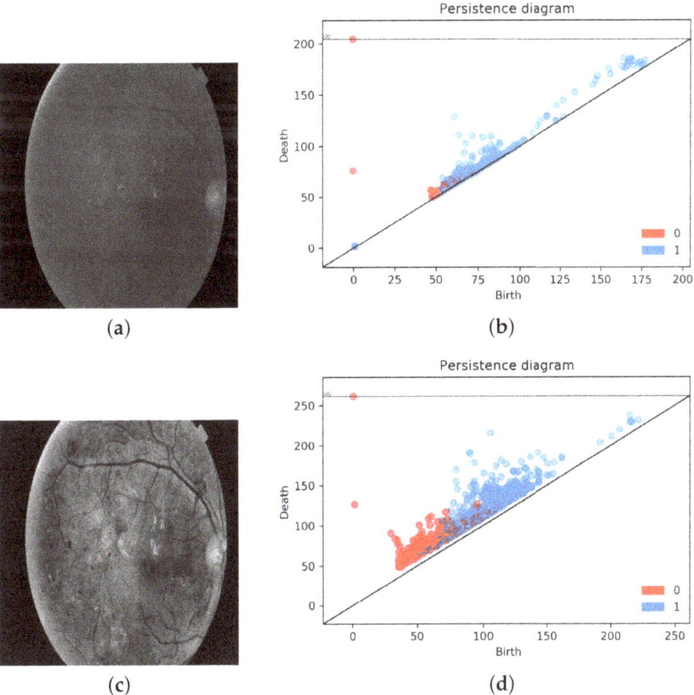

Figure 12. Comparison of the original image and the CLAHE-filtered image. (**a**) Original image, (**b**) persistence diagram of the original image (**a**), (**c**) CLAHE-filtered image, (**d**) persistence diagram of the filtered image (**c**).

The texture of an image can be described by its statistical properties and this information is useful to classify images [42]. For extracting texture features, we used the well-known gray-level co-occurrence matrix (GLCM) [43]. GLCM extracts texture information regarding the structural arrangement of surfaces by using a displacement vector defined by its radius and orientation. We used three distances $(1, 2, 3)$ and four directions $(0°, 45°, 90°, 135°)$ to obtain the GLCM features. From each of the co-occurrence matrices, two global statistics were extracted, energy and homogeneity, resulting in $3 \times 4 \times 2 = 24$ texture features for each image.

Table 2 gives a list of extracted features from the APTOS dataset during the filtration process. In total, 144 features were extracted for each dimension ($fdim_0$ and $fdim_1$) from the CLAHE-filtered image in descending order of persistence. Similarly, 100 (dim_0 and dim_1) topological features for each dimension were extracted. Note that dim_0 represents β_0 and dim_1 represents β_1 Betti numbers, respectively. A total of 24 GLCM features were extracted from the original gray-level image.

In this paper, feature engineering and learning algorithms were implemented with the following Python libraries: Gudhi [44,45] for calculating persistent homology, PyTorch [46] for modeling and execution of ResNet 1D, and scikit-learn [47] for implementation of other machine learning algorithms. Moreover, libraries such as NumPy [48] and pandas [49] were used for computing matrices and analyzing the data structure. All tests were conducted using a desktop workstation with Intel i7-9700K at 3.6 GHz, 8 CPU cores, 16 GB RAM, and Gigabyte GeForce RTX 2080 GPU. The following algorithms were used.

Deep Residual Network, suggested by [50], is an ensemble of VGG-19 [51], plain network, and residual network as a solution to the network depth-accuracy degradation problem. This is done by a residual learning framework, which is a feedforward network with a shortcut. Multi-scale 1D ResNet is used in this work, where multi-scale refers to flexible convolutional kernels rather than flexible strides [52]. The authors use different sizes of kernels so that the network can learn features from original signals with different views with multiple scales. The structure of the model is described in Figure 13. The 1D ResNet model consists of a number of subblocks of the basic CNN blocks. A *basic CNN block* computes batch normalization after convolution as follows: $y = W \otimes x + b, s = \text{BN}(y)$, and $h = \text{ReLU}(s)$, where $\otimes$ denotes the convolution operator and BN is a batch normalization operator. Moreover, stacking two basic CNN blocks forms a *subblock* of the basic CNN blocks as follows: $h_1 = \text{Basic}(x)$, $h_2 = \text{Basic}(h_1)$, $y = h_2 + x$, and $\hat{h} = \text{ReLU}(y)$, where the *Basic* operator denotes the basic block described above. Using the above method, it is possible to construct multiple sub-blocks of the CNN with different kernel sizes. For training the network, we used an early stopping option if there was no improvement in the validation loss after 20 epochs. Using the early stopping option and a learning rate of 0.01, the network was trained over 100 epochs, since the average training time was around 50 epochs.

For the other machine learning models, the *random forest* algorithm with 200 trees, *gini* as a criterion, and unlimited depth was used. For the $K - nearest\ neighbors\ (kNN)$, the following parameters were used: $k = 5$ and *Minkowski* as the distance metric. While the random forest algorithm is an ensemble method based on the concept of bagging, GBM [53] uses the concept of boosting, iteratively training the model by adding new weak models consecutively with the negative gradient from the loss function. Both extreme gradient boosting ($XGBoost$) [54,55] and *lightGBM* [56] are advanced models of the gradient boosting machines. LightGBM combines two techniques: *Gradient-based One-Side Sampling* and *Exclusive Feature Bundling* [57]. Since our dataset is tabular, XGBoost, which is a more efficient implementation of GBM, was used. For the XGBoost implementation, the following training parameters were used: 1000 *n_estimators* for creating weak learners, *learning rate* = 0.3 (*Eta*), and *max_depth* = 6. For the LightGBM implementation, the following training parameters were used: 1000 *n_estimators* for creating weak learners, *learning rate* = 0.1 (*Eta*), and *num_leaves* = 31.

Table 2. Results of feature engineering process applied to the APTOS dataset.

img	label	glcm1	glcm2	⋯	glcm24	dim0_0	dim0_1	⋯	dim0_99	dim1_0	dim1_1	⋯	dim1_99	fdim0_0	fdim0_1	⋯	fdim0_143	fdim1_0	fdim1_1	⋯	fdim1_143
0	2	0.1603	0.1571	⋯	0.4639	1	0.0366	⋯	0	1.2054	1.1815	⋯	0	0.9999	0.2060	⋯	0.0319	1.7698	1.6339	⋯	1.1067
⋮																					
3661	2	0.1196	0.1160	⋯	0.5387	0.9999	0.0020	⋯	0	1.4787	1.0636	⋯	0	0.9999	0.1295	⋯	0.0042	1.9493	1.3658	⋯	1.10478

For all the datasets (except the Indian fruits dataset), 80% of the dataset was used for training and 20% was used for testing. The Indian fruits dataset was already separated into 90% for training and 10% for testing. This ratio was used in our experiments.

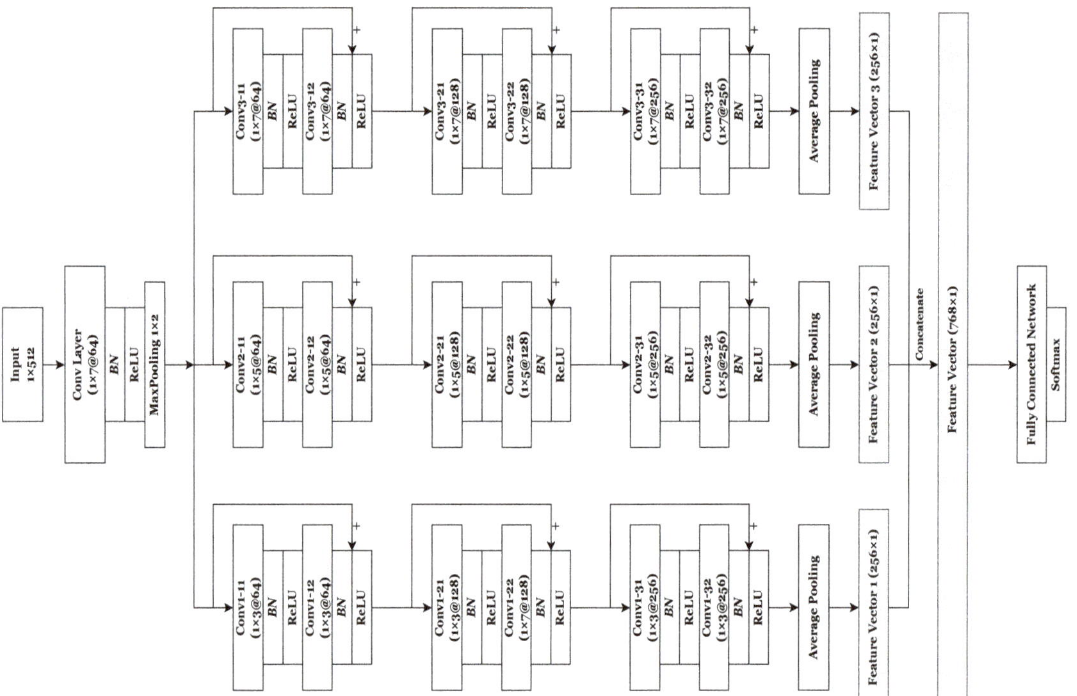

Figure 13. Structure of the multi-scale 1D ResNet.

4. Results and Discussion

Table 3 gives the accuracy, weighted F1 score, and run time information for each of the datasets. The accuracy score reported with the benchmark datasets is given in the last column. The best result is indicated in blue. Overall, ResNet 1D outperforms other ML models, while different types of gradient boosting machines show fairly good accuracy and weighted F1 scores. In terms of binary classification image datasets, as in the concrete dataset, most of the algorithms achieve 0.99 accuracy and F1 score. However, for the multi-class image datasets, the SVM and kNN perform poorly, mainly due to the inherent difficulty of finding the best parameters. All machine learning models perform significantly worse than the benchmark results with the Fashion MNIST and APTOS datasets with the extracted topological features. This is because it is hard to obtain good trainable topological signatures from images that have low resolution, even though Fashion MNIST was resized (please see Table 1).

In the case of the APTOS dataset, imbalanced training data were the main cause of the poor results. For example, Label 0 indicates the absence of diabetic retinopathy and has the highest number of images (See Figure 6). However, the presence of diabetic retinopathy can be found in four classes, of which Label 2 (severity level 2.0) has the largest number of cases. As a result, more than half of the examples were classified as Label 2 (see Figure 14c).

Table 3. Accuracy and weighted F1 score for each dataset.

		ResNet 1D	CART	GBM	LightGBM	Random Forest	SVM	XGBoost	kNN	Related Works—Benchmark
Concrete	Accuracy	0.994	0.989	0.991	**0.9945**	0.993	0.956	0.9935	0.890	**0.999** with CNN [35]
	Weighted F1	0.994	0.988	0.989	0.994	0.992	0.955	0.993	0.884	
	Run time	465.87	9.08	252.15	11.25	7.63	214.05	59.15	1.93	
Mangopest	Accuracy	0.931	0.764	0.681	0.898	0.869	0.474	0.889	0.666	0.76 with CNN [37]
	Weighted F1	0.931	0.764	0.676	0.898	0.869	0.439	0.889	0.663	
	Run time	760.94	17.17	5562.09	260.62	13.94	662.45	2041.22	2.33	
Indian fruits	Accuracy	1.000	0.9608	0.9608	0.9608	0.9608	0.7313	0.9608	0.676	0.999 SVM with deep features [38]
	Weighted F1	1.000	0.9608	0.9608	0.9608	0.9608	0.7236	0.9608	0.656	
	Run time	271.21	4.44	4265.09	82.55	4.13	72.73	451.65	1.18	
Fashion MNIST	Accuracy	0.7427	0.567	0.696	0.749	0.693	0.535	0.746	0.397	**0.99** with CNN [58]
	Weighted F1	0.7414	0.569	0.694	0.749	0.692	0.529	0.746	0.390	
	Run time	467.12	8.36	1808.07	89.37	7.66	935.21	1108.20	3.38	
APTOS	Accuracy	0.7326	0.698	0.760	0.787	0.782	0.674	0.775	0.655	**0.971** with CNN [59]
	Weighted F1	0.667	0.695	0.737	0.771	0.757	0.591	0.764	0.637	
	Run time	61.81	0.63	86.02	13.16	0.70	3.49	42.34	0.08	
Colorectal histology	Accuracy	0.892	0.75	0.842	0.869	0.855	0.679	0.874	0.759	0.874 with SVM [39]
	Weighted F1	0.89	0.727	0.832	0.850	0.834	0.686	0.843	0.743	
	Run time	86.23	1.18	255.08	12.52	1.10	4.06	44.63	0.14	
	Accuracy	0.882 ± 0.109	0.789 ± 0.147	0.822 ± 0.121	0.876 ± 0.087	0.856 ± 0.102	0.675 ± 0.154	0.873 ± 0.90	0.674 ± 0.148	
	Weighted F1	0.871 ± 0.125	0.784 ± 0.148	0.815 ± 0.124	0.870 ± 0.091	0.851 ± 0.105	0.654 ± 0.164	0.866 ± 0.092	0.662 ± 0.147	

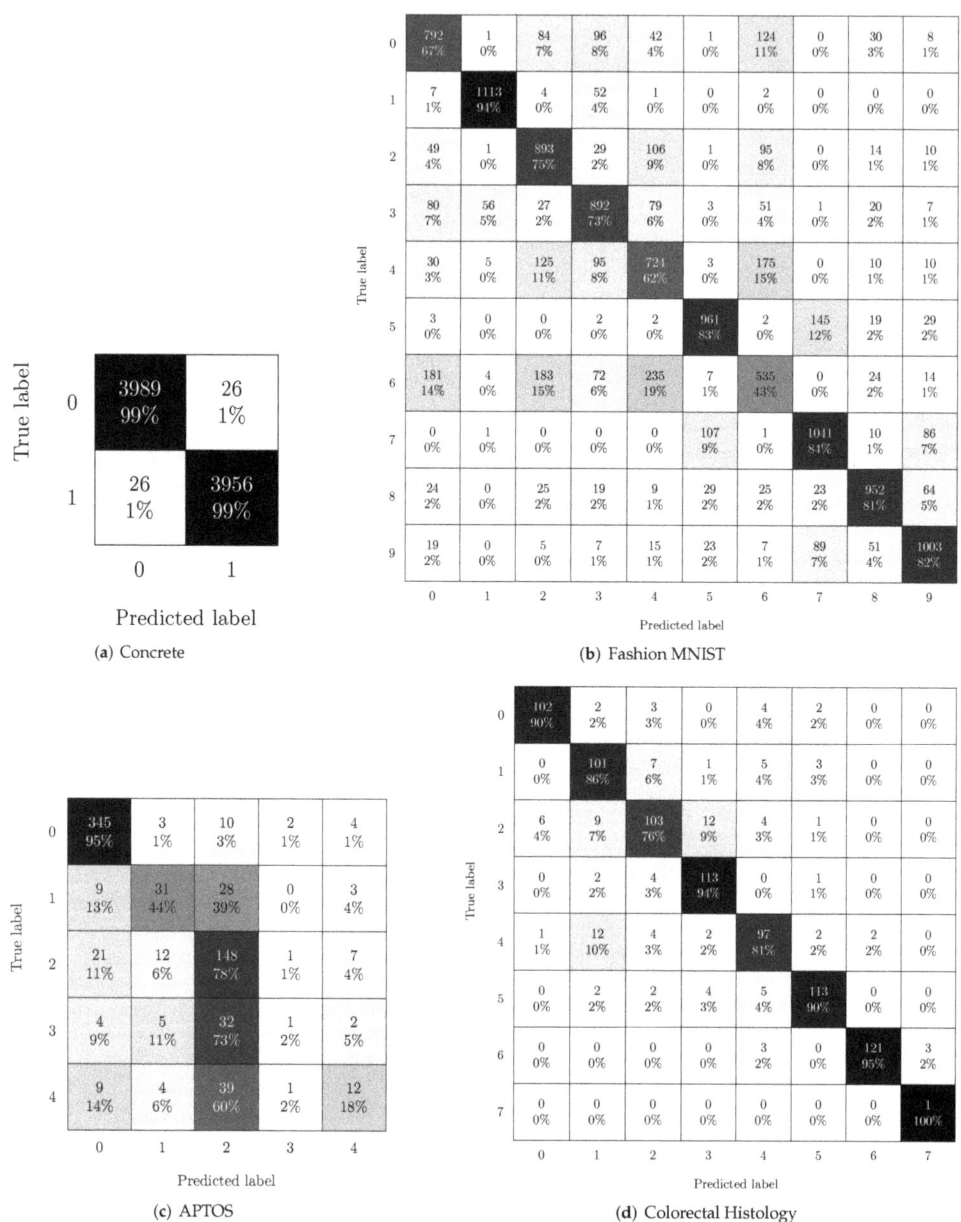

Figure 14. Confusion matrices of implementation with ResNet 1D. (**a**) Concrete, (**b**) Fashion MNIST, (**c**) APTOS, (**d**) Colorectal Histology.

Imbalanced data such as Mangopest and Indian fruits were classified well because there were sufficient training examples. In summary, the best classification performance using cubical homology with the ResNet 1D classifier was obtained for 2 out of 6 datasets using our proposed feature extraction method and score measure. However, these topological signatures were not helpful in the classification of the Fashion MNIST and APTOS images. For the Indian fruits dataset, the model classifies with an accuracy of 1.00, which is comparable since there is just 0.001 improvement. Similarly, with the concrete dataset, the result is comparable, with only a slight difference ($\leq$0.005) with the benchmark result.

In Table 4, we give an illustration of the performance of classifiers using features derived from cubical homology (TDA), GLCM, and combined TDA-GLCM for two datasets. It is clear from the results that the combined feature set using topological features and GLCM features results in better classifier accuracy.

Confusion matrices for experiments with the 1D ResNet model are given in Figures 14–16. It is noteworthy that, for these datasets, the application of cubical homology has led to meaningful results in 4 out of 6 datasets.

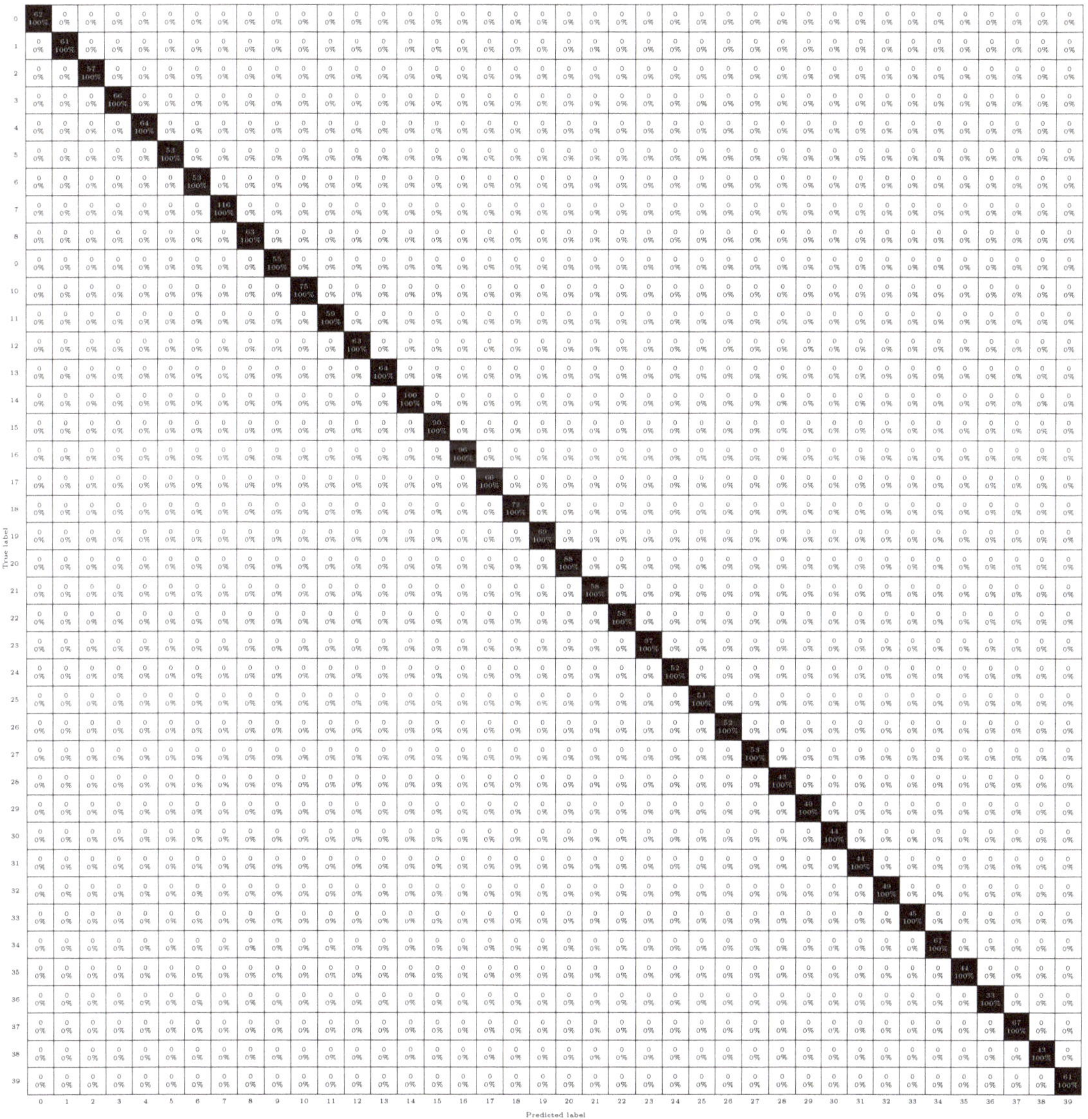

Figure 15. Confusion matrix for the Indian fruits dataset with ResNet 1D implementation.

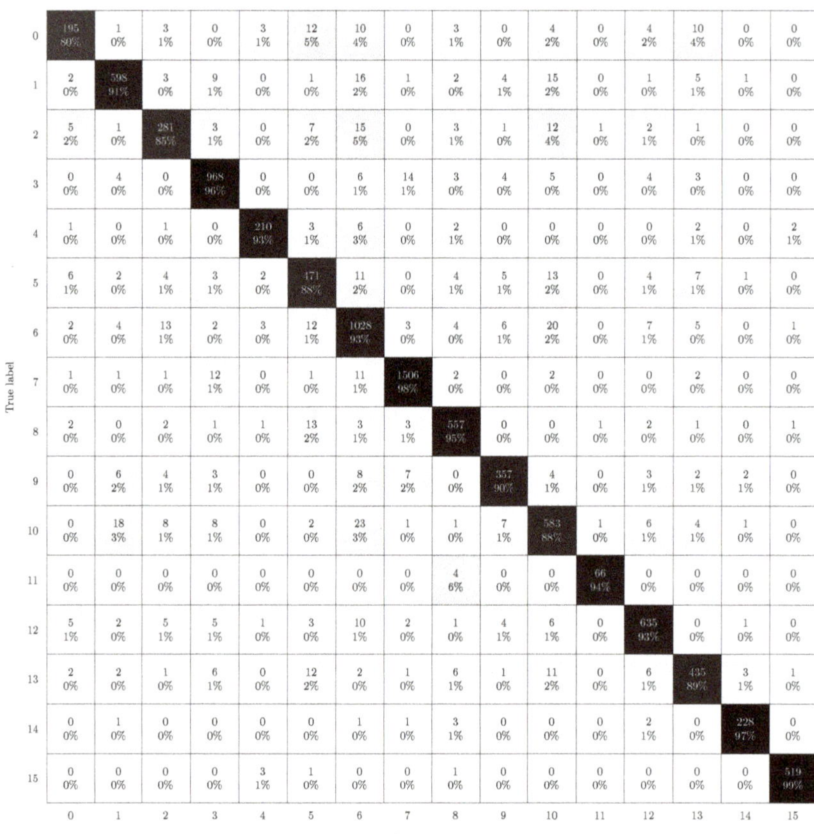

Figure 16. Confusion matrix for the Mangopest dataset with ResNet 1D implementation.

Table 4. Comparison of performance of GLCM+TDA, TDA feature-only, and GLCM-only implemented by 1D ResNet model on two datasets.

	Colorectal Histology Dataset	APTOS Dataset
GLCM+TDA	0.892	0.7326
TDA	0.7697	0.6739
GLCM	0.694	0.7252

5. Conclusions

The focus of this paper was on feature extraction from 2D image datasets using a specific topological method (cubical homology) and a novel score measure. These features were then used as input to well-known classification algorithms to study the efficacy of the proposed feature extraction process. We proposed a novel scoring method to transform the 2D input images into a one-dimensional array to vectorize the topological features. In this study, six published datasets were used as benchmarks. ResNet 1D, LightGBM, XGBoost, and five well-known machine learning models were trained on these datasets. Our experiments demonstrated that, in three out of six datasets, our proposed topological feature method is comparable to (or better than) the benchmark results in terms of accuracy. However, with two datasets, the performance of our proposed topological feature method is poor, due to either low resolution or an imbalanced dataset. We also demonstrate that topological features combined with GLCM features result in better classification accuracy in two of the datasets. This study reveals that the application of cubical homology to image classification shows promise. Since the conversion of input images to 2D data is very time-consuming, future work will involve (i) seeking more efficient ways to reduce the time for pre-processing and (ii) experimentation with more varied datasets. The problem of poor accuracy with imbalanced datasets needs further exploration.

Author Contributions: S.C.: Conceptualization, Investigation, Formal Analysis, Supervision, Resources, Original Draft. S.R.: Methodology, Validation, Software, Editing. All authors have read and agreed to the publisher version of the manuscript.

Funding: This research was funded by NSERC Discovery Grant#194376 and University of Winnipeg Major Research Grant#14977.

Institutional Review Board Statement: Not applicable.

Informed Consent Statement: Not applicable.

Data Availability Statement: Not applicable.

Conflicts of Interest: The authors declare no conflicts of interest.

References

1. Edelsbrunner, H.; Letscher, D.; Zomorodian, A. Topological persistence and simplification. In Proceedings of the 41st Annual Symposium on Foundations of Computer Science, Redondo Beach, CA, USA, 12–14 November 2000; IEEE Computer Society Press: Los Alamitos, CA, USA, 2000; pp. 454–463.
2. Chazal, F.; Michel, B. An Introduction to Topological Data Analysis: Fundamental and Practical aspects for Data Scientists. *arXiv* **2017**, arXiv:1710.04019.
3. Zomorodian, A.; Carlsson, G. Computing persistent homology. *Discr. Comput. Geom.* **2005**, *33*, 249–274. [CrossRef]
4. Carlsson, G. Topology and data. *Bull. Am. Math. Soc.* **2009**, *46*, 255–308. [CrossRef]
5. Edelsbrunner, H.; Harer, J. Persistent homology. A survey. *Contemp. Math.* **2008**, *453*, 257–282.
6. Zomorodian, A.F. Computing and Comprehending Topology: Persistence and Hierarchical Morse Complexes. Ph.D. Thesis, University of Illinois at Urbana-Champaign, Urbana, IL, USA, 2001.
7. Aktas, M.E.; Akbas, E.; Fatmaoui, A.E. Persistence Homology of Networks: Methods and Applications. *arXiv* **2019**, arXiv:math.AT/1907.08708.
8. Garin, A.; Tauzin, G. A topological "reading" lesson: Classification of MNIST using TDA. In Proceedings of the 2019 18th IEEE International Conference On Machine Learning And Applications (ICMLA), Boca Raton, FL, USA, 16–19 December 2019; pp. 1551–1556.
9. Adams, H.; Chepushtanova, S.; Emerson, T.; Hanson, E.; Kirby, M.; Motta, F.; Neville, R.; Peterson, C.; Shipman, P.; Ziegelmeier, L. Persistence Images: A Stable Vector Representation of Persistent Homology. *arXiv* **2016**, arXiv:cs.CG/1507.06217.
10. Kramár, M.; Goullet, A.; Kondic, L.; Mischaikow, K. Persistence of force networks in compressed granular media. *Phys. Rev. E* **2013**, *87*, 042207. [CrossRef]
11. Nakamura, T.; Hiraoka, Y.; Hirata, A.; Escolar, E.G.; Nishiura, Y. Persistent homology and many-body atomic structure for medium-range order in the glass. *Nanotechnology* **2015**, *26*, 304001. [CrossRef]
12. Dunaeva, O.; Edelsbrunner, H.; Lukyanov, A.; Machin, M.; Malkova, D.; Kuvaev, R.; Kashin, S. The classification of endoscopy images with persistent homology. *Pattern Recognit. Lett.* **2016**, *83*, 13–22. [CrossRef]
13. Reininghaus, J.; Huber, S.; Bauer, U.; Kwitt, R. A stable multi-scale kernel for topological machine learning. In Proceedings of the IEEE Conference on Computer Vision and Pattern Recognition, Boston, MA, USA, 7–12 June 2015; pp. 4741–4748.
14. Iijima, T. Basic theory on the normalization of pattern (in case of typical one-dimensional pattern). *Bull. Electro-Tech. Lab.* **1962**, *26*, 368–388.
15. Bonis, T.; Ovsjanikov, M.; Oudot, S.; Chazal, F. Persistence-based pooling for shape pose recognition. In Proceedings of the International Workshop on Computational Topology in Image Context, Marseille, France, 15–17 June 2016; pp. 19–29.
16. Dey, T.; Mandal, S.; Varcho, W. Improved image classification using topological persistence. In Proceedings of the Conference on Vision, Modeling and Visualization, Bonn, Germany, 25–27 September 2017; pp. 161–168.
17. Kindelan, R.; Frías, J.; Cerda, M.; Hitschfeld, N. Classification based on Topological Data Analysis. *arXiv* **2021**, arXiv:cs.LG/2102.03709.
18. Carrière, M.; Chazal, F.; Ike, Y.; Lacombe, T.; Royer, M.; Umeda, Y. Perslay: A neural network layer for persistence diagrams and new graph topological signatures. In Proceedings of the International Conference on Artificial Intelligence and Statistics (PMLR), Online, 26–28 August 2020; pp. 2786–2796.
19. Chung, M.K.; Lee, H.; DiChristofano, A.; Ombao, H.; Solo, V. Exact topological inference of the resting-state brain networks in twins. *Netw. Neurosci.* **2019**, *3*, 674–694. [CrossRef]

20. Don, A.P.H.; Peters, J.F.; Ramanna, S.; Tozzi, A. Topological View of Flows Inside the BOLD Spontaneous Activity of the Human Brain. *Front. Comput. Neurosci.* **2020**, *14*, 34. [CrossRef]
21. Don, A.P.; Peters, J.F.; Ramanna, S.; Tozzi, A. Quaternionic views of rs-fMRI hierarchical brain activation regions. Discovery of multilevel brain activation region intensities in rs-fMRI video frames. *Chaos Solitons Fractals* **2021**, *152*, 111351. [CrossRef]
22. Krizhevsky, A.; Sutskever, I.; Hinton, G.E. Imagenet classification with deep convolutional neural networks. *Adv. Neural Inf. Process. Syst.* **2012**, *25*, 1097–1105. [CrossRef]
23. Hofer, C.D.; Kwitt, R.; Niethammer, M. Learning Representations of Persistence Barcodes. *J. Mach. Learn. Res.* **2019**, *20*, 1–45.
24. Umeda, Y. Time series classification via topological data analysis. *Inf. Media Technol.* **2017**, *12*, 228–239. [CrossRef]
25. Pun, C.S.; Xia, K.; Lee, S.X. Persistent-Homology-Based Machine Learning and its Applications—A Survey. *arXiv* **2018**, arXiv:math.AT/1811.00252.
26. Allili, M.; Mischaikow, K.; Tannenbaum, A. Cubical homology and the topological classification of 2D and 3D imagery. In Proceedings of the 2001 International Conference on Image Processing (Cat. No. 01CH37205), Thessaloniki, Greece, 7–10 October 2001; Volume 2, pp. 173–176.
27. Kot, P. Homology calculation of cubical complexes in Rn. *Comput. Methods Sci. Technol.* **2006**, *12*, 115–121. [CrossRef]
28. Strömbom, D. Persistent Homology in the Cubical Setting: Theory, Implementations and Applications. Master's Thesis, Lulea University of Technology, Lulea, Sweden, 2007.
29. Choe, S. Cubical homology-based Image Classification-A Comparative Study. Master's Thesis, University of Winnipeg, Winnipeg, MB, Canada, 2021.
30. Fisher, M.; Springborn, B.; Schröder, P.; Bobenko, A.I. An algorithm for the construction of intrinsic Delaunay triangulations with applications to digital geometry processing. *Computing* **2007**, *81*, 199–213. [CrossRef]
31. Otter, N.; Porter, M.A.; Tillmann, U.; Grindrod, P.; Harrington, H.A. A roadmap for the computation of persistent homology. *EPJ Data Sci.* **2017**, *6*, 1–38. [CrossRef]
32. Kaczynski, T.; Mischaikow, K.M.; Mrozek, M. *Computational Homology*; Springer: Berlin/Heidelberg, Germany, 2004.
33. Kalies, W.D.; Mischaikow, K.; Watson, G. Cubical approximation and computation of homology. *Banach Cent. Publ.* **1999**, *47*, 115–131. [CrossRef]
34. Marchese, A. Data Analysis Methods Using Persistence Diagrams. Ph.D. Thesis, University of Tennessee, Knoxville, TN, USA, 2017.
35. Özgenel, Ç.F.; Sorguç, A.G. Performance comparison of pretrained convolutional neural networks on crack detection in buildings. In Proceedings of the International Symposium on Automation and Robotics in Construction (ISARC), Berlin, Germany, 20–25 July 2018; Volume 35, pp. 1–8.
36. Avilés-Rodríguez, G.J.; Nieto-Hipólito, J.I.; Cosío-León, M.d.l.Á.; Romo-Cárdenas, G.S.; Sánchez-López, J.d.D.; Radilla-Chávez, P.; Vázquez-Briseño, M. Topological Data Analysis for Eye Fundus Image Quality Assessment. *Diagnostics* **2021**, *11*, 1322. [CrossRef]
37. Kusrini, K.; Suputa, S.; Setyanto, A.; Agastya, I.M.A.; Priantoro, H.; Chandramouli, K.; Izquierdo, E. Data augmentation for automated pest classification in Mango farms. *Comput. Electron. Agric.* **2020**, *179*, 105842. [CrossRef]
38. Behera, S.K.; Rath, A.K.; Sethy, P.K. Fruit Recognition using Support Vector Machine based on Deep Features. *Karbala Int. J. Mod. Sci.* **2020**, *6*, 16. [CrossRef]
39. Kather, J.N.; Weis, C.A.; Bianconi, F.; Melchers, S.M.; Schad, L.R.; Gaiser, T.; Marx, A.; Zöllner, F.G. Multi-class texture analysis in colorectal cancer histology. *Sci. Rep.* **2016**, *6*, 27988. [CrossRef]
40. Xiao, H.; Rasul, K.; Vollgraf, R. Fashion-MNIST: A Novel Image Dataset for Benchmarking Machine Learning Algorithms. *arXiv* **2017**, arXiv:1708.07747.
41. Pizer, S.M.; Amburn, E.P.; Austin, J.D.; Cromartie, R.; Geselowitz, A.; Greer, T.; ter Haar Romeny, B.; Zimmerman, J.B.; Zuiderveld, K. Adaptive histogram equalization and its variations. *Comput. Vision Graph. Image Process.* **1987**, *39*, 355–368. [CrossRef]
42. Gadkari, D. Image Quality Analysis Using GLCM. Master's Thesis, University of Central Florida, Orlando, FL, USA, 2004.
43. Mohanaiah, P.; Sathyanarayana, P.; GuruKumar, L. Image texture feature extraction using GLCM approach. *Int. J. Sci. Res. Publ.* **2013**, *3*, 1–5.
44. The GUDHI Project. *GUDHI User and Reference Manual*, 3.4.1 ed.; GUDHI Editorial Board; GUDHI: Nice, France, 2021.
45. Dlotko, P. Cubical complex. In *GUDHI User and Reference Manual*, 3.4.1 ed.; GUDHI Editorial Board; GUDHI: Nice, France, 2021.
46. Paszke, A.; Gross, S.; Massa, F.; Lerer, A.; Bradbury, J.; Chanan, G.; Killeen, T.; Lin, Z.; Gimelshein, N.; Antiga, L.; et al. PyTorch: An Imperative Style, High-Performance Deep Learning Library. In *Advances in Neural Information Processing Systems 32*; Wallach, H., Larochelle, H., Beygelzimer, A., d'Alché-Buc, F., Fox, E., Garnett, R., Eds.; Curran Associates, Inc.: North Adams, MA, USA, 2019; pp. 8024–8035.
47. Pedregosa, F.; Varoquaux, G.; Gramfort, A.; Michel, V.; Thirion, B.; Grisel, O.; Blondel, M.; Prettenhofer, P.; Weiss, R.; Dubourg, V.; et al. Scikit-learn: Machine Learning in Python. *J. Mach. Learn. Res.* **2011**, *12*, 2825–2830.
48. Harris, C.R.; Millman, K.J.; van der Walt, S.J.; Gommers, R.; Virtanen, P.; Cournapeau, D.; Wieser, E.; Taylor, J.; Berg, S.; Smith, N.J.; et al. Array programming with NumPy. *Nature* **2020**, *585*, 357–362. [CrossRef] [PubMed]
49. McKinney, W. Data structures for statistical computing in python. In Proceedings of the 9th Python in Science Conference, Austin, TX, USA, 28 June–3 July 2010; Volume 445, pp. 51–56.
50. He, K.; Zhang, X.; Ren, S.; Sun, J. Deep residual learning for image recognition. In Proceedings of the IEEE Conference on Computer Vision and Pattern Recognition, Las Vegas, NV, USA, 27–30 June 2016; pp. 770–778.

51. Mateen, M.; Wen, J.; Song, S.; Huang, Z. Fundus image classification using VGG-19 architecture with PCA and SVD. *Symmetry* **2019**, *11*, 1. [CrossRef]
52. Liu, R.; Wang, F.; Yang, B.; Qin, S.J. Multiscale kernel based residual convolutional neural network for motor fault diagnosis under nonstationary conditions. *IEEE Trans. Ind. Inform.* **2019**, *16*, 3797–3806. [CrossRef]
53. Friedman, J.H. Greedy function approximation: A gradient boosting machine. *Ann. Stat.* **2001**, *29*, 1189–1232. [CrossRef]
54. Chen, T.; He, T.; Benesty, M.; Khotilovich, V.; Tang, Y.; Cho, H. Xgboost: Extreme gradient boosting. *R Package Version 0.4-2* **2015**, *1*, 1–4.
55. Chen, T.; Guestrin, C. Xgboost: A scalable tree boosting system. In Proceedings of the 22nd ACM SIGKDD International Conference on Knowledge Discovery and Data Mining, San Francisco, CA, USA, 13–17 August 2016; pp. 785–794.
56. Ke, G.; Meng, Q.; Finley, T.; Wang, T.; Chen, W.; Ma, W.; Ye, Q.; Liu, T.Y. Lightgbm: A highly efficient gradient boosting decision tree. *Adv. Neural Inf. Process. Syst.* **2017**, *30*, 3146–3154.
57. Patel, V.; Choe, S.; Halabi, T. Predicting Future Malware Attacks on Cloud Systems using Machine Learning. In Proceedings of the 2020 IEEE 6th International Conference on Big Data Security on Cloud (BigDataSecurity), High Performance and Smart Computing, (HPSC) and Intelligent Data and Security (IDS), Baltimore, MD, USA, 25–27 May 2020; pp. 151–156.
58. Kayed, M.; Anter, A.; Mohamed, H. Classification of garments from fashion MNIST dataset using CNN LeNet-5 architecture. In Proceedings of the 2020 International Conference on Innovative Trends in Communication and Computer Engineering (ITCE), Aswan, Egypt, 8–9 February 2020; pp. 238–243.
59. Tymchenko, B.; Marchenko, P.; Spodarets, D. Deep learning approach to diabetic retinopathy detection. *arXiv* **2020**, arXiv:2003.02261.

Article

Towards Predictive Vietnamese Human Resource Migration by Machine Learning: A Case Study in Northeast Asian Countries

Nguyen Hong Giang [1,2], Tien-Thinh Nguyen [3,*], Chac Cau Tay [4,5,*], Le Anh Phuong [6,*] and Thanh-Tuan Dang [7,*]

[1] Faculty of Architecture, ThuDauMot University, ThuDauMot 820000, Vietnam; giangnh@tdmu.edu.vn
[2] Department of Civil Engineering, National Kaohsiung University of Science and Technology, Kaohsiung 807618, Taiwan
[3] Department of International Business, National Kaohsiung University of Science and Technology, Kaohsiung 82445, Taiwan
[4] Department of Business Administration, Southern Taiwan University of Science and Technology, Tainan 71005, Taiwan
[5] Shenghua Marketing Co., Ltd., Kaohsiung 814026, Taiwan
[6] Department of Computer Science, Hue University of Education, Hue University, Hue 49118, Vietnam
[7] Department of Industrial Engineering and Management, National Kaohsiung University of Science and Technology, Kaohsiung 807618, Taiwan
* Correspondence: ngtienthinh101096@gmail.com (T.-T.N.); chaccautay@gmail.com (C.C.T.); leanhphuong@dhsphue.edu.vn (L.A.P.); tuandang.ise@gmail.com (T.-T.D.)

Abstract: Labor exports are currently considered among the most important foreign economic sectors, implying that they contribute to a country's economic development and serve as a strategic solution for employment creation. Therefore, with the support of data collected between 1992 and 2020, this paper proposes that labor exports contribute significantly to Vietnam's socio-economic development. This study also aims to employ the Backpropagation Neural Network (BPNN), k-Nearest Neighbor (kNN), and Random Forest Regression (RFR) models to analyze labor migration forecasting in Taiwan, Korea, and Japan. The study results indicate that the BPNN model was able to achieve the highest accuracy regarding the actual labor exports. In terms of these accuracy metrics, this study will aid the Vietnamese government in establishing new legislation for Vietnamese migrant workers in order to improve the nation's economic development.

Keywords: labor exports; Northeast Asian Countries; backpropagation neural network (BPNN); k-Nearest Neighbor (kNN); random forest regression (RFR); decision making

Citation: Giang, N.H.; Nguyen, T.-T.; Tay, C.C.; Phuong, L.A.; Dang, T.-T. Towards Predictive Vietnamese Human Resource Migration by Machine Learning: A Case Study in Northeast Asian Countries. *Axioms* **2022**, *11*, 151. https://doi.org/10.3390/axioms11040151

Academic Editor: Oscar Humberto Montiel Ross

Received: 16 February 2022
Accepted: 18 March 2022
Published: 24 March 2022

Publisher's Note: MDPI stays neutral with regard to jurisdictional claims in published maps and institutional affiliations.

1. Introduction

Human resources play a pivotal role in the prosperity of a country, and global development in general [1,2]. Globalization has resulted in people having more opportunities to work in many different countries [3,4]. With the rapid development of the economy in developed countries, labor resources are needed to meet requirements of the economy's growth rate [5,6]. In addition, some developed countries face the dilemma of an aging population structure [7,8]. Therefore, there is an urgent need for foreign human resources to make up for the shortage of domestic human resources [9]. In terms of solving these difficulties, many policies have been implemented with the aim of attracting human resources from overseas. Therefore, many countries have developed labor export development plans to create jobs, thus increasing income for each individual and contributing to the growth of the economy in terms of Gross Domestic Product (GDP) [10,11].

Currently, many countries in the Asia-Pacific region participate in labor exports, with the Philippines, Indonesia, Thailand, and China being direct competitors to Vietnam in the international labor market, as they share similarities [12]. The experiences of these countries in managing, operating, and developing labor export activities can serve as valuable lessons for Vietnam. It is estimated that around 20 million Southeast Asians work outside of their home countries, with approximately half of those being in the Middle East [13]. Additionally, inheriting the results and experiences gained through the era of international labor cooperation with socialist nations from 1980 to 1990, the Party and the State of Vietnam changed its level of awareness and promoted labor export activities [14]. On 9 November 1991, the government issued Decree No. 370/HDBT, which clearly stated: "Putting migrant workers to working abroad for a definite time is a method to generate jobs, produce more incomes for workers and boost foreign currency revenue for the country, resulting in the strengthening of economic, cultural, scientific and technical collaboration relations between Vietnam and other countries that use labor on the principle of equality, mutual benefit, respect for each other's laws and national traditions" [15]. In 2018, 500,000 Vietnamese contract labor migrants worked abroad, according to the latest statistics [16]. Taiwan, Korea, and Japan are still three important labor export markets for Vietnam, accounting for more than 70% of the total number of laborers working abroad in 2008 and 2009, and are three markets that continue to have a high demand for Vietnamese workers [17]. Additionally, several studies on labor exports in East and Southeast Asia have reported workers migrating from Vietnam to Singapore, South Korea, Taiwan, and Japan [18,19].

When it comes to Backpropagation Neural Network (BPNN), k-Nearest Neighbor (kNN), and Random Forest Regression (RFR) models, many studies have been conducted by applying these models for prediction [20–26]. Chen, Lai [27] used BPNN and ANN algorithms to forecast that the majority of international visitors to Taiwan would be from the main markets of Hong Kong, Japan, and Macau from January 1971 to August 2009. Likewise, Vihikan, Putra [28] employed BPNN to predict in the arrival of international tourists in Bali to assist the government in constructing the nation's tourism strategies. Wang, Wu [29] showed that the monthly inward tourism flow to China between 2000 and 2013 could be forecasted by deploying an Enhanced Backpropagation Neural Network. In addition, Lin, Malyscheff [30] used ANN models to predict engineering students' future retention, consistently achieving an overall prediction accuracy of around 70% or higher. Moreover, the research used RFR and Support Vector Regression (SVR) models to estimate global foreign tourist arrivals, achieving prediction accuracies of 99.4 percent using SVR and 84.7 percent using RFR [31]. Additionally, the research used the kNN technique to investigate the unique nature of campground management, testing several forecasting approaches in order to determine which was best suited to the unique behavior of camping tourists and the unique nature of campsites [32].

This paper presents a forecast of Vietnamese labor migration using the kNN, RFR, and BPNN models. The data collected in this study include 29 observations from 1992 to 2020. With the support of the collected data, the paper analyzes the role of labor exports in Vietnam's socio-economic development and compares this export labor with that of other Asian countries. After that, the three algorithms are compared on the basis of the results of their statistical accuracy indicators. More importantly, this research highlights the possible future contexts of labor migration between Vietnam and East Asia, including Korea, Republic of China (Taiwan), and Japan. Some assume that the migration corridor extending between Vietnam and other East Asian nations has progressed to the point where labor migration is only one element in a more varied movement pattern. Nevertheless, there will continue to be significant demand for Vietnamese workers to work in Japan, Korea, and Taiwan for the foreseeable future. More importantly, this research could assist the government of Vietnam in enacting new regulations for Vietnamese migrant workers in order to boost the socio-economic situation. Consequently, the Vietnamese government could determine a different approach for our country's labor export activities in the context of integration.

The following is a description of the paper's structure. The first section of the paper is an introduction. The BPNN, kNN, and RFR models' approach is introduced in Section 2. The study results and discussions are described in Sections 3 and 4. Finally, the conclusions are presented in Section 5.

2. Materials and Methods

The process of the following experimental stages in this study is summarized in Figure 1. Firstly, the database is preprocessed and tested by statistical methods, and it is also divided into training and testing sets. Secondly, the BPNN, RFR, and kNN models are used to learn the training samples and obtain the optimal network parameters. Finally, the three models' performances are compared using metrics from the accuracy measurement indicators as Mean Square Error (MSE), Root-Mean-Square Error (RMSE), Mean Absolute Error (MAE), Correlation Coefficient (R), and Correlation of Determination (R^2) in the possible result stage. At the same time, seeking the most suitable prediction model for the study if the RMSE, MAE, and MSE gain the lowest values, and R and R^2 approach the highest values; on the other hand, the experiments are adjusted by component parameters of models or training data size to look for the optimal accuracy measurement indicators.

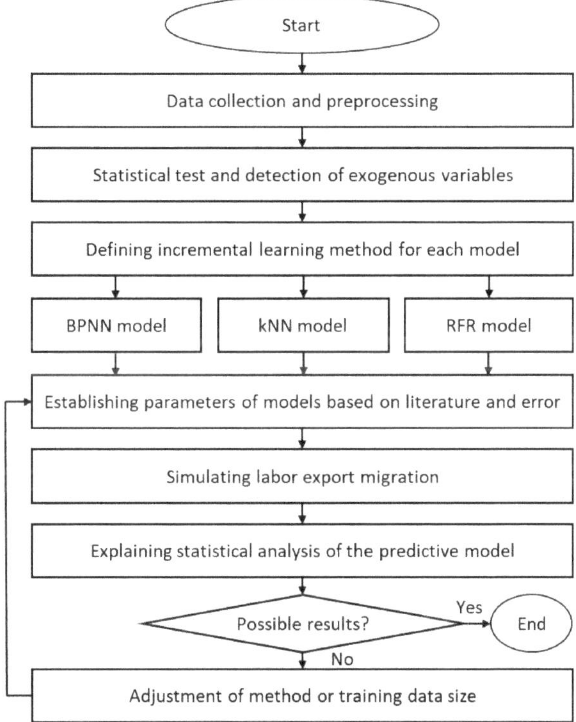

Figure 1. The process of the experimental stages in this study.

2.1. Database

This paper uses a database containing Vietnam labor immigration to Korea, Japan, and Taiwan from 1992 to 2020 obtained by the Department of Overseas Labor (DOLAB). The Python 3.9 software is deployed for data analysis. The data in Figure 2 show that the number of Vietnamese workers migrating to Northeast Asian Countries has witnessed a fluctuation from 1992 to 2020. The total of Vietnamese laborers who immigrated to Northeast Asian Countries was 1,373,712 people for 29 years, in which Japan, Taiwan, and Korea occupied about 811,138,367,967, and 194,607 people, respectively. At the same time, more than 100,000 employees per yearly migrated to the region during the period of 2015–2019. However, the total number of Vietnamese workers in 2020 going to work in Japan, South Korea, and Taiwan is 76,355 persons, implying a decrease in the number of 55% compared with that in 2019 and a decline in total labor exports of nearly 56% compared with the data in 2018. Moreover, the highest total number of Vietnamese workers working in Taiwan, Japan, and South Korea was 68,737 employees in 2018, then 82,703 employees in 2019, and 18,141 in 2008.

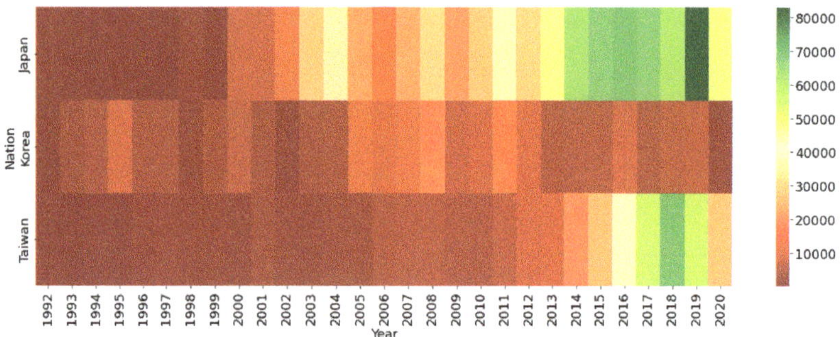

Figure 2. The heatmap of annual labor exports to Japan, Taiwan, and Korea.

The expected statistical results as the Mean, Min, and Max values, St Dev, Skew, and Kurt are also pointed out in Table 1 to quantitatively analyze the yearly labor export characteristics to interpret their significance further. The mean is the crucial popular measure of central trend, and the indicator may be used with continuous data. The Minimum and Maximum values indicate the margin of the time series. The Standard Deviation is used to measure the degree of data dispersion. Finally, Skewness and Kurtosis are deployed to estimate whether the distribution of the sample is normal or not. The statistics result in Table 1 indicate as below: The Mean and Standard Deviation of the labor exports to North-East Asian countries were 25,422 persons and 28,962 persons in Japan, 18,786 persons and 12,689 persons in Taiwan, 4132 persons and 6711 persons in Korea, respectively. The skewness for a normal distribution is zero, and any symmetric data should have a skewness near zero. Negative values for the skewness indicate skewed left data, and positive values for the skewness indicate skewed right data [33,34]. Skewness coefficients were low for the data sets. This approach is appropriate for modeling because a high skewness coefficient has a considerable negative effect on ANN performance [35]. Hence, the values of Kurt and Skew of the data fluctuating from −8.87 to 2.64 could be accepted for prediction through these models. The labor exports to the region using Augmented Dickey-Fuller (ADF) test also point out that the database is separated two parts such as the data of labor exports to Korea (*p*-value = 0.02, test statistic = −3.16) has a unit root, this data is stationary, and the data of labor exports to Japan (*p*-value = 0.62, test statistic = −1.32) and Taiwan (*p*-value = 0.99, Test Statistic = 2.22) do not have a unit root, these data are non-stationary. Furthermore, the time series data are stationary; they can be easily modeled with higher accuracy than their non-stationary counterparts [36].

Table 1. Descriptive statistics and formal hypothesis test of labor exports to Northeast Asian Countries from 1992 to 2020.

Item	Japan	Taiwan	Korea
Mean	25,422	18,786	4132
SD	28,962	12,689	6711
Min	5	15	216
Max	82,703	68,737	18,141
Kurt	−0.87	2.64	1.05
Skew	0.53	1.89	0.91
Results of Augmented Dickey-Fuller Test:			
Test Statistic	−1.32	2.22	−3.16
p-value	0.62	0.99	0.02
Critical Value (1%)	−3.70	−3.83	−3.69
Critical Value (5%)	−2.98	−3.03	−2.97
Critical Value (10%)	−2.63	−2.65	−2.62

The input data patterns of the three nations were randomly selected with two sections. The first section was used for the training phase, and it contained roughly 70% of the total data. The remaining 30% of the labor export data was contained in the second component, which was used for the test phase.

2.2. Backpropagation Neural Network (BPNN)

BNPP can learn and store various data, akin to the human brain. Therefore, a single hidden layer is represented in Figure 3. Furthermore, BPNN has three layers: an input layer, at least one hidden layer, and an output layer. Weights connect adjacent layers, always scattered among 0 and 1. Although several heuristic techniques have been deployed by many researchers [37,38], there is no systematic theory for verifying the number of input nodes and hidden layer nodes. Experiments or trial and error based on the test data's least mean square error are the most fitted methods for determining the appropriate indicators of input and hidden nodes. The following Equations (1) and (2) indicate the relationships of McCulloch–Pitts (M–P) neurons in hidden layer and the output layer [39].

$$b_n = f\left(\sum_{i=1}^{d} v_{in}x_i + \gamma_n\right) \tag{1}$$

$$y_j = f\left(\sum_{n=1}^{m} w_{nj}b_n + p_j\right) \tag{2}$$

where b_n is the result of nth hidden neuron, x_i is the ith valuable input out of d inputs, y_j is the jth output value, m is the total number of hidden neurons, f is the activation function of the rectified linear unit (ReLU), v_{in} and w_{nj} are the weight terms, γ_n and p_j are the bias terms.

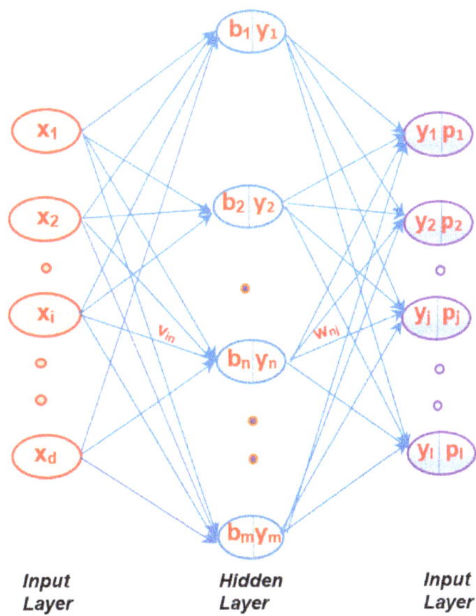

Figure 3. Backpropagation neural network structure.

Figure 3 demonstrates the BPNN structure below, showing 29 input nodes from x1 to xd (which d is also the 29th year) for the input layer; the 29 neurons for the output layer are also the values of the labor exports. At the same time, one hidden layer contains neurons from b1 to bm (m is the 200th node). Each neuron of the hidden layer and the output layer have in charge of Weight and Bias, such as v_{in}, γ_n and w_{nj}, p_j to correspond for neuron b_n and neuron y_j, respectively. Each hidden layer neuron takes the output from input layer neurons and converts these values with a weighted linear sum into the output layer. The output layer obtains the values from the hidden layer. Furthermore, the activation function for the hidden layer is ReLU function. Adam's method is stochastic optimizations to the solver of weight optimization. The training method for the models is regression.

2.3. K-Nearest Neighbor (kNN)

In machine learning, kNN is one of the most fundamental supervised learning algorithms. The kNN model can be used for regression as well as classification. Moreover, because the observed data has a numeric label to forecast, kNN is employed as a predictive method with a multi-input multi-output (MIMO) strategy for forecasting [32]. kNN begins by initialing the number of k. Hence, the distance between the training and testing instances is calculated. Then, the prediction is made by calculating the test data's average k nearest distance neighbor. Subsequently, the Euclidean Distance is deployed as witnessed in Equation (3).

$$d(x,y) = \sqrt{\sum_{i=1}^{k}(x-y)^2} \tag{3}$$

where d symbolizes the distance among data points, x represents the testing data, y is the training data.

2.4. Random Forest Regression (RFR)

Random forest is a regression method for classifying or forecasting the numerical value of a variable that incorporates the outputs of various decision tree algorithms [40–42]. When an (x) input vector containing the values of the various evidentiary features analyzed for a certain training area is received by a random forest. RFR creates a total of K regression trees and then averages the output. The random forest regression forecaster as shown in Equation (4) when K grows the $\{T(x)\}_1^K$ tree [43].

$$\hat{f}_{rf}^{K}(x) = \frac{1}{K}\sum_{k=1}^{K} T(x) \tag{4}$$

To avoid different trees correlations, RFR promotes tree variety by allowing them to grow from diverse training data subsets obtained through a method known as bagging [43]. Bagging is a method for generating training data that randomly resample the original dataset with replacement data. Therefore, certain data may be used many times during training, whereas others may never be used. Hence, better stability is gained, as it becomes more resistant when faced with minor deviations in input data while also increasing forecast accuracy [40].

2.5. Performance Metrics

Estimating results are based on calculating and comparing the actual values to the forecasted values. These metrics of the accuracy measurement parameters include the Mean Square Error (MSE), Root-Mean-Square Error ($RMSE$), Mean Absolute Error (MAE), Correlation Coefficient (R), and Correlation of Determination (R^2). Furthermore, the error metrics are defined as follows [44–46]:

$$MSE = \frac{\sum_{t=1}^{n}(x_t - x_t')^2}{n} \tag{5}$$

$$MAE = \frac{\sum_{t=1}^{n}|x_t - x_t'|}{n} \tag{6}$$

$$RMSE = \sqrt{\frac{\sum_{t=1}^{n}(x_t - x_t')^2}{n}} \tag{7}$$

$$R^2 = 1 - \frac{\sum_{t=1}^{n}(x_t - x_t')^2}{\sum_{t=1}^{n}\left(x_t - \frac{1}{n}\sum_{t=1}^{n}x_t\right)^2} \tag{8}$$

$$R = \frac{\sum_{t=1}^{n}(x_t - \overline{x})\left(x_t' - \overline{x'}\right)}{\sqrt{\sum_{t=1}^{n}(x_t - \overline{x})^2}\sqrt{\sum_{t=1}^{n}\left(x_t' - \overline{x'}\right)^2}} \tag{9}$$

where x_t, x_t' are the observed and estimated values in the period time t, and n is the number of the observed values in the testing data. $\overline{x}$, $\overline{x'}$ are mean of the observed and estimated value. The R^2 and R should be approaching *1* to indicate strong model performance, and the MSE, MAE, and RMSE should be as close to *zero* as possible.

3. Results Analysis

Three models were established to predict labor exports in three countries (Taiwan, Japan, and Korea). For comparing the models based on the estimating performance of labor exports, simulating experiments have pointed out the optimum models with the main indicators in Table 2.

Table 2. The optimum component for three models.

kNN		RFR		BPNN	
Item	*Configuration*	*Item*	*Configuration*	*Item*	*Configuration*
leaf_size	40	n_estimators	30	Number of inputs	29
metric	euclidean	max_depth	None	Number of hidden layers	1
n_jobs	2	max_features	1	Hidden layer sizes	200
n_neighbors	3	min_samples_leaf	1	Number of outputs	29
p	2	min_samples_split	2	activation	ReLU
		bootstrap	False	optimize	adam

Line graphs in Figure 4 illustrate the RFR (red lines), kNN (grey lines), and BPNN (green lines) algorithms using testing data that were compared with the actual line graphs (blue lines). As far as the predicted models of labor immigration in the three nations are concerned, the BPNN lines are the nearest ones compared with the real data in Taiwan, Korea, and Japan. As a result, the grey lines show that kNN algorithms achieve the second-most accurate level in Taiwan and Japan, whereas kNN only achieves the lowest level in Korea in a selection of random years from 1994 to 2016. Therefore, the accuracy parameters for the labor immigration forecast are calculated (as shown in Table 3) to assess the three forecasting models' accuracy levels. The outperform of the simulation in Table 3 indicates the estimation accuracy using kNN, RFR, and BPNN algorithms has a significant reliability.

Additionally, when deploying kNN, RFR, and BPNN models in three nations with the accuracy parameters (MAPE, MAE, RMSE, R-squared, and NSE), the BPNN algorithm has the greatest accuracy indicators in Taiwan, Japan, and Korea, whereas the RFR method has the lowest. Furthermore, BPNN is the best machine learning algorithm in Taiwan, Japan, and Korea with $BPNN_{Taiwan}$ $BPNN_{Japan}$, and $BPNN_{Korea}$ achieved higher estimation accuracy (with $MAPE_{Japan} = 0.006$, $MAPE_{Taiwan} = 0.006$, and $MAPE_{Korea} = 0.07$) than RFR (with $MAPE_{Japan} = 0.145$, $MAPE_{Taiwan} = 0.137$, and $MAPE_{Korea} = 0.051$) and kNN (with $MAPE_{Japan} = 0.073$, $MAPE_{Taiwan} = 0.066$, and $MAPE_{Korea} = 0.051$). Similarly, based on MAE, RMSE, and NSE indicators, the BPNN algorithm in Taiwan, Japan, and Korea acquired greater accuracy parameters. In terms of forecast models for Taiwan, Japan, and Korea, the parameters of BPNN achieve a similar level of accuracy; however, $BPNN_{Korea}$ earns the highest one compared with others (with $MAE_{Korea} = 57$, and $RMSE_{Korea} = 67$), RFR_{Korea} gained the second-highest one (with $MAE_{Korea} = 488$, and $RMSE_{Korea} = 1002$), and kNN_{Korea} is the lowest one (with $MAE_{Korea} = 638$, and $RMSE_{Korea} = 1570$). In addition, the scatter plots and three models in Figure 5 calculate the simulation results. As indicated, the dispersion of points around the diagonal of the BPNN model earns the highest accuracy parameters in three countries; the kNN machine learning algorithm is the second-lowest, and the RFR algorithm achieves the lowest accurate level.

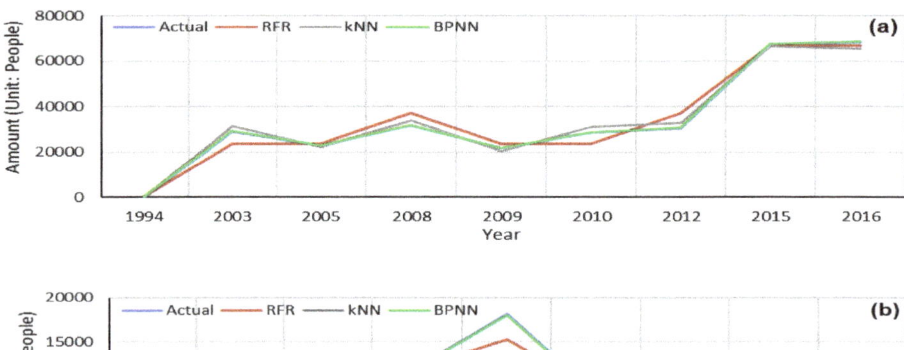
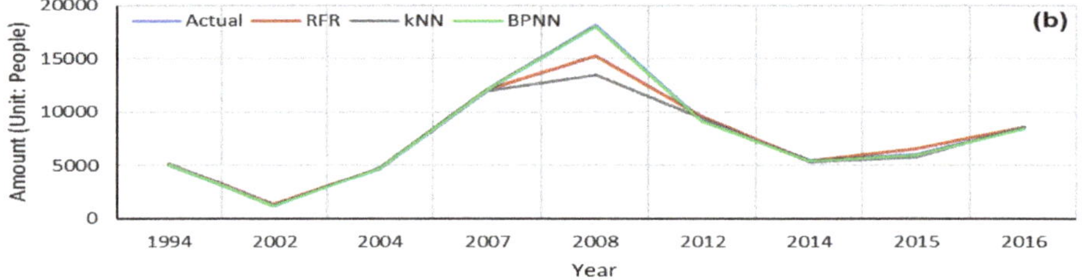
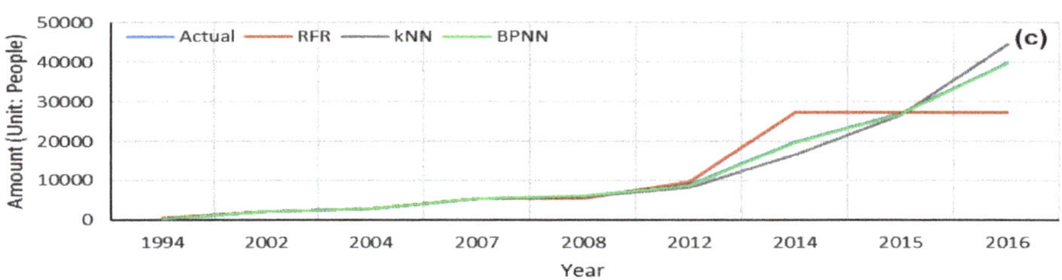

Figure 4. Vietnam labor immigration prediction for (**a**) Japan, (**b**) Korea, and (**c**) Taiwan. (**a**) labor immigration prediction for Japan; (**b**) labor immigration prediction for Korea; (**c**) labor immigration prediction for Taiwan.

Table 3. Accuracy parameters for labor immigration prediction.

Parameters	Japan			Taiwan			Korea		
	kNN	RFR	BPNN	kNN	RFR	BPNN	kNN	RFR	BPNN
MAPE	0.073	0.145	0.006	0.066	0.137	0.006	0.051	0.051	0.007
MAE	1657	2976	191	1023	2492	71	638	488	57
RMSE	1919	3858	225	1890	4917	102	1570	1002	67
R-squared	0.991	0.982	0.999	0.993	0.803	0.999	0.814	0.937	0.999
NSE	0.991	0.964	0.999	0.978	0.853	0.999	0.887	0.954	0.999

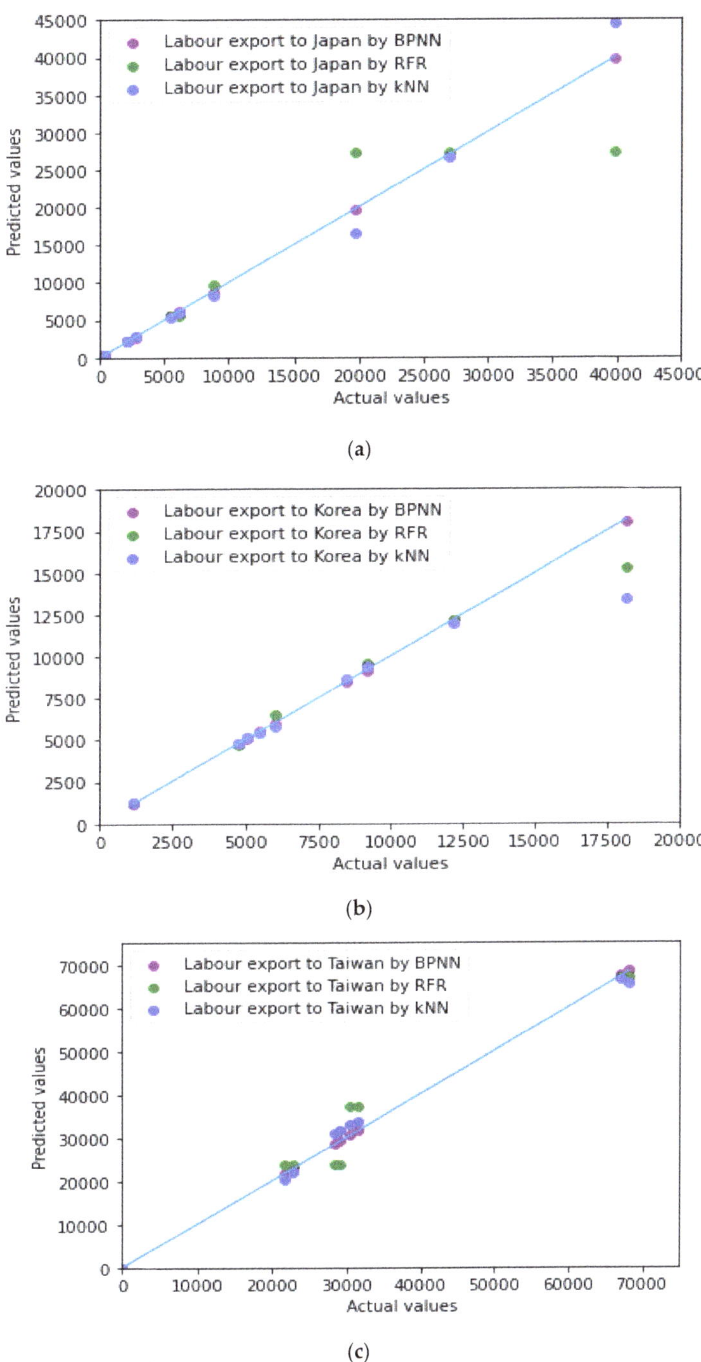

Figure 5. The best performance parameters for prediction: (**a**) labor exports to Japan, (**b**) labor exports to Korea, (**c**) labor exports to Taiwan (unit: people).

The Taylor diagrams examine the performance of estimated and real values using the standard deviation and correlation used in evaluating the models. Moreover, the Taylor chart shows the standard deviation and correlation between the actual and anticipated datasets for the models and general consistency between observed and estimated values when the correlation value approaches 1, as shown in Figure 6. This study can be considered for the kNN, RFR, and BPNN algorithms in Japan, which have SD (testing phase) = 20,649, 20,463, and 19,847, respectively, and are near to the actual data (with SD = 28,962), resulting in the three models achieving the same accurate level in predicting; meanwhile, the BPNN model's standard deviations in Japan, and Korea implied that the predicted values of BPNN model are nearest with the real values (with SD_{Japan} = 11,089, and SD_{Korea} = 3637). Hence, the smaller the standard deviation value, the stronger the relationship. The Taylor plot indicates that kNN, RFR, and BPNN algorithms are the most accurate and optimal forecasting.

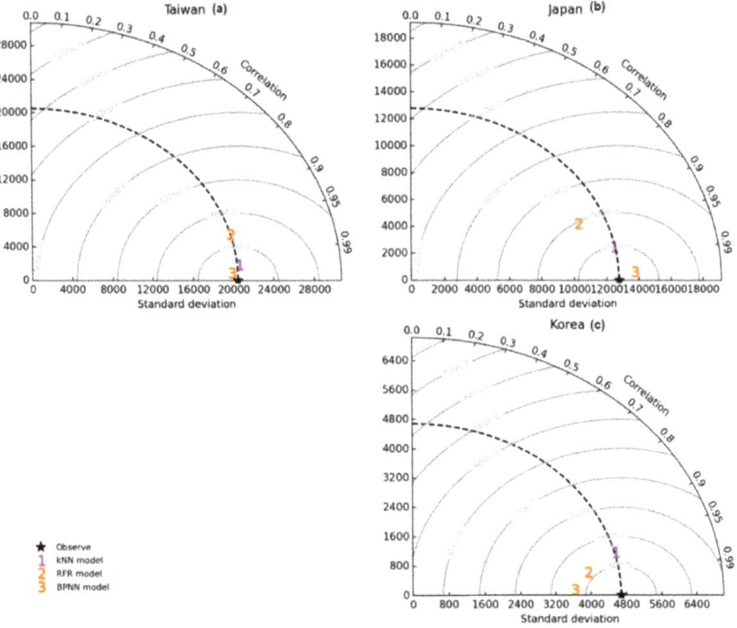

Figure 6. Taylor diagram representing the best performance of BPNN, RFR, kNN models of (**a**) labor exports to Japan, (**b**) labor exports to Taiwan, (**c**) labor exports to Korea.

4. Discussions

Regarding labor exports of Asian countries, Indonesia was the second-largest sending country of labor migrants in Asia after the Philippines, with a long history of immigration and emigration. The Philippines, for example, sent about one million employees abroad in 2005, Indonesia about 400,000, and Bangladesh and Sri Lanka each more than 200,000; Meanwhile, Vietnam only deployed 70,000 to 80,000 people in the same year. According to the most recent published estimates, there are around 500,000 Vietnamese contract workers globally [47]. According to the International Organization for Migration, Indonesian labor migrants significantly raised from 517,619 to 696,746 between 1996 and 2007, with the top five destinations being Saudi Arabia, Malaysia, Taiwan, Singapore, and Hong Kong [48]. Furthermore, although Japan's openness approves highly qualified migrants through open policies, highly skilled migrants in the targeted nations are still relatively small. In 2010, 198,000 highly skilled migrants in Japan accounted for barely 9% of the 2.1 million migrants [49]. This export labor also compared with other Asian countries pointed out that with amount more than 100,000 employees per yearly were migrated to Japan, Taiwan, and Korea, Vietnam has third ranking of Asia during the period of 2015–2019.

Furthermore, foreign worker exports help improve employees' lives, but they also substantially impact the country's economy. The number of remittances from overseas to Vietnam remitted to the country is expected to be between three and four billion dollars per year. Hence, this is encouraging news for the Vietnamese economy [50].

Human resource immigration has played a crucial role in Vietnam's socio-economic development. Firstly, it can improve income and transform workers' perceptions. Many workers, for instance, have returned to Vietnam to establish small and medium-sized businesses, contributing to the eradication of large-scale enterprises and using practical experiences gained from many areas of the world in Vietnam. Secondly, labor migration can help to alleviate poverty and advance Vietnam's socio-economic growth. Then, labor migration provides a substantial source of foreign currency while lowering investment costs to alleviate the issue of domestic employment. Finally, labor exports are also a mechanism for transferring an innovative technology from other nations, assisting in training a quality workforce, and strengthening international cooperation links between Vietnam and other countries in the Northeast Asian Countries. Therefore, forecasting labor exports is significantly important for Vietnam's labor exporting. The forecast results will assist labor export policymakers in assessing and considering foreign countries' potential and labor demand. Simultaneously, governments can develop vocational training programs and acquire human resources that fit those countries' needs.

This paper used kNN, BPNN, and RFR models to analyze and estimate Vietnamese human resources working for the Northeast Asian Countries. The different influential factors and parameters have been described in the simulations. The following key findings are as three models of BPNN, RFR, and kNN indicating forecast results were fairy accuracy for three countries Taiwan, Korea, and Japan. Regarding the error index, the BPNN model predicting Vietnamese labor immigration to Northeast Asian Countries obtained the best R^2, MAPE, RMSE, and MAE value. At the same time, the RFR and kNN models showed better forecasting performances. The points indicate that BPNN was the highlight compared with two RFR and kNN algorithms. In addition, the forecasting errors in the case of the models increased if the testing data increased.

The model was demonstrated excellently by a training accuracy of 89.98% and validation accuracy of 84.05%. In terms of estimating tourism demand in Japan, Chen, Lai [27] used a novel forecasting model based on empirical mode decomposition (EMD) and a BPNN, which revealed that the MAPE, RMSE of the proposed EMD-BPNN algorithm are, respectively, 0.958%, 1443, implying that these output values are higher than this study parameter. Moreover, Mishra et al. (2021) [31] study results witnessed that for predicting international tourists with tiny datasets, the random forest regression works well (with an R-square of 0.847), pointing out higher values than this study outputs in terms of R-square indicators.

In addition, the method using three models was the most classical one; however, the models were deployed for the time series data give performance comparable to deep learning models (as Long short-term memory (LSTM) models), but faster training speed and less resource-intensive (as the computer's memory). For instance, Wang et al. (2020) [29] deployed BPNN and LSTM models to forecast the monthly inbound tourism to mainland China (2000–2013), the monthly tourist arrival Turkey from different countries (2000–2011), and the monthly inbound tourism to The United States of America (June 2006–May 2018). The research result showed that R of BPNN model with 0.999 to approximate the R of LMST model with 0.998.

Although this study used annual data to anticipate labor force migration, the results showed that three models in three specific countries produced reliable results. As a result, it can predict labor migration in other countries around the globe.

5. Conclusions

Labor exports have been among of the most important tasks in Vietnam's socio-economic development over the past three decades. Labor exports will help increase income and improve skills for workers. In addition, labor exports also contribute to reducing unemployment in the country. Hence, the study aimed to examine the adjustment in the Vietnamese labor migration to Northeast Asian Countries. Based on annual data from 1992 to 2020, this study focused on the role of labor exports on Vietnam's socio-economic factors and compared them with the movement of human resources to other Asian countries. Afterward, the study implemented three machine learning models for estimation. At the same time, the result indicated that three models earned high accurate results for predicting labor migration, in which the BPNN model showed a more accurate level than other algorithms. Hence, the BPNN model can be effectively applied for labor migration prediction. In the context of integration, forecasting human resource export variation may find a separate direction for Northeast Asian Countries' human resource activities. Although the MAPE, MAE, and RMSE parameters of three models also attained the high level of accuracy which were compared with the other studies' parameters values, the BPNN model earned the highest accuracy of the actual labor exports with $MAPE_{Japan,\_BPNN} = 0.06$, $MAE_{Japan,\_BPNN} = 191$, $RMSE_{Japan\_BPNN} = 225$, $MAPE_{Korea,\_BPNN} = 0.006$, $MAE_{Korea\_BPNN} = 71$, $RMSE_{Korea,\_BPNN} = 102$, $MAPE_{Taiwan\_BPNN} = 0.007$, $MAE_{Taiwan\_BPNN} = 57$, $RMSE_{Taiwan\_BPNN} = 67$, respectively. In addition, this study proved that machine learning models play a key role in the decision-making progress for conducting an effect of labor exports. Using Augmented Dickey-Fuller (ADF) test indicates that stationary data will attain more forecast results' accurate parameters compared with non-stationary data. The study results support stakeholders in establishing projects related to constructing and training high-quality human resources migration, increasing the laborers' surplus value. Moreover, human resource migration problems help governments enforce suitable policies to appeal to labor from overseas. Another possible future work is to deeply analyze the GDP contribution of human resource immigration for Vietnam's economy.

Author Contributions: Conceptualization, N.H.G.; Data curation, T.-T.N.; Formal analysis, L.A.P.; Funding acquisition, N.H.G.; Investigation, T.-T.N.; Methodology, L.A.P.; Project administration, C.C.T.; Software, N.H.G.; Validation, C.C.T.; Writing—original draft, N.H.G., T.-T.N. and C.C.T.; Writing—review and editing, C.C.T. and T.-T.D. All authors have read and agreed to the published version of the manuscript.

Funding: This research received no external funding.

Institutional Review Board Statement: Not applicable.

Informed Consent Statement: Not applicable.

Data Availability Statement: Not applicable.

Acknowledgments: The authors appreciate the support from the National Kaohsiung University of Science and Technology, Taiwan; Southern Taiwan University of Science and Technology, Taiwan; Shenghua Marketing Co., Ltd., Taiwan; ThuDauMot University, Vietnam; and Hue University, Vietnam.

Conflicts of Interest: The authors declare no conflict of interest.

References

1. Crouch, G.I.; Ritchie, J.B. Tourism, Competitiveness, and Societal Prosperity. *J. Bus. Res.* **1999**, *44*, 137–152.
2. Rockström, J.; Williams, J.; Daily, G.; Noble, A.; Matthews, N.; Gordon, L.; Wetterstrand, H.; Declerck, F.; Shah, M.; Steduto, P.; et al. Sustainable Intensification of Agriculture for Human Prosperity and Global Sustainability. *Ambio* **2017**, *46*, 4–17. [PubMed]
3. Iversen, T.; Cusack, T.R. The Causes of Welfare State Expansion: Deindustrialization or Globalization? *World Politics* **2000**, *52*, 313–349.
4. Friedman, B.A. Globalization Implications for Human Resource Management Roles. *Empl. Responsib. Rights J.* **2007**, *19*, 157–171.
5. Awan, A.G. Relationship between Environment and Sustainable Economic Development: A Theoretical Approach to Environmental Problems. *Int. J. Asian Soc. Sci.* **2013**, *3*, 741–761.

6. Litvinenko, V.S. Digital Economy as a Factor in the Technological Development of the Mineral Sector. *Nat. Resour. Res.* **2020**, *29*, 1521–1541.
7. Ince Yenilmez, M. Economic and Social Consequences of Population Aging the Dilemmas and Opportunities in the Twenty-First Century. *Appl. Res. Qual. Life* **2015**, *10*, 735–752.
8. Yu, M. Study on the Development Dilemma and Breakthrough of the Traditional Sports of Ethnic Minority in the Northwest of Guangxi under the Background of Urbanization. In Proceedings of the 2017 International Conference on Innovations in Economic Management and Social Science (IEMSS 2017), Hangzhou, China, 15–16 April 2017; Atlantis Press: Paris, France, 2017.
9. Hughes, J.C.; Rog, E. Talent Management: A Strategy for Improving Employee Recruitment, Retention and Engagement within Hospitality Organizations. *Int. J. Contemp. Hosp. Manag.* **2008**, *20*, 743–757.
10. Goldsmith, P.D.; Gunjal, K.; Ndarishikanye, B. Rural–Urban Migration and Agricultural Productivity: The Case of Senegal. *Agric. Econ.* **2004**, *31*, 33–45.
11. Kapstein, E.B. Workers and the World Economy. *Foreign Aff.* **1996**, *75*, 16.
12. Chaponnière, J.-R.; Cling, J.-P.; Zhou, B. Vietnam following in China's footsteps: The third wave of emerging Asian economies. *South. Engines Glob. Growth* **2010**, 114–140.
13. Hugo, G. International labour migration and migration policies in Southeast Asia. *Asian J. Soc. Sci.* **2012**, *40*, 392–418.
14. Tuấn, T.M. Labor export management in some countries and practice in Vietnam. *J. Econ. Dev.* **2019**, 52–60.
15. Nguyen, T. Vietnam and Its Diaspora: An Evolving Relationship. In *Emigration and Diaspora Policies in the Age of Mobility*; Springer: Berlin/Heidelberg, Germany, 2017; pp. 239–255.
16. Hoang, L.A. Debt and (un) freedoms: The case of transnational labour migration from Vietnam. *Geoforum* **2020**, *116*, 33–41.
17. Bélanger, D.; Ueno, K.; Hong, K.T.; Ochiai, E. From foreign trainees to unauthorized workers: Vietnamese migrant workers in Japan. *Asian Pac. Migr. J.* **2011**, *20*, 31–53.
18. Ishizuka, F. International labor migration in Vietnam and the impact of receiving countries' policies. In *Institute of Developing Economies*; Japan External Trade Organization: Chiba, Japan, 2013.
19. Ho, T.T.N. *Gender and Transnational Migration: Vietnamese Women Workers and Negotiation of Self-Identity in Multicultural Taiwan*; Young Scholar Workshop; European Research Center on Contemporary: Tuebingen, Germany, 2012.
20. Chen, T.; Yin, X.; Peng, L.; Rong, J.; Yang, J.; Cong, G. Monitoring and Recognizing Enterprise Public Opinion from High-Risk Users Based on User Portrait and Random Forest Algorithm. *Axioms* **2021**, *10*, 106.
21. Hu, X.; Wang, J.; Wang, L.; Yu, K. K-Nearest Neighbor Estimation of Functional Nonparametric Regression Model under Na Samples. *Axioms* **2022**, *11*, 102.
22. Gaxiola, F.; Melin, P.; Valdez, F.; Castro, J.R.; Manzo-Martínez, A. Pso with Dynamic Adaptation of Parameters for Optimization in Neural Networks with Interval Type-2 Fuzzy Numbers Weights. *Axioms* **2019**, *8*, 14.
23. Shi, C.; Zhuang, X. A Study Concerning Soft Computing Approaches for Stock Price Forecasting. *Axioms* **2019**, *8*, 116.
24. Juraev, D.A.; Noeiaghdam, S. Modern Problems of Mathematical Physics and Their Applications. *Axioms* **2022**, *11*, 45.
25. Too, J.; Abdullah, A.R.; Saad, N.M. Hybrid Binary Particle Swarm Optimization Differential Evolution-Based Feature Selection for Emg Signals Classification. *Axioms* **2019**, *8*, 79.
26. Bretti, G.J.A. Differential Models, Numerical Simulations and Applications. *Axioms* **2021**, *10*, 260.
27. Chen, C.-F.; Lai, M.-C.; Yeh, C.-C. Forecasting tourism demand based on empirical mode decomposition and neural network. *Knowl.-Based Syst.* **2012**, *26*, 281–287.
28. Vihikan, W.O.; Putra, I.K.G.D.; Dharmaadi, I.P.A. Foreign Tourist Arrivals Forecasting Using Recurrent Neural Network Backpropagation through Time. *TELKOMNIKA (Telecommun. Comput. Electron. Control)* **2017**, *15*, 1257–1264.
29. Wang, L.; Wu, B.; Zhu, Q.; Zeng, Y.-R. Forecasting Monthly Tourism Demand Using Enhanced Backpropagation Neural Network. *Neural Process. Lett.* **2020**, *52*, 2607–2636.
30. Imbrie, P.; Lin, J.J.-J.; Malyscheff, A. Artificial Intelligence Methods to Forecast Engineering Students' Retention Based on Cognitive and Non Cognitive Factors. In Proceedings of the 2008 Annual Conference & Exposition, Pittsburgh, PA, USA, 22–25 June 2008.
31. Mishra, R.K.; Urolagin, S.; Jothi, J.; Nawaz, N.; Haywantee, R. Machine Learning based Forecasting Systems for Worldwide International Tourists Arrival. *Int. J. Adv. Comput. Sci. Appl.* **2021**, *12*, 55–64.
32. Rice, W.L.; Park, S.Y.; Pan, B.; Newman, P. Forecasting campground demand in US national parks. *Ann. Tour. Res.* **2019**, *75*, 424–438.
33. Sahu, S.K.; Dey, D.K.; Branco, M.D. A new class of multivariate skew distributions with applications to bayesian regression models. *Can. J. Stat.* **2003**, *31*, 129–150.
34. Brys, G.; Hubert, M.; Struyf, A. A Robust Measure of Skewness. *J. Comput. Graph. Stat.* **2004**, *13*, 996–1017.
35. Olyaie, E.; Abyaneh, H.Z.; Mehr, A.D. A comparative analysis among computational intelligence techniques for dissolved oxygen prediction in Delaware River. *Geosci. Front.* **2017**, *8*, 517–527.
36. Aiello, M.; Yang, Y.; Zou, Y.; Zhang, L.J. *Artificial Intelligence and Mobile Services-Aims 2018*; Springer: Berlin/Heidelberg, Germany, 2018.
37. Nawi, N.M.; Khan, A.; Rehman, M.Z. A new back-propagation neural network optimized with cuckoo search algorithm. In Proceedings of the International Conference on Computational Science and Its Applications, Ho Chi Minh City, Vietnam, 24–27 June 2013; Springer: Berlin/Heidelberg, Germany, 2013; pp. 413–426.

38. Wang, L.; Zeng, Y.; Chen, T. Back propagation neural network with adaptive differential evolution algorithm for time series forecasting. *Expert Syst. Appl.* **2015**, *42*, 855–863.
39. Zhao, H.-Liu, W.; Guan, H.; Fu, C. Analysis of Diaphragm Wall Deflection Induced by Excavation Based on Machine Learning. *Math. Probl. Eng.* **2021**, *2021*, 6664409.
40. Breiman, L. Random forests. *Mach. Learn.* **2001**, *45*, 5–32.
41. Guo, L.; Chehata, N.; Mallet, C.; Boukir, S. Relevance of airborne lidar and multispectral image data for urban scene classification using Random Forests. *ISPRS J. Photogramm. Remote Sens.* **2011**, *66*, 56–66.
42. Rodriguez-Galiano, V.F.; Ghimire, B.; Rogan, J.; Chica-Olmo, M.; Rigol-Sanchez, J.P. An assessment of the effectiveness of a random forest classifier for land-cover classification. *ISPRS J. Photogramm. Remote Sens.* **2012**, *67*, 93–104.
43. Rodriguez-Galiano, V.; Sanchez-Castillo, M.; Chica-Olmo, M.; Chica-Rivas, M.J.O.G.R. Machine learning predictive models for mineral prospectivity: An evaluation of neural networks, random forest, regression trees and support vector machines. *Ore Geol. Rev.* **2015**, *71*, 804–818.
44. Kardani, N.; Zhou, A.; Nazem, M.; Shen, S.L. Estimation of bearing capacity of piles in cohesionless soil using optimised machine learning approaches. *Geotech. Geol. Eng.* **2020**, *38*, 2271–2291.
45. Touzani, S.; Granderson, J.; Fernandes, S. Gradient boosting machine for modeling the energy consumption of commercial buildings. *Energy Build.* **2018**, *158*, 1533–1543.
46. Yang, J.-H.; Yang, M.-S. Engineering. A control chart pattern recognition system using a statistical correlation coefficient method. *Comput. Ind. Eng.* **2005**, *48*, 205–221.
47. Bélanger, D. Labor migration and trafficking among Vietnamese migrants in Asia. *Ann. Am. Acad. Political Soc. Sci.* **2014**, *653*, 87–106.
48. Deng, J.B.; Wahyuni, H.I.; Yulianto, V.I. Labor migration from Southeast Asia to Taiwan: Issues, public responses and future development. *Asian Educ. Dev. Stud.* **2020**, *10*, 69–81.
49. Oishi, N. The limits of immigration policies: The challenges of highly skilled migration in Japan. *Am. Behav. Sci.* **2012**, *56*, 1080–1100.
50. Nguyen, P.H. Remittances and competitiveness: A case study of Vietnam. *J. Econ. Bus. Manag.* **2017**, *5*, 79–83.

Article

Lexicon-Enhanced Multi-Task Convolutional Neural Network for Emotion Distribution Learning

Yuchang Dong and Xueqiang Zeng *

School of Computer & Information Engineering, Jiangxi Normal University, Ziyang Road 99, Nanchang 330022, China; yixuxi_dong@jxnu.edu.cn
* Correspondence: xqzeng@jxnu.edu.cn

Abstract: Emotion distribution learning (EDL) handles emotion fuzziness by means of the emotion distribution, which is an emotion vector that quantitatively represents a set of emotion categories with their intensity of a given instance. Despite successful applications of EDL in many practical emotion analysis tasks, existing EDL methods have seldom considered the linguistic prior knowledge of affective words specific to the text mining task. To address the problem, this paper proposes a text emotion distribution learning model based on a lexicon-enhanced multi-task convolutional neural network (LMT-CNN) to jointly solve the tasks of text emotion distribution prediction and emotion label classification. The LMT-CNN model designs an end-to-end multi-module deep neural network to utilize both semantic information and linguistic knowledge. Specifically, the architecture of the LMT-CNN model consists of a semantic information module, an emotion knowledge module based on affective words, and a multi-task prediction module to predict emotion distributions and labels. Extensive comparative experiments on nine commonly used emotional text datasets showed that the proposed LMT-CNN model is superior to the compared EDL methods for both emotion distribution prediction and emotion recognition tasks.

Keywords: emotion distribution learning; text-based emotion analysis; affective words; multi-task CNN

Citation: Dong, Y.; Zeng, X. Lexicon-Enhanced Multi-Task Convolutional Neural Network for Emotion Distribution Learning. *Axioms* **2022**, *11*, 181. https://doi.org/10.3390/axioms11040181

Academic Editor: Oscar Humberto Montiel Ross

Received: 26 February 2022
Accepted: 14 April 2022
Published: 17 April 2022

Publisher's Note: MDPI stays neutral with regard to jurisdictional claims in published maps and institutional affiliations.

1. Introduction

Emotion analysis aims to recognize and analyze the emotions behind massive human behavior data such as text, pictures, movies, and music [1–3]. With the rapid development of Internet-based social media, text-based emotion classification shows promising application prospects in many emerging artificial intelligence fields such as public opinion analysis, commodity recommendation, and business decision making [4]. In recent years, the text-oriented emotion classification model has become a research hotspot in the field of natural language processing and machine learning [5].

Many traditional emotion classification models adopt the multi-label learning paradigm and assume that an example is associated with some emotion labels. On this technical line, many scholars have proposed various effective works to solve the problem of emotion recognition. Multi-label learning can deal with multi-emotion recognition tasks, but it cannot quantitatively model a variety of emotions with different expression intensities [6]. To solve this problem, Zhou et al. proposed emotion distribution learning (EDL) [7] motived by label distribution learning (LDL) [6]. Different from the traditional emotion classification model, EDL associates an emotion distribution vector with each example (e.g., facial image or text sentence). The emotion distribution vector records the expression degree of a given example of each emotion label, and its dimension is the number of all emotions. In recent years, many EDL research works have been published in top conferences and journals in the field of machine learning. For example, Yang et al. proposed a circular-structured representation for visual emotion distribution learning by exploiting the intrinsic relationship between emotions based on psychological models [8]. Xu and Wang proposed a method for learning emotion distribution based on an attention mechanism in 2021, using an emotional graph-based network to explore the correlation between various regions in the image and emotion distribution [9]. Fei et al. proposed a latent emotional memory network that can learn latent emotional distributions without external knowledge, and the model has been used well for classification tasks [10]. Zhao et al. proposed a small sample text EDL model of the meta-learning method [11]; Jia et al. proposed a facial EDL method using local low-rank label correlation [12]; Pang et al. proposed a basic model of an acceleration algorithm to predict the emotion distribution of unlabeled files [13]. These EDL methods can effectively record the intensity of examples of different emotion labels and show better performance than the traditional emotion classification model. However, most of the existing EDL methods do not introduce the unique affective word information containing prior emotion knowledge into the prediction model.

Affective words are words with different emotional tendencies [14], which have generally been manually labeled based on emotional linguistic knowledge. Different affective words are usually used to describe different emotional characteristics, and different combinations of affective words can also express different emotional tendencies. At present, some scholars have used affective word information in the field of emotion analysis. Teng et al. showed that affective words have a significant effect in predicting emotion in 2016 [14]. Zhang et al. proposed lexicon-based emotion distribution label enhancement (LLE) in 2018 [15]. Tong et al. annotated affective words manually and used them for emotion analysis [16]. These studies show that affective words can significantly improve the performance of the emotion analysis model. However, so far, there is no EDL method to use affective word information for emotion distribution prediction.

To address this problem, we propose a text emotion distribution learning model based on a lexicon-enhanced multi-task convolutional neural network (LMT-CNN). The overall architecture of the LMT-CNN model has three major modules: semantic information, emotion knowledge, and multi-task prediction. The semantic information module uses the sliding window convolution neural network to extract the semantic information of the text from the word embedding space of the input text. Based on the affective words extracted from the text, the emotion knowledge module uses the lexicon to introduce the corresponding emotional prior knowledge to synthesize an emotion knowledge vector. The input of the multi-task prediction module is constructed from the outputs of the first two modules. Then, the final emotion distribution is predicted through a full connection layer. The two prediction tasks of the emotion distribution output layer are emotion distribution prediction based on Kullback–Leibler (KL) loss [17] and emotion classification based on cross-entropy loss. The emotion with the highest score in the emotion distribution output layer is used as the dominant emotion output for emotion classification. The existing EDL research work shows that the multi-task convolution neural network model combined with KL loss and cross-entropy loss can achieve better performance by simultaneously training emotion distribution prediction and emotion classification tasks in an end-to-end manner [15].

Different from the existing EDL work based on neural networks, the proposed LMT-CNN method considers the linguistic prior knowledge of affective words unique to the text mining task and combines it with text-based semantic information to construct an end-to-end deep neural network. We evaluated the LMT-CNN method on the English emotion distribution dataset Semeval [18], Chinese emotion distribution dataset Ren-CECps [19], four English single-label emotion datasets (Fairy Tales [20], TEC [21], CBET, ISEAR [22], and Affect in Tweets [23]), and Chinese single-label emotion datasets NLP&CC 2013 and NLP&CC 2014 [24], comparing LMT-CNN methods with multiple baseline methods. Compared with the baseline methods, our LMT-CNN method achieved the best performance in almost all measurements.

The rest of the paper is organized as follows. First, Section 2 briefly reviews some related works. Then, Section 3 describes the proposed lexicon-enhanced multi-task convolutional neural network for emotion distribution learning. Section 4 describes in detail the datasets used for the experiments, experimental setup, and experimental results and analysis. Section 5 mainly discusses the limitations of our proposed method. Finally, Section 6 concludes the entire paper and provides an outlook on future work.

2. Related Work

Emotions are ubiquitous in our daily life and play an important role in our lives, influencing our decisions and judgments. For a long time, people have expected a variety of ways to recognize the diverse emotions that humans generate. Many researchers have proposed a variety of algorithms aiming at making machines that have emotional intelligence through artificial intelligence algorithms, meaning machines that have the ability to recognize, interpret, and process emotions. For this reason, machines should first learn to recognize human emotions from their external and implicit emotional cues.

Emotion models are the basis of emotion recognition systems, which define the expression of emotions. To measure emotions quantitatively, psychologists considered that emotions exist in multiple states and therefore proposed various emotion models to distinguish different emotional states. Two of the most prominent emotion representation models are the categorical emotional state (CES) and the dimensional emotional space (DES) [25]. CES classifies emotions into different basic categories and considers each type of basic emotion independent. Among the popular CES models are binary emotions [26], Ekman's six emotions [27], and Plutchik's eight emotions [28]. Binary emotions contain positive, negative, and sometimes neutral emotions. In this case, "emotion" is often referred to as "sentiment". In the Paul Ekman model, emotions are independent and can be distinguished into six basic categories depending on how they are perceived by the experiencer. These basic emotions are joy, sadness, anger, disgust, surprise, and fear. Plutchik's eight emotions consist of amusement, anger, awe, contentment, disgust, excitement, fear, and sadness. The DES model assumes that emotions do not exist independently of each other and that there are interactions between emotions. Therefore, some scholars consider locating emotions in a multidimensional space, such as valence–arousal–dominance (VAD) [29]. In our work, Ekman's six emotions were used as the target emotion label set.

Early work in emotion recognition focused on combining machine learning methods to learn emotional features to recognize the emotions embedded in text, speech, or images. Examples include the methods of Naive Bayes, maximum entropy, and support vector machines [30]. Later, Vrysis et al. proposed a method for emotion recognition combined with lexicon-based and rule-based algorithms [31]. The lexicon-based approach relies on the semantic direction of the text with the polarity of the words and phrases appearing in it [32]. Rule-based algorithms design extraction rules based on syntactic dependencies [33], which need to be used within the controllable range of the rules and may lead to incorrect emotion judgments if they are beyond the range of the rules. We believe that the performance of this approach largely depends on the quality of the lexicon and the number and quality of the rules formulated. It is more suitable when the data are lacking, and there is a performance bottleneck when dealing with large-scale data, combined with the fact that now the path of the method based on deep learning language models is more attractive in the field of emotion classification [34].

Deep learning changes in feature engineering and feature learning both make problem solving easier. Traditionally, the efficiency of machine learning algorithms is highly dependent on how well the input data are represented. For this reason, feature engineering has been the most critical step in the machine learning workflow. In contrast, deep learning algorithms can automate feature extraction, which allows researchers to extract features with minimal domain knowledge and manpower. Another transformative aspect of deep learning is that models can learn all representation layers together at the same time. Through common feature learning, once the model modifies an internal feature, all other features that depend on that feature will automatically adapt accordingly without human intervention.

three modules: semantic information module, emotion knowledge module, and multi-task prediction module. The specific network structure is shown in Figure 2.

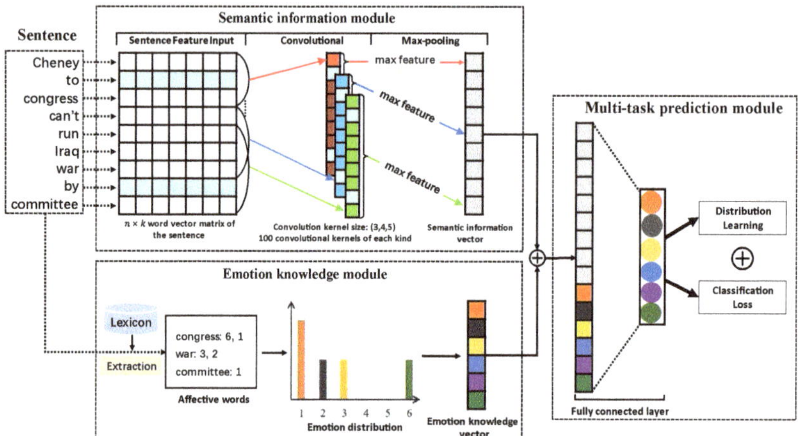

Figure 2. Overall framework of the LMT-CNN model (emotion labels 1: anger, 2: disgust, 3: sadness, 4: surprise, 5: fear, 6: joy).

3.1. Semantic Information Module

The semantic information module is a convolution neural network (CNN) built on word embedding for extracting semantic information from text. The CNN model was originally created in the domain of image recognition and classification, and its recent application in NLP has proven to be successful with excellent results [15]. Another competitive approach for semantic extraction is the long short-term memory (LSTM) network, which has been shown to have a good ability to learn sequential data [36]. LSTM controls the transmission state by gating the states, stores information that needs to be remembered for a long time, and forgets unimportant information. To the best of our knowledge, there is no evidence to prove which approach is definitely better for emotion recognition. However, because of the many contents introduced, LSTM leads to more parameters and makes the training more time-consuming. Since our aim was to verify the influence of prior knowledge of affective words on the model, we did not want to involve unnecessarily complex computations. Hence, we followed the previous successful EDL work by Zhang et al. [15] and used CNN to extract semantic information.

The semantic information module is composed of the input layer, convolution layer, and max-pooling layer. The specific workflow is as follows: Firstly, each sentence of the text dataset is used as the input of the module. Then, each word is transformed into a word embedding based on the pretrained word embedding model. Finally, the word embedding matrix representing the original sentence is processed by the convolution layer and max-pool layer to output a semantic information vector.

We use $S = \{(s_1, D_1), (s_2, D_2), \ldots, (s_n, D_n)\}$, which represents the training text dataset, where n is the number of sentences in the dataset, s_i ($i \in \{1, 2, \ldots, n\}$) is the i-th sentence, $D_i = \{d_{s_i}^{a_1}, d_{s_i}^{a_2}, \ldots, d_{s_i}^{a_N}\}$ represents for sentence s_i corresponding emotion distribution, $A = \{a_1, a_2, \ldots, a_N\}$ is a finite set of emotion labels, and N is the number of emotion labels. Next, we specifically describe the functions of the input layer, convolution layer, and max-pooling layer of this module.

Input Layer: The length of the input sentence is M, and $x_i \in R^k$ is the k-dimensional word2vec word embedding representation of the i-th word in the sentence. In this way, the input sentence is represented as a $M \times k$ word embedding matrix x as

$$x = x_1 \oplus x_2 \oplus \ldots \oplus x_M, \tag{1}$$

where $\oplus$ is the concatenation operator. Note that, if the sentence length is less than M, the end of the word embedding matrix is filled with 0.

Convolution layer: A set of filters $w \in \mathbb{R}^{h \times k}$ with a sliding window size of h are adopted to generate new features. The width of the filter is the same as the width of the word embedding matrix, and thus the filter can only move in the height direction to conduct the convolution on several adjacent words. For example, set $x_{p:p+q}$ denotes the concatenation of word $x_p, x_{p+1}, \ldots, x_{p+q}$, and a new feature v_p is generated from the word window $x_{p:p+h-1}$ as

$$v_p = f\left(w \cdot x_{p:p+h-1} + b\right), \tag{2}$$

where $f(\cdot)$ is a nonlinear active function, and $b \in \mathbb{R}$ is a bias term. The filter is used for each possible word window in the sentence $x_{1:h}, x_{2:h+1}, \ldots, x_{M-h+1:M}$ to produce the feature mapping v by

$$v = [v_1, v_2, \ldots, v_{M-h+1}]. \tag{3}$$

Max-pooling layer: We use the standard max-pooling operation to obtain the generated feature map and take the maximum value of v as the characteristic of this particular filter:

$$v_{semantic} = \max(v), \tag{4}$$

this operation can capture the feature with the highest value as the most important feature for each feature map.

Through the above operations, a semantic feature can be extracted by each filter. The module uses several filters of different sizes to extract multiple semantic features from a given sentence.

3.2. Emotion Knowledge Module

The emotion knowledge module uses the lexicon to extract affective words from the sentence text and then synthesizes an emotion knowledge vector based on the prior emotion labels corresponding to affective words.

Given sentence s_i, we extract all affective words in s_i from the lexicon and obtain all the emotion labels corresponding to each emotion word. For each emotion, we set the number of emotion labels of corresponding affective words as m_{ij} and the total number of all emotion labels as C_i. Then, we set the emotion knowledge vector to $r_i = [r_{i1}, r_{i2}, \ldots, r_{iN}]$, where

$$r_{ij} = \begin{cases} \frac{m_{ij}}{C_i}, & \text{if } C_i \neq 0 \\ \frac{1}{N}, & \text{else} \end{cases}, \ j = 1, 2, \ldots, N. \tag{5}$$

Taking the input sentence in Figure 2 as an example, we extract three emotional words—congress, war, and committee—by looking up the lexicon. Among them, the affective word congress has two emotion labels (anger and joy), war has two emotion labels (disgust and sadness), and committee has one emotion label (anger). The total number of emotion labels of all affective words in the sentence is 5, the number of categories of emotional labels is 6, and the corresponding emotion knowledge vector is $\left[\frac{2}{5}, \frac{1}{5}, \frac{1}{5}, 0, 0, \frac{1}{5}\right]$.

In addition, when a sentence does not contain any affective words, the emotion knowledge weight of the sentence is the same on the six emotion labels, and its corresponding emotion knowledge vector is $\left[\frac{1}{6}, \frac{1}{6}, \frac{1}{6}, \frac{1}{6}, \frac{1}{6}, \frac{1}{6}\right]$.

3.3. Multi-Task Prediction Module

The multi-task prediction module supports two tasks: emotion distribution prediction and emotion classification, including the semantic synthesis layer and emotion distribution output layer.

Semantic synthesis layer: We splice the output $v_{semantic}$ of the semantic information module and the output r_i of the emotion knowledge module to generate the semantic synthesis vector $\hat{v}$ by

$$\hat{v} = v_{semantic} \oplus \alpha \cdot r_i. \tag{6}$$

where $\oplus$ is the vector splicing operation, and the parameter α is used to control the ratio of adding emotional word information. When $\alpha = 0$, it means that the semantic synthesis layer only contains semantic information and does not consider the prior knowledge of affective words; when $\alpha = 1$, it means that all affective word information is considered in the semantic synthesis layer, and the semantic information is combined with all affective words prior to knowledge splicing. We believe that the importance of semantic information in the LMT-CNN model should be greater than affective word knowledge, that is, the value of parameters α should not be too large. This is because, in general, manually labeled emotion labels have high reliability, while the prior information of general affective words has more noise. This study analyzed the value of the parameter α in the experimental part where it is further discussed.

Emotion distribution output layer: Semantic synthesis vector $\hat{v}$ after a full connection layer transformation and the output result of the emotion distribution layer are obtained. The dimension of the emotion distribution layer is the number of categories of emotion labels.

The LMT-CNN model uses an end-to-end manner to train the emotion distribution prediction and emotion classification tasks simultaneously. For the distribution prediction task, we use KL loss [13] to measure the distance between the real distribution and the predicted one. The KL loss is defined as

$$E_{edl}(s, D) = -\frac{1}{N} \sum_{i=1}^{d} \sum_{j=1}^{N} d_{s_i}^{j} \ln p_j^{(i)}, \tag{7}$$

where $p_j^{(i)} = \frac{e_j^{(i)}}{\sum_{t=1}^{N} e_t^{(i)}}$, $d_{s_i}^{j}$ represents the sum of each emotion label loss in sentence s_i.

The optimization objective of the emotion classification task is cross-entropy loss $E_{cls}(s, y)$, and its formula is

$$E_{cls}(s, y) = -\frac{1}{N} \sum_{i=1}^{d} \sum_{j=1}^{N} \mathbf{1}(y_i = j) ln p_j^{(i)}, \tag{8}$$

where $\mathbf{1}(y_i = j) = 1$ means when y_i is correctly divided into j classes; otherwise, $\mathbf{1}(y_i = j) = 0$.

The total loss of the emotion distribution output layer is a weighted combination of KL loss and cross-entropy loss as

$$E = (1 - \lambda)E_{cls}(s, y) + \lambda E_{edl}(s, D), \tag{9}$$

where λ is a weight parameter to control the importance of two kinds of losses. According to the experimental results of the existing related work, we set $\lambda = 0.7$ [15].

Finally, we use the stochastic gradient descent (SGD) algorithm [48] to optimize the LMT-CNN model. When the prediction task is emotion classification, we take the emotion with the highest expression degree in the emotion distribution output from the full connection layer as the real emotion of the sentence.

Recent research work shows that the emotion distribution model based on a multi-task convolutional neural network (MT-CNN) [15] can achieve good results by simultaneously training emotion distribution prediction and emotion classification tasks in an end-to-end manner. However, the MT-CNN model does not consider the linguistic prior knowledge of affective words, which is very important in the task of text analysis. In the experimental part, the performance of LMT-CNN and MT-CNN models is compared and analyzed.

The source code of the LMT-CNN model is released at our website [49].

4. Experiments

To investigate the performance of the LMT-CNN model proposed in this paper, we conducted three groups of experiments on two emotion distribution datasets and five single-label emotion datasets. First, to verify the effectiveness of adding affective word information to the LMT-CNN model, we changed the value of α and recorded the performance of LMT-CNN. Second, to test the performance of the emotion distribution prediction of the LMT-CNN model, we used 8 different emotion distribution learning methods compared with the LMT-CNN method. Third, to examine the performance of the LMT-CNN model on emotion classification tasks, we used the deep-learning-based TextCNN, MT-CNN, and LMT-CNN emotion distribution learning models for comparative experiments.

4.1. Dataset

There were 9 commonly used text emotional datasets adopted in the experiments, namely Semeval [18], Ren-CECps [19], Fair Tales [20], TEC [21], CBET, ISEAR [22], Affect in Tweets [23], NLP&CC 2013, and NLP&CC 2014 [24]. Among them, Semeval and Ren-CECps are emotion distribution datasets, and the other 7 datasets are traditional single-label datasets. We list the details of all experimental datasets in Table 1.

Table 1. Experimental datasets (the sentence number of each emotion in the SemEval and Ren-CECps datasets is the sentence number whose dominant emotion is the corresponding emotion).

Dataset	Anger	Disgust	Sadness	Surprise	Fear	Joy	All
SemEval	91	42	192	435	262	228	1250
Ren-CECps	1117	2293	5841	536	-	4137	13,924
Fairy Tales	216	-	264	114	166	444	1204
TEC	1555	761	3830	3849	2816	8240	21,051
CBET	8540	8540	8540	8540	8540	8540	51,240
ISEAR	1087	1081	1083	-	1090	1090	5431
Affect in Tweets	844	-	747	-	1105	793	3489
NLP&CC 2013	1146	2075	1562	482	186	2140	7581
NLP&CC 2014	1899	3130	2478	820	299	2805	11,431

SemEval is an English text dataset containing multiple emotions, including 1250 news headlines labeled with 6 emotion labels (anger, disgust, sadness, surprise, fear, and joy). Each headline has a percentile score for each emotion. The multiple emotions associated with each sample can be regarded as an emotion distribution vector, and the length of the vector is 1 through normalization.

Ren-CECps is a Chinese text emotion distribution dataset. The sentence corpus comes from Chinese blogs. Each sentence is labeled with 8 emotion labels, and each sentence has a score for each emotion. In order to be consistent with the types of emotions in other datasets, we selected 13,924 sentences of 5 emotions (anger, disgust, sadness, surprise, and joy) for experiments. The emotion distribution label retains the scores of the corresponding five emotions, and the multiple emotion labels corresponding to each sentence can be regarded as an emotion distribution vector, and the length of the vector is 1 by normalization.

Fairy Tales contains 1204 sentences from 185 children's stories, each sentence labeled with one of five emotions (anger, sadness, surprise, fear, and joy). TEC contains 21,051 emotional tweets, each marked with one of six emotions (anger, disgust, sadness, surprise, fear, and joy). CBET is marked with 76,860 tweets of 9 emotions, and 8540 tweets are collected for each emotion. We retained a total of 51,240 tweets of 6 emotions. ISEAR contains 7666 English sentences. Each English sentence describes the situation and experience of different people when they experience seven main emotions. We retained five emotions, a total of 5431 sentences. Affect in Tweets is created from tweets. There are four emotions contained in dataset: anger, fear, joy, and sadness. The dataset contains 3489 English comments from tweets.

NLP&CC 2013 is a Chinese Weibo corpus containing 7 emotion labels released by the Natural Language Processing and Chinese Computing Conference in 2013, containing 10,552 emotional sentences and 29,633 non-emotional sentences. In this experiment, 7581 emotional sentences marked with six emotions of anger, disgust, sadness, surprise, fear, and joy were retained. Table 1 respectively lists the detailed information of the dataset. NLP&CC 2014 Chinese microblog emotion analysis dataset is from Sina Weibo, and all tweets in the dataset are divided into two categories with or without emotion; those without emotion are labeled as none, and those with emotion are divided into 7 emotion categories. In this paper, six categories of emotions were selected: happiness, anger, sadness, fear, disgust, and surprise; a total of 11,431 Chinese tweets were selected.

The English lexicon used in the experiment was formed by the merger of Emosenticnet [50] and NRC [51]. The Emosenticnet lexicon contains 13,189 English affective words, and each affective word is labeled with 6 emotion labels. Each affective word in Emosenticnet and NRC lexicons is marked with one or more emotion labels. When the two lexicons were merged, we retained the 6 emotion labels (anger, disgust, sadness, surprise, fear, and joy) and deleted the affective words that were not labeled with these 6 emotions. For an affective word shared by the two dictionaries, the emotion label is the union of the emotion label of the affective word in the two lexicons. In the end, we obtained an English affective lexicon consisting of 15,603 affective words and 6 emotion labels. The Chinese affective lexicon uses the emotion vocabulary text database of the Dalian University of Technology [52], which has 27,466 affective words and 11 emotion labels. We kept 5 emotion labels (anger, disgust, sadness, surprise, and joy) corresponding to the Chinese dataset and deleted the affective words that were not labeled with these 5 emotions. Finally, 15,179 Chinese affective words were retained, and each word was labeled with 1 label.

4.2. Experimental Setup

The experimental process adopted the standard stratified 10-fold cross-validation. We partitioned the dataset before feeding the data into the network. All samples were divided into ten categories evenly under the constraint of keeping the category proportions essentially the same. Each time, one of the samples was selected as the test set, and the remaining data were merged as the training dataset. In total, 90% of the training set was used for neural network training, and the remaining 10% was used as the validation set. Each fold cross-validation was a separate emotion prediction task, repeated ten times, and the average evaluation indicators of the ten cross-validations were used to evaluate the final performance of the model. The data division of all models participating in the comparison experiment remained the same.

Since the output of the LDL model is a distribution vector, a single indicator can reflect only one aspect of the algorithm on a particular data, and it is difficult for us to determine which indicator is the best. Geng et al. proposed that when comparing different LDL algorithms in the same dataset, multiple indicators can be used to evaluate and compare the algorithms [6]. Following four principles, six indicators were finally selected to evaluate the LDL algorithms. Based on this suggestion, Zhou et al. chose six indicators to measure the average similarity or distance between the actual and predicted emotion distributions in the EDL model, respectively [7]. Similarly, we used six evaluation indicators for our emotion distribution prediction task evaluation, namely *Euclidean, Sørensen, Squaredχ^2, KL Divergence, Cosine,* and *Intersection* [7]. The calculation formulas of the 6 evaluation indicators are as follows:

$$Euclidean(P, Q) = \sqrt{\sum_{i=1}^{N}(P_i - Q_i)^2} \tag{10}$$

$$Sorensen(P, Q) = \frac{\sum_{i=1}^{N}|P_i - Q_i|}{\sum_{i=1}^{N}|P_i + Q_i|} \tag{11}$$

$$Squared\chi^2(P, Q) = \sum_{i=1}^{N}\frac{(P_i - Q_i)^2}{P_i + Q_i} \tag{12}$$

$$K - L_{(P,Q)} = \sum_{i=1}^{N} P_i ln \frac{P_i}{Q_i} \tag{13}$$

$$Cosine(P,Q) = \frac{\sum_{i=1}^{N}(P_i \times Q_i)}{\sqrt{\sum_{i=1}^{N} P_i{}^2} \times \sqrt{\sum_{i=1}^{C} Q_i{}^2}} \tag{14}$$

$$Intersection(P,Q) = \sum_{i=1}^{N} min(P_i, Q_i) \tag{15}$$

where P is the true emotion distribution of the text, Q is the predicted emotion distribution, and N is the number of emotion types contained in the sentence. These 6 indicators are used to measure the similarity between the predicted emotion distribution Q and the real emotion distribution P.

The emotion classification task uses 4 commonly used classification task evaluation indicators, namely *Precision, Recall, F1-score* and *Accuracy*. The specific calculation formulas of the 4 evaluation indicators are as follows:

$$Precision = \frac{|TP|}{|TP + FP|} \tag{16}$$

$$Recall = \frac{|TP|}{|TP + FN|} \tag{17}$$

$$F1 - score = \frac{2 \times precision \times recall}{precision + recall} \tag{18}$$

$$Accuracy = \frac{|TP + TN|}{|A|} \tag{19}$$

where TP represents the number of positive samples with correct predictions, FP represents the number of negative samples with incorrect predictions, FN represents the number of positive samples with incorrect predictions, TN represents the number of negative samples with correct predictions, and A represents the total number of all samples. *Precision* represents the proportion of all samples predicted to be positive that are actually positive. *Recall* indicates the proportion of all positive examples in the sample that are correctly predicted. The *F1-score* is the weighted average (or harmonic mean) of *Precision* and *Recall*. Therefore, this score takes both false positives and false negatives into account to strike a balance between *Precision* and *Recall*. *Accuracy* is the most intuitive measure of performance and reflects the proportion of samples correctly classified by the prediction model.

For all the datasets we used, we set the filter windows of CNN to 3, 4, 5 in our experiments, each with 100 feature maps; the dropout rate was 0.5, the learning rate was set to 0.05, and the mini-batch size was 50 [53]. We used the word embedding method to convert each instance into a word embedding matrix. We counted the average length of Chinese dataset samples as 100 and the average length of English dataset samples as 15, and thus we set the word vector matrix shape as 100×300 and 15×300, respectively. We used publicly available word2vec vectors. The English word vector is trained from 100 billion words in Google News [54], and the Chinese word vector is trained from Sogou News [55]. The dimension of all these word vectors is 300. In particular, the parameters of the LMT-CNN model were set as shown in Table 2.

For the computational complexity of the LMT-CNN model, we chose Params and FLOPs to measure the complexity. Params refers to the total number of parameters to be trained in the model training and is only related to the defined network structure. FLOPs is the number of floating-point operations, which can be regarded as the calculation amount of the model [56]. We regard the multi-add combination as a floating-point operation. FLOPs are related to different layer operation structures, and the maximum length of the input sentences influences the size of the computation. We calculated the parameter quantities of the convolutional layer and the fully connected layer that use different height convolution kernels; we also calculated the calculation amount of convolutional layer and fully connected layer at different sentence maximum length M. The results are shown in Table 3.

Table 2. LMT-CNN model parameter settings.

Parameter	Set Value
Word vector dimension	300
Sliding window	(3, 4, 5)
Optimizer	SGD
Learning rate	0.05
Dropout	0.5
Batch size	50
Word embedding matrix dimension in Chinese text	100×300
Word embedding matrix dimension in English text	15×300

Table 3. The computational complexity of the LMT-CNN model.

Layers	Params	MFLOPs	
		$M = 15$	$M = 100$
Conv2d ($k_h = 3$)	90,100	234.260	1765.960
Conv2d ($k_h = 4$)	120,100	312.260	2353.960
Conv2d ($k_h = 5$)	150,100	390.260	2941.960
Fully connected	1842	0.004	0.004
Total	362,142	936.784	7061.884

4.3. The Influence of the Weight Coefficient α of Affective Words on the Performance of LMT-CNN Model

The affective word weight coefficient α is an important parameter of the LMT-CNN model, which is used to control the weight of affective word information and semantic information. To verify the effectiveness of adding affective word information to the LMT-CNN model, we changed the value of α from 0 to 1.5 (every 0.1) and recorded the scores of the *Euclidean, KL Divergence, Cosine,* and *Accuracy* indicators of the LMT-CNN model on the Semeval and Ren-CECps datasets. The detailed experimental results are shown in Figures 3 and 4.

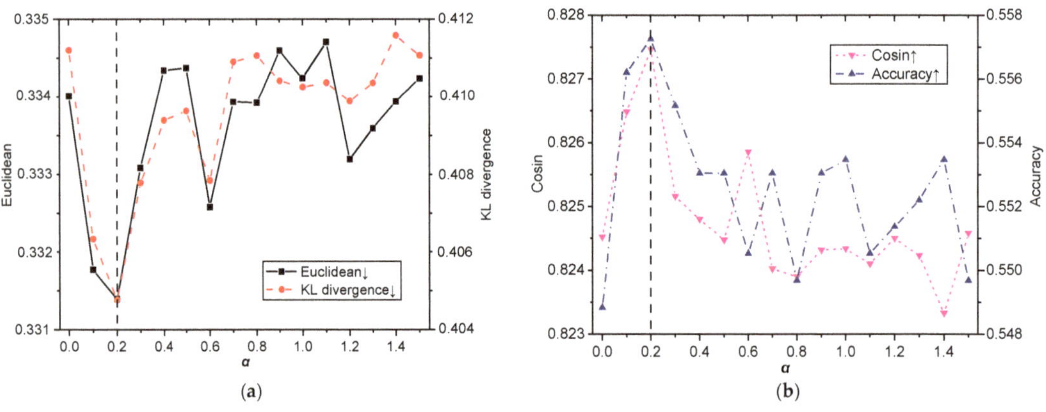

Figure 3. The influence of the weight coefficient α on the performances of the LMT-CNN model on the Semeval dataset ($\uparrow$ means that the larger the indicator is, the better, and $\downarrow$ means that the smaller the indicator is, the better). (**a**) *Euclidean* and *KL Divergence*; (**b**) *Cosine* and *Accuracy*.

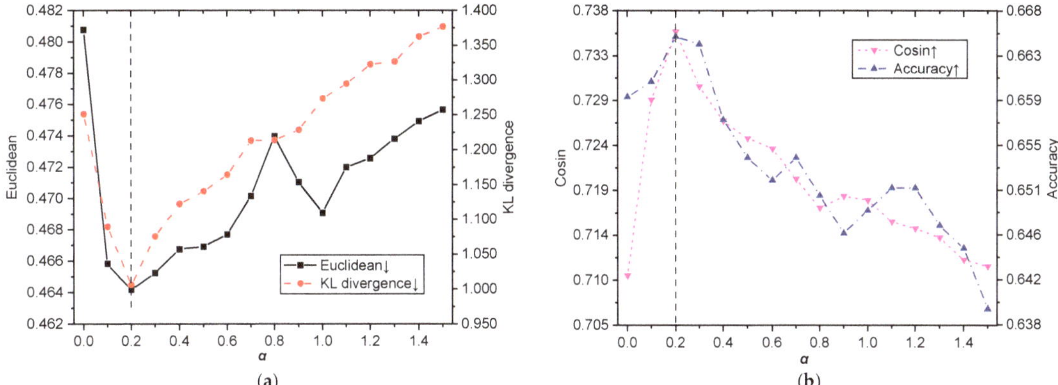

Figure 4. The Influence of the weight coefficient α on the performances of the LMT-CNN model on the Ren-CECps dataset ($\uparrow$ means that the larger the indicator is, the better, and $\downarrow$ means that the smaller the indicator is, the better). (**a**) *Euclidean* and *KL Divergence*; (**b**) *Cosine* and *Accuracy*.

As can be seen from the experimental results of the English dataset in Figure 3, although the scores of the four evaluation indicators vary greatly with the change in weight α, the optimal value is reached when $\alpha = 0.2$. When α increases from 0 to 0.2, the two indicators in Figure 3a gradually decrease, and the two indicators in Figure 3b gradually increase, which shows that it is beneficial to increase the linguistic knowledge of affective words. When $\alpha = 0.2$, the four indicators in Figure 3 reach the best value, and the performance of the LMT-CNN model is the best, which shows that emotion knowledge and semantic information reach a balance. When $\alpha > 0.2$, the indicators in Figure 3a,b show an unstable upward and downward trend, respectively, and the performance of the model begins to decrease as a whole, which shows that when the weight of affective words is too large, the noise of affective word information affects the performance of the model.

For the Chinese dataset Ren-CECps, it can be seen from the experimental results in Figure 4 that the changing trend of the four evaluation indicators with the increase in weight is similar to that of the English dataset. The *Euclidean* and *KL Divergence* indicators in Figure 4a rise first and then fall; *Cosine* and *Accuracy* indicators in Figure 4b fall first and then rise. The four evaluation indicators are in $\alpha = 0.2$, indicating that the affective word information and semantic information reach a balance.

The above experimental results show that affective words contain effective emotional information. Adding affective word information to the EDL model can help improve the performance of emotion distribution prediction. However, since affective words have more noise, affective word information weight α should not be too large. Considering the balance of text semantic information and affective word information, in the following experiments of this paper, we set $\alpha = 0.2$ for the dataset for the LMT-CNN model.

4.4. Contrast Experiment of Multiple Text Emotion Distribution Prediction Methods

To verify the performance of the emotion distribution prediction of the LMT-CNN model, we selected 8 different EDL methods as baseline methods and conducted a comprehensive comparative experiment on the Semeval and Ren-CECps datasets. The compared EDL methods included AA-KNN, AA-BP, SA-LDSVR, SA-IIS, SA-BFGS, SA-CPNN [6], TextCNN [53], and MT-CNN [15]. The detailed model configurations are described as follows.

Specifically, AA-KNN and AA-BP are extended versions of the classical KNN algorithm and BP (backpropagation) neural network to solve LDL tasks [6]. In the AA-KNN model, we set the number of neighbor samples as 4, used *Euclidean* distance to calculate the distance from the sample point to be classified to each other sample, and predicted the category of the sample to be tested through the nearest 4 neighboring samples. In the AA-BP model, we created a three-layer network; the number of neurons in the first layer was 24, and the transfer function was *"tansig"*. The number of neurons in the second layer was 60, and the transfer function was *"tansig"*. The number of neurons in the output layer was 6, and the transfer function was *"purelin"*. We set the learning rate to 0.05 and the momentum factor to the default value of 0.9.

Additionally, SA-LDSVR, SA-IIS, SA-BFGS, and SA-CPNN are algorithms specifically designed for LDL tasks [6]. In the SA-LDSVR model, we chose radial basis function (RBF) as the core function, the penalty parameter was 1.0, and the other parameters were the default values. The SA-IIS method establishes the maximum entropy model and uses the IIS algorithm to estimate the parameters. SA-BFGS follows the idea of an effective quasi-Newton method BFGS to further improve IIS-LLD. SA-CPNN has a network structure similar to Modha's neural network, where the only difference is its training manner is supervised.

Both TextCNN and MT-CNN are deep-neural-network-based models. TextCNN is a convolutional neural network model for text emotion classification [53]. The height of convolutional kernel size was divided into three groups (3, 4, 5), and the width was 300, which was equal to the dimension of the word vectors. There were 32 channels in each group. Batch size and learning rate were set to 16 and 0.001. MT-CNN is a multi-task convolutional neural network model that simultaneously predicts the distribution of text emotion and the dominant emotion of the text [15]. The height of convolutional kernel size was divided into three groups (3, 4, 5), and the width was 300, which was equal to the dimension of the word vectors. We combined the cross-entropy loss with the KL loss by setting the weight (λ). Therefore, considering the balance between distribution prediction and classification performance in the experiment, $\lambda = 0.7$ [15] was set for this model, which means that the weight of the cross-entropy loss was 0.3, and the weight of the KL loss was 0.7.

The detailed experimental results of all emotion distribution learning methods participating in the comparison of the datasets are shown in Tables 4 and 5. In the table, ↑ means that the larger the indicator is, the better, and ↓ means that the smaller the indicator is, the better. The optimal results of each indicator are marked in bold.

Table 4. Experimental results comparing the performance of 9 emotion distribution learning methods for emotion distribution prediction and emotion classification on the SemEval dataset (↑ indicates that the larger the indicator is, the better; ↓ indicates that the smaller the indicator is, the better.). The best performances of each indicator are marked in bold.

EDL Method	Emotion Distribution Prediction						Emotion Classification			
	Euc (↓)	Sør (↓)	Squ (↓)	KL (↓)	Cos (↑)	Int (↑)	Pre (%)	Rec (%)	F1 (%)	Acc (%)
AA-KNN	0.5249	0.4883	0.5649	0.8729	0.6064	0.5117	26.14	19.52	15.33	36.80
AA-BP	0.4677	0.4483	0.4775	0.7858	0.6915	0.5517	38.83	32.13	33.32	43.20
SA-LDSVR	0.4489	0.4360	0.4262	0.5982	0.7367	0.5640	23.77	25.87	21.16	37.60
SA-IIS	0.4804	0.4719	0.4922	0.6871	0.6954	0.5281	26.50	24.08	21.99	39.20
SA-BFGS	0.5573	0.4839	0.5245	1.4149	0.6285	0.5161	32.22	31.55	31.41	40.00
SA-CPNN	0.6688	0.6003	0.7849	2.3375	0.4780	0.3997	21.94	19.25	19.53	23.20
TextCNN	0.7100	0.6068	0.9746	2.0047	0.5353	0.3932	37.29	32.01	33.16	42.65
MT-CNN	0.3344	0.3200	0.3636	0.4118	0.8239	0.6800	44.81	**42.05**	41.27	54.96
LMT-CNN	**0.3324**	**0.3177**	**0.3586**	**0.4057**	**0.8261**	**0.6823**	**45.20**	42.00	**41.68**	**55.76**

Table 5. Experimental results comparing the performance of 9 emotion distribution learning methods for emotion distribution prediction and emotion classification on the Ren-CECps dataset (↑ indicates that the larger the indicator is, the better; ↓ indicates that the smaller the indicator is, the better). The best performances of each indicator are marked in bold.

EDL Method	Emotion Distribution Prediction						Emotion Classification			
	Euc (↓)	Sør (↓)	Squ (↓)	KL (↓)	Cos (↑)	Int (↑)	Pre (%)	Rec (%)	F1 (%)	Acc (%)
AA-KNN	0.7918	0.7826	1.1809	**0.7575**	0.3546	0.2115	12.53	12.43	11.44	16.47
AA-BP	0.7042	0.7465	0.9998	1.3841	0.5049	0.2535	8.37	12.42	5.76	24.83
SA-LDSVR	0.7050	0.7579	1.0166	1.3969	0.5006	0.2421	8.79	12.55	7.39	24.00
SA-IIS	0.7073	0.7546	1.0132	1.4031	0.4967	0.2454	12.16	12.65	8.97	22.80
SA-BFGS	0.7079	0.7495	1.0069	1.4081	0.4966	0.2504	12.01	.12.93	9.91	23.08
SA-CPNN	0.7450	0.8015	1.1136	1.6538	0.4159	0.1984	9.88	12.41	4.96	19.14
TextCNN	0.5033	0.3945	0.6410	1.0831	0.7043	0.6055	54.65	47.32	48.86	62.57
MT-CNN	0.4990	0.3883	0.6293	1.0629	0.7076	0.6117	56.07	47.54	49.71	63.72
LMT-CNN	**0.4682**	**0.3657**	**0.5865**	1.0309	**0.7330**	**0.6343**	**58.72**	**50.27**	**52.22**	**66.95**

In the emotion classification task on Semeval dataset, except that MT-CNN *Recall* is higher than LMT-CNN, LMT-CNN model is 0.39%, 0.41%, and 0.80% higher than MT-CNN model in *Precision, F1-score*, and *Accuracy*. On Ren-CECps dataset, LMT-CNN is 2.65%, 3.73%, 2.51%, and 3.23% higher than MT-CNN in *Precision, Recall, F1-score*, and *Accuracy*. The experiment results show that the emotion knowledge vector extracted from the emotion lexicon can effectively increase the dominant emotion information of sentences and help to improve the emotion classification performance of the LMT-CNN model.

4.5. Comparison of Emotion Classification Performance on Single-Label Datasets

To examine the performance of the LMT-CNN model on the emotion classification task, we conducted comparative experiments using deep-learning-based TextCNN, MT-CNN, and LMT-CNN emotion distribution learning models on seven single-label datasets. Note that the LLE label enhancement method [12] was used to transform the single-label dataset into the emotion distribution dataset.

When the classification task is multi-category, the performance of the classification model can be evaluated by comparing the scores of the classification indicators for each individual category. If using the arithmetic mean of all categories on different indicators as a criterion for evaluating the model, in the case of data imbalance, the categories with fewer data will affect the scores of the indicators more. For example, the *Precision* indicator may score relatively high on categories with small sample sizes. This improves the average of *Precision* indicators on the overall data to some extent, while in reality, not so many samples are correctly classified. Therefore, we compared the performance of the three CNN-based EDL models on each emotion label. The detailed comparative experimental results for each individual emotion category are shown in Table 6, where the last column shows the macro averaged score of the corresponding indicators. The optimal results of each indicator are marked in bold.

As can be seen from Table 6, LMT-CNN achieves the overall best results on all seven datasets. Taking the macro averaged *F1-score* as an example, on the Fairy Tales, TEC, CBET, ISEAR, and Affect in Tweets datasets, the macro averaged F1 value of the LMT-CNN model is 0.40%, 1.32%, 0.91%, 2.17%, and 3.36% higher than that of the MT-CNN model and 0.64%, 2.37%, 1.88%, 3.54%, and 4.78% higher than that of TextCNN. Comparing the performance of LMT-CNN and the baseline model on each emotion, we can see that our model gives better results on almost all emotions. Moreover, the classification of Anger, Sadness and Fear emotions is more challenging, and the performance in all models is relatively low. For example, in the experimental results of Chinese datasets NLP&CC 2013 and NLP&CC 2014, the scores of *Recall* and *F1-score* on the three emotion labels Anger, Sadness, and Fear were relatively low, while the classification on the two labels Joy and Surprise achieved better results.

Compared to the two baseline methods, LMT-CNN shows different degrees of performance score improvement on different datasets. Overall, on all seven datasets, the LMT-CNN method almost always scored higher than the baseline method being compared for the classification indicators in each class. This shows that the classification performance is improved more effectively by combining the emotion knowledge module with semantic connotations for the same input text features. Adding the consideration of prior knowledge of affective words clearly helps in emotion recognition, and our particular method learns more information in the sentence while considering the information of affective words.

In addition, consistent with the experimental results of Zhang et al. [15], the MT-CNN model scored higher than the TextCNN model on each evaluation indicator in the experiments on the seven datasets. This result shows that the performance of the multi-task neural network is significantly better than that of the traditional single-task networks in the emotion classification task. MT-CNN and LMT-CNN use the combination of cross-entropy loss and KL loss to train multi-task neural networks. In the training process, emotion distribution prediction and emotion classification tasks improve each other and can achieve better performance than single-task CNN.

Table 6. Comparative performance of three CNN-based EDL models for emotion classification. The best performances of each indicator on each dataset are marked in bold.

Dataset	Model	Indicator	Emotion						Averaged
			Anger	Disgust	Sadness	Surprise	Fear	Joy	
Fairy Tales	TextCNN	Pre (%)	82.05%	71.24%	66.05%	-	66.15%	76.06%	72.31%
		Rec (%)	65.53%	76.46%	71.20%	-	65.58%	73.55%	70.46%
		F1 (%)	63.14%	68.35%	68.87%	-	72.87%	76.39%	69.92%
	MT-CNN	Pre (%)	82.13%	71.51%	66.17%	-	66.85%	76.56%	72.64%
		Rec (%)	65.83%	76.67%	71.23%	-	66.10%	73.48%	70.66%
		F1 (%)	63.75%	68.35%	69.42%	-	73.17%	76.09%	70.16%
	LMT-CNN	Pre (%)	**82.31%**	**71.67%**	**66.67%**	-	**67.14%**	**76.92%**	**72.94%**
		Rec (%)	**66.96%**	**77.29%**	**71.50%**	-	**66.32%**	**73.77%**	**71.17%**
		F1 (%)	**64.48%**	**68.89%**	**69.80%**	-	**73.23%**	**76.42%**	**70.56%**
TEC	TextCNN	Pre (%)	52.38%	48.74%	49.13%	57.12%	65.70%	55.66%	54.79%
		Rec (%)	31.67%	30.74%	76.57%	43.57%	48.49%	52.91%	47.33%
		F1 (%)	38.36%	54.05%	72.25%	46.04%	30.93%	57.33%	49.83%
	MT-CNN	Pre (%)	55.59%	49.44%	51.57%	58.51%	67.56%	55.56%	56.37%
		Rec (%)	30.60%	30.97%	78.56%	44.87%	48.60%	53.85%	47.91%
		F1 (%)	38.47%	53.05%	72.98%	45.17%	35.77%	59.86%	50.88%
	LMT-CNN	Pre (%)	**56.36%**	**50.38%**	**52.74%**	**58.77%**	**68.33%**	**55.78%**	**57.06%**
		Rec (%)	**35.30%**	**30.81%**	**78.90%**	**45.73%**	**50.31%**	**58.79%**	**49.97%**
		F1 (%)	**38.99%**	**55.00%**	**73.08%**	**48.45%**	**37.65%**	**60.04%**	**52.20%**
CBET	TextCNN	Pre (%)	57.82%	54.78%	53.19%	57.67%	63.87%	66.80%	59.02%
		Rec (%)	50.63%	61.57%	61.90%	52.21%	63.53%	72.48%	60.39%
		F1 (%)	54.56%	59.09%	68.50%	50.03%	62.68%	70.89%	60.96%
	MT-CNN	Pre (%)	59.01%	57.70%	53.43%	59.77%	73.19%	69.39%	62.08%
		Rec (%)	51.73%	64.23%	67.22%	53.63%	64.35%	72.45%	62.27%
		F1 (%)	54.66%	60.10%	69.29%	50.88%	65.60%	71.06%	61.93%
	LMT-CNN	Pre (%)	**62.09%**	**60.40%**	**54.78%**	**62.80%**	**75.19%**	**67.81%**	**63.85%**
		Rec (%)	**53.09%**	**65.51%**	**68.32%**	**54.58%**	**65.44%**	**72.90%**	**63.31%**
		F1 (%)	**55.92%**	**60.42%**	**70.74%**	**51.46%**	**66.60%**	**71.91%**	**62.84%**

Table 6. *Cont.*

Dataset	Model	Indicator	Emotion						Averaged
			Anger	Disgust	Sadness	Surprise	Fear	Joy	
ISEAR	TextCNN	Pre (%)	61.36%	63.50%	76.64%	-	70.67%	79.30%	70.29%
		Rec (%)	70.84%	64.24%	74.21%	-	71.66%	64.59%	69.11%
		F1 (%)	62.14%	65.22%	76.39%	-	72.09%	73.97%	69.96%
	MT-CNN	Pre (%)	61.31%	64.68%	80.27%	-	72.16%	81.13%	71.91%
		Rec (%)	71.62%	64.46%	77.37%	-	73.66%	69.36%	71.29%
		F1 (%)	65.68%	67.63%	77.00%	-	74.25%	72.09%	71.33%
	LMT-CNN	Pre (%)	**62.28%**	**66.00%**	**82.07%**	-	**72.50%**	**82.15%**	**73.00%**
		Rec (%)	**72.38%**	**65.10%**	**79.34%**	-	**74.40%**	**71.64%**	**72.57%**
		F1 (%)	**66.54%**	**70.64%**	**80.68%**	-	**74.95%**	**74.69%**	**73.50%**
Affect in Tweets	TextCNN	Pre (%)	60.60%	55.26%	-	-	34.69%	28.57%	44.78%
		Rec (%)	43.59%	39.60%	-	-	53.12%	21.11%	39.36%
		F1 (%)	41.02%	49.12%	-	-	40.98%	29.05%	40.04%
	MT-CNN	Pre (%)	60.83%	55.57%	-	-	33.57%	32.67%	45.66%
		Rec (%)	49.33%	40.18%	-	-	53.42%	24.29%	41.80%
		F1 (%)	43.97%	49.38%	-	-	43.04%	29.05%	41.36%
	LMT-CNN	Pre (%)	62.71%	56.18%	-	-	35.39%	36.51%	47.70%
		Rec (%)	50.70%	41.21%	-	-	53.61%	25.26%	42.70%
		F1 (%)	45.22%	50.07%	-	-	44.25%	30.00%	42.39%
NLP&CC 2013	TextCNN	Pre (%)	44.86%	49.12%	52.37%	58.72%	50.00%	73.03%	54.68%
		Rec (%)	32.97%	42.08%	30.59%	52.59%	38.38%	67.00%	43.93%
		F1 (%)	34.84%	55.38%	36.32%	47.15%	33.76%	71.83%	46.55%
	MT-CNN	Pre (%)	46.55%	49.77%	57.06%	63.27%	53.11%	78.07%	57.97%
		Rec (%)	34.01%	48.26%	32.86%	53.92%	39.26%	**69.86%**	46.36%
		F1 (%)	37.94%	53.98%	36.65%	51.38%	37.82%	73.45%	48.54%
	LMT-CNN	Pre (%)	**51.55%**	**51.18%**	**60.00%**	**66.97%**	**54.03%**	**78.91%**	**60.44%**
		Rec (%)	**36.56%**	**50.90%**	**32.68%**	**58.13%**	**38.68%**	69.80%	**47.79%**
		F1 (%)	**38.45%**	**56.38%**	**38.82%**	**52.82%**	**38.78%**	**73.55%**	**49.80%**
NLP&CC 2014	TextCNN	Pre (%)	40.60%	45.03%	62.20%	59.63%	51.11%	70.04%	54.77%
		Rec (%)	36.36%	60.14%	25.18%	58.42%	24.39%	78.13%	47.10%
		F1 (%)	38.37%	51.50%	35.85%	59.02%	33.02%	73.86%	48.60%
	MT-CNN	Pre (%)	40.22%	49.63%	70.59%	62.87%	52.78%	72.43%	58.09%
		Rec (%)	36.65%	63.83%	26.36%	60.31%	25.27%	77.81%	48.37%
		F1 (%)	38.36%	55.84%	38.39%	61.56%	34.18%	75.02%	50.56%
	LMT-CNN	Pre (%)	**43.54%**	**50.57%**	**72.92%**	**63.10%**	**53.54%**	**75.91%**	**61.60%**
		Rec (%)	**39.56%**	**64.45%**	**26.47%**	**66.28%**	**28.57%**	**79.32%**	**50.78%**
		F1 (%)	**41.46%**	**56.67%**	**38.84%**	**64.65%**	**37.26%**	**77.58%**	**52.74%**

5. Limitation

In the extensive comparative experiments, the proposed LMT-CNN method shows superior performance on both emotion distribution prediction tasks and emotion classification tasks to the baseline methods. However, our method has some limitations in the following aspects that can be improved in follow-up work.

- The used prior linguistic knowledge is static. Affective words, as a priori knowledge in linguistics, need to build the affective lexicon in advance to associate affective words with emotion labels. However, once the lexicon is established, the affective words have static emotion properties and cannot be updated in real time according to the changes in external emotion semantics. We suppose that dynamic updating of the lexicon is a possible solution.
- The LMT-CNN model does not consider cross-language performance differences. The prior knowledge of affective words is not always present in all languages, especially in

some minor languages. In the future, this deficiency can be addressed through transfer learning, where the rich linguistic knowledge in English is used to boost that of other minor languages.

- Not all state-of-the-art EDL algorithms were directly compared in our experiments. In recent years, many effective EDL models have been proposed, and the algorithms implemented on different datasets are not directly comparable. Re-implementing all state-of-the-art EDL algorithms on their datasets is too time-consuming to illustrate the validity of prior linguistic knowledge. We re-implemented some representative EDL algorithms and performed experimental comparisons on several popular emotional datasets. Although a direct comparison with other algorithms is not possible on different data, classes, and applications, empirical results show that our proposed LMT-CNN method achieves optimal results on all the evaluation metrics.

6. Conclusions

This paper proposes text emotion distribution learning based on the lexicon-enhanced multi-task convolutional neural network (LMT-CNN) model. The LMT-CNN model consists of a semantic information module, emotion knowledge module, and multi-task prediction module. Different from the existing EDL model, LMT-CNN comprehensively considers the linguistic prior knowledge of affective words and text semantic information. A number of comparative experimental results show that the performance of the LMT-CNN model is better than the existing EDL methods in emotion distribution prediction and emotion classification.

In future work, we will consider making more effective use of prior knowledge and try some different emotion modeling methods to predict the emotion distribution.

Author Contributions: Y.D. participated in designing the study, wrote the paper's first draft, performed the computer coding implementation, and performed experiments. X.Z. conceptualized and designed the study, conducted the literature review, and rewrote the paper. All authors have read and agreed to the published version of the manuscript.

Funding: This research was supported by the Natural Science Foundation of China under grant no. 61866017.

Data Availability Statement: The data that support the findings of this study are included in the corresponding referenced articles.

Conflicts of Interest: The funder had no role in study design, data collection and analysis, decision to publish, or preparation of the manuscript.

References

1. Liu, T.; Du, Y.; Zhou, Q. Text emotion recognition using GRU neural network with attention mechanism and emoticon emotions. In Proceedings of the 2020 2nd International Conference on Robotics, Intelligent Control and Artificial Intelligence, Shanghai, China, 17–19 October 2020; pp. 278–282.
2. Yadollahi, A.; Shahraki, A.G.; Zaiane, O.R. Current state of text sentiment analysis from opinion to emotion mining. *ACM Comput. Surv. (CSUR)* **2017**, *50*, 1–33. [CrossRef]
3. Saraswat, M.; Chakraverty, S.; Kala, A. Analyzing emotion based movie recommender system using fuzzy emotion features. *Int. J. Inf. Technol.* **2020**, *12*, 467–472. [CrossRef]
4. Wang, W.; Zhu, K.; Wang, H.; Wu, Y.-C.J. The impact of sentiment orientations on successful crowdfunding campaigns through text analytics. *IET Softw.* **2017**, *11*, 229–238. [CrossRef]
5. Zhang, L.; Wang, S.; Liu, B. Deep learning for sentiment analysis: A survey. *Wiley Interdiscip. Rev. Data Min. Knowl. Discov.* **2018**, *8*, e1253. [CrossRef]
6. Geng, X. Label distribution learning. *IEEE Trans. Knowl. Data Eng.* **2016**, *28*, 1734–1748. [CrossRef]
7. Zhou, D.; Zhang, X.; Zhou, Y.; Zhao, Q.; Geng, X. Emotion Distribution Learning from Texts. In Proceedings of the 2016 Conference on Empirical Methods in Natural Language Processing, Austin, TX, USA, 1–4 November 2016; pp. 638–647.
8. Yang, J.; Li, J.; Li, L.; Wang, X.; Gao, X. A circular-structured representation for visual emotion distribution learning. In Proceedings of the IEEE/CVF Conference on Computer Vision and Pattern Recognition, Nashville, TN, USA, 20–25 June 2021; pp. 4237–4246.
9. Xu, Z.; Wang, S. Emotional attention detection and correlation exploration for image emotion distribution learning. *IEEE Trans. Affect. Comput.* **2021**, *1*, 1. [CrossRef]

10. Fei, H.; Zhang, Y.; Ren, Y.; Ji, D. Latent emotion memory for multi-label emotion classification. In Proceedings of the AAAI Conference on Artificial Intelligence, New York, NY, USA, 7–12 February 2020; pp. 7692–7699.
11. Zhao, Z.; Ma, X. Text emotion distribution learning from small sample: A meta-learning approach. In Proceedings of the 2019 Conference on Empirical Methods in Natural Language Processing and the 9th International Joint Conference on Natural Language Processing (EMNLP-IJCNLP), Hong Kong, China, 3–7 November 2019; pp. 3957–3967.
12. Jia, X.; Zheng, X.; Li, W.; Zhang, C.; Li, Z. Facial emotion distribution learning by exploiting low-rank label correlations locally. In Proceedings of the IEEE/CVF Conference on Computer Vision and Pattern Recognition, Long Beach, CA, USA, 15–20 June 2019; pp. 9841–9850.
13. Pang, J.; Rao, Y.; Xie, H.; Wang, X.; Wang, F.L.; Wong, T.-L.; Li, Q. Fast supervised topic models for short text emotion detection. *IEEE Trans. Cybern.* **2019**, *51*, 815–828. [CrossRef]
14. Teng, Z.; Vo, D.-T.; Zhang, Y. Context-sensitive lexicon features for neural sentiment analysis. In Proceedings of the 2016 conference on Empirical Methods in Natural Language Processing, Austin, TX, USA, 1–5 November 2016; pp. 1629–1638.
15. Zhang, Y.; Fu, J.; She, D.; Zhang, Y.; Wang, S.; Yang, J. Text emotion distribution learning via multi-task convolutional neural network. In Proceedings of the 27th International Joint Conference on Artificial Intelligence, Stockholm, Sweden, 13–19 July 2018; pp. 4595–4601.
16. Tong, R.M. An operational system for detecting and tracking opinions in on-line discussion. In Proceedings of the Working Notes of the ACM SIGIR 2001 Workshop on Operational Text Classification, New York, NY, USA; 2001; pp. 1–6.
17. Gao, B.-B.; Xing, C.; Xie, C.-W.; Wu, J.; Geng, X. Deep label distribution learning with label ambiguity. *IEEE Trans. Image Process.* **2017**, *26*, 2825–2838. [CrossRef]
18. Strapparava, C.; Mihalcea, R. Semeval-2007 task 14: Affective text. In Proceedings of the Fourth International Workshop on Semantic Evaluations (SemEval-2007), Prague, Czech Republic, 23–24 June 2007; pp. 70–74.
19. Quan, C.; Ren, F. A blog emotion corpus for emotional expression analysis in Chinese. *Comput. Speech Lang.* **2010**, *24*, 726–749. [CrossRef]
20. Alm, C.O.; Sproat, R. Emotional sequencing and development in fairy tales. In Proceedings of the International Conference on Affective Computing and Intelligent Interaction, Beijing, China, 22–24 October 2005; pp. 668–674.
21. Mohammad, S. # Emotional tweets. In Proceedings of the * SEM 2012: The First Joint Conference on Lexical and Computational Semantics–Volume 1: Proceedings of the main conference and the shared task, and Volume 2: Proceedings of the Sixth International Workshop on Semantic Evaluation (SemEval 2012), Montréal, QC, Canada, 7–8 June 2012; pp. 246–255.
22. Scherer, K.R.; Wallbott, H.G. Evidence for universality and cultural variation of differential emotion response patterning. *J. Personal. Soc. Psychol.* **1994**, *66*, 310. [CrossRef]
23. Mohammad, S.; Bravo-Marquez, F.; Salameh, M.; Kiritchenko, S. Semeval-2018 task 1: Affect in tweets. In Proceedings of the 12th International Workshop on Semantic Evaluation, New Orleans, LA, USA, 5–6 June 2018; pp. 1–17.
24. Yao, Y.; Wang, S.; Xu, R.; Liu, B.; Gui, L.; Lu, Q.; Wang, X. The construction of an emotion annotated corpus on microblog text. *J. Chin. Inf. Process.* **2014**, *28*, 83–91.
25. Zhao, S.; Ding, G.; Huang, Q.; Chua, T.-S.; Schuller, B.W.; Keutzer, K. Affective image content analysis: A comprehensive survey. In Proceedings of the 27th International Joint Conference on Artificial Intelligence, Stockholm, Sweden, 13–19 July 2018; pp. 5534–5541.
26. Zhao, S.; Ding, G.; Han, J.; Gao, Y. Personality-aware personalized emotion recognition from physiological signals. In Proceedings of the 27th International Joint Conference on Artificial Intelligence, Stockholm, Sweden, 13–19 July 2018; pp. 1660–1667.
27. Ekman, P. Basic emotions. *Handb. Cogn. Emot.* **1999**, *98*, 16.
28. Mikels, J.A.; Fredrickson, B.L.; Larkin, G.R.; Lindberg, C.M.; Maglio, S.J.; Reuter-Lorenz, P.A. Emotional category data on images from the International Affective Picture System. *Behav. Res. Methods* **2005**, *37*, 626–630. [CrossRef]
29. Schlosberg, H. Three dimensions of emotion. *Psychol. Rev.* **1954**, *61*, 81. [CrossRef]
30. Perikos, I.; Hatzilygeroudis, I. Aspect based sentiment analysis in social media with classifier ensembles. In Proceedings of the 2017 IEEE/ACIS 16th International Conference on Computer and Information Science, Wuhan, China, 24–26 May 2017; pp. 273–278.
31. Vrysis, L.; Vryzas, N.; Kotsakis, R.; Saridou, T.; Matsiola, M.; Veglis, A.; Arcila-Calderón, C.; Dimoulas, C. A Web Interface for Analyzing Hate Speech. *Future Internet* **2021**, *13*, 80. [CrossRef]
32. Machova, K.; Mach, M.; Vasilko, M. Comparison of Machine Learning and Sentiment Analysis in Detection of Suspicious Online Reviewers on Different Type of Data. *Sensors* **2021**, *22*, 155. [CrossRef] [PubMed]
33. Khan, I.U.; Khan, A.; Khan, W.; Su'ud, M.M.; Alam, M.M.; Subhan, F.; Asghar, M.Z. A Review of Urdu Sentiment Analysis with Multilingual Perspective: A Case of Urdu and Roman Urdu Language. *Computers* **2021**, *11*, 3. [CrossRef]
34. Catelli, R.; Pelosi, S.; Esposito, M. Lexicon-based vs. Bert-based sentiment analysis: A comparative study in Italian. *Electronics* **2022**, *11*, 374. [CrossRef]
35. Franzoni, V.; Milani, A.; Biondi, G. SEMO: A semantic model for emotion recognition in web objects. In Proceedings of the International Conference on Web Intelligence, Leipzig, Germany, 23–26 August 2017; pp. 953–958.
36. Dashtipour, K.; Gogate, M.; Cambria, E.; Hussain, A. A novel context-aware multimodal framework for Persian sentiment analysis. *Neurocomputing* **2021**, *457*, 377–388. [CrossRef]

37. Rahman, W.; Hasan, M.K.; Lee, S.; Zadeh, A.; Mao, C.; Morency, L.-P.; Hoque, E. Integrating multimodal information in large pretrained transformers. In Proceedings of the Conference of Association for Computational Linguistics. Meeting, Seattle, WA, USA, 5–10 July 2020; p. 2359.

38. Kydros, D.; Argyropoulou, M.; Vrana, V. A content and sentiment analysis of Greek tweets during the pandemic. *Sustainability* **2021**, *13*, 6150. [CrossRef]

39. Chen, T.; Yin, X.; Peng, L.; Rong, J.; Yang, J.; Cong, G. Monitoring and recognizing enterprise public opinion from high-risk users based on user portrait and random forest algorithm. *Axioms* **2021**, *10*, 106. [CrossRef]

40. Zou, H.; Xiang, K. Sentiment classification method based on blending of emoticons and short texts. *Entropy* **2022**, *24*, 398. [CrossRef] [PubMed]

41. Socher, R.; Perelygin, A.; Wu, J.; Chuang, J.; Manning, C.D.; Ng, A.Y.; Potts, C. Recursive deep models for semantic compositionality over a sentiment treebank. In Proceedings of the 2013 Conference on Empirical Methods in Natural Language Processing, Seattle, WA, USA, 18–21 October 2013; pp. 1631–1642.

42. Zheng, X.; Jia, X.; Li, W. Label distribution learning by exploiting sample correlations locally. In Proceedings of the AAAI Conference on Artificial Intelligence, New Orleans, LA, USA, 2–7 February 2018.

43. Batbaatar, E.; Li, M.; Ryu, K.H. Semantic-emotion neural network for emotion recognition from text. *IEEE Access* **2019**, *7*, 111866–111878. [CrossRef]

44. Qin, X.; Chen, Y.; Rao, Y.; Xie, H.; Wong, M.L.; Wang, F.L. A constrained optimization approach for cross-domain emotion distribution learning. *Knowl.-Based Syst.* **2021**, *227*, 107160. [CrossRef]

45. Wang, Y.; Pal, A. Detecting emotions in social media: A constrained optimization approach. In Proceedings of the Twenty-fourth International Joint Conference on Artificial Intelligence, Buenos Aires, Argentina, 25–31 July 2015.

46. Abdi, A.; Shamsuddin, S.M.; Hasan, S.; Piran, J. Deep learning-based sentiment classification of evaluative text based on Multi-feature fusion. *Inf. Process. Manag.* **2019**, *56*, 1245–1259. [CrossRef]

47. Ke, P.; Ji, H.; Liu, S.; Zhu, X.; Huang, M. SentiLARE: Sentiment-aware language representation learning with linguistic knowledge. In Proceedings of the 2020 Conference on Empirical Methods in Natural Language Processing, Online, 16–20 November 2020; pp. 6975–6988.

48. Robbins, H.; Monro, S. A stochastic approximation method. *Ann. Math. Stat.* **1951**, *22*, 400–407. [CrossRef]

49. Lexicon-Enhanced Multi-Task Convolutional Neural Network for Emotion Distribution Learning. Available online: https://github.com/yc-Dong/Lexicon-enhanced-multi-task-CNN (accessed on 19 December 2021).

50. Poria, S.; Gelbukh, A.; Cambria, E.; Hussain, A.; Huang, G.-B. EmoSenticSpace: A novel framework for affective common-sense reasoning. *Knowl. -Based Syst.* **2014**, *69*, 108–123. [CrossRef]

51. Mohammad, S.; Turney, P. *Nrc Emotion Lexicon*; National Research Council: Ottawa, ON, Canada, 2020; Volume 2013, p. 2.

52. Xu, L.; Lin, H.; Pan, Y.; Ren, H.; Chen, J. Construction of emotional vocabulary ontology. *J. China Soc. Sci. Tech. Inf.* **2008**, *27*, 180–185.

53. Kim, Y. Convolutional neural networks for sentence classification. In Proceedings of the 2014 Conference on Empirical Methods in Natural Language Processing, Doha, Qatar, 25–29 October 2014; pp. 1746–1751.

54. Mikolov, T.; Sutskever, I.; Chen, K.; Corrado, G.S.; Dean, J. Distributed representations of words and phrases and their compositionality. *Adv. Neural Inf. Process. Syst.* **2013**, *26*, 3111–3119.

55. Li, S.; Zhao, Z.; Hu, R.; Li, W.; Liu, T.; Du, X. Analogical reasoning on Chinese morphological and semantic relations. In Proceedings of the 56th Annual Meeting of the Association for Computational Linguistics (Volume 2: Short Papers), Melbourne, Australia, 15–20 July 2018; pp. 138–143.

56. Molchanov, P.; Tyree, S.; Karras, T.; Aila, T.; Kautz, J. Pruning convolutional neural networks for resource efficient inference. In Proceedings of the 5th International Conference on Learning Representations, ICLR 2017-Conference Track Proceedings, Toulon, France, 24–26 April 2017.

Article

A Schelling Extended Model in Networks—Characterization of Ghettos in Washington D.C.

Diego Ortega [1,*] and **Elka Korutcheva** [1,2]

[1] Dto. Física Fundamental, Universidad Nacional de Educación a Distancia (UNED), E-28040 Madrid, Spain; ekorutch@gmail.com
[2] G. Nadjakov Institute of Solid State Physics, Bulgarian Academy of Sciences, 1784 Sofia, Bulgaria
* Correspondence: dortega144@alumno.uned.es

Abstract: Segregation affects millions of urban dwellers. The main expression of this reality is the creation of ghettos which are city parts characterized by a combination of features: low income, poor cultural level... Segregation models have been usually defined over regular lattices. However, in recent years, the focus has shifted from these unrealistic frameworks to other environments defined via geographic information systems (GIS) or networks. Nevertheless, each one of them has its drawbacks: GIS demands high-resolution data, that are not always available, and networks tend to have limited real-world applications. Our work tries to fill the gap between them. First, we use some basic GIS information to define the network, and then, run an extended Schelling model on it. As a result, we obtain the location of ghettos. After that, we analyze which parts of the city are segregated, via spatial analysis and machine learning and compare our results. For the case study of Washington D.C., we obtain an 80% accuracy.

Keywords: ghettos; GIS; networks; Washington D.C.; machine learning; segregation

MSC: 91D10

Citation: Ortega, D.; Korutcheva, E. A Schelling Extended Model in Networks—Characterization of Ghettos in Washington D.C. *Axioms* **2022**, *11*, 457. https://doi.org/10.3390/axioms11090457

Academic Editors: Joao Paulo Carvalho and Oscar Humberto Montiel Ross

Received: 5 July 2022
Accepted: 1 September 2022
Published: 6 September 2022

Publisher's Note: MDPI stays neutral with regard to jurisdictional claims in published maps and institutional affiliations.

1. Introduction

A ghetto is a part of a city in which members of a minority group live, especially as a result of political, social, legal, environmental, or economic pressure [1]. Regrettably, millions of people live in ghettos, bringing out the magnitude of the matter. Although their common feature is the impoverishment of the zone, different kinds of ghettos can be found across the world.

Ghettos constitute the main expression of segregation. Segregation can be understood as the practice of separating groups of people with differing characteristics, often connoting a condition of inequality [2]. Even though segregation can have its origin in economic, cultural, or religious motives, we focus on the racial one. Racial segregation restricts people of a different race to areas whose facilities have lower standards than the rest of the city, i.e., ghettos. An actual example of these is some black neighborhoods in the city of Washington D.C., our case study.

The understanding and measuring of segregation can be complex. On one hand, models of segregation have been studied from the sociophysics field [3–13]. Whereas these contributions provide deep insights into different aspects of segregation, they take place in simple square lattices. On the other hand, GIS allows us to define more realistic frameworks. However, the measure of segregation itself has drawn substantial attention during the last thirty years [14–21], but a consistent and generally agreeable definition of segregation has not been formulated yet. The ultimate trend in the field consists in the use of microdata to minimize the modifiable areal unit problem (MAUP) [22]. Nevertheless, this approach requires the use of data on the personal level which is not always available.

In this article, we propose a method to bridge the gap between Sociophysics models and the measure of segregation by GIS techniques. Our procedure relies on capturing the ghettos' location with a segregation model and basic data from the city in situations where detailed data can be not completely reliable. This is a novel approach to the segregation field, as we can see in Section 2.3. We start by using basic GIS information from our region of interest to define a network and then, run on this framework an extended Schelling model [3]. Our enhanced version takes into account the economic contribution to segregation, including terms for the housing market and the financial gap between the population. The final result is the location of ghettos on the network, which can be mapped into their corresponding city areas. Finally, these predicted ostracized regions are compared to the ones characterized as ghettos by spatial analysis (SA) and machine learning (ML) algorithms. For the real case study of Washington D.C., the obtained accuracy is $80 \pm 7\%$.

The paper is organized as follows. In Section 2 the segregation literature previously cited is reviewed. Section 3 explains how to define the network and discusses the system dynamics. In Section 4 we describe our findings, accentuating the connection between the simple network model and the real segregation phenomena. Discussion of our results is found in Section 5. Finally, our conclusions and proposals for further work are explored in Section 6. A guideline is illustrated in Figure 1.

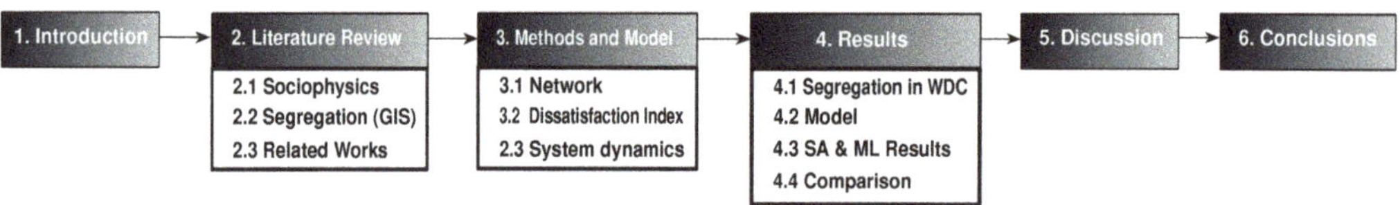

Figure 1. Scheme of the paper organization. WDC, SA, and ML are the abbreviations of Washington D.C, Spatial Analysis, and Machine Learning, respectively.

2. Literature Review

As our work involves the study of segregation from the Sociophsyics and Geographic fields we try to cover their evolution in Sections 2.1 and 2.2, respectively. Papers related to our work are discussed in Section 2.3.

2.1. The Schelling Model and Other Related Models from Sociophysics

One of the first approaches to the segregation field was put forward by T. C. Schelling [3]. His work considered two different social groups (*red* and *blues*) distributed over a square lattice with some vacancies. As people tend to seek out neighbors who are similar to themselves, large clusters of the same type of agents are typically created. Each agent has an *state* and an *action*. An agent is in a state of happiness when the fraction of different neighbors in her/his neighborhood, f_d, does not exceed the tolerance value T. However, if $f_d > T$, the agent is unhappy. Regarding actions, an unhappy agent can relocate in a vacancy if happiness can be attained in the new location. Otherwise, the agent is characterized as happy and no action is performed. Several extensions of this model have been developed and can be classified into two broad categories: more physically oriented or socioeconomically centered.

The Schelling model is related to the Blume-Emery-Griffiths model (BEG), which was originally used to study He^3-He^4 mixtures [4]. This framework can be linked to the Schelling model by associating the type of agents with the spin values considered in their work: *red* and *blue* agents with $s = \pm 1$, respectively, while vacancies imply a null spin. For certain parameter values, the Hamiltonian of the system is reduced when similar agents group together and clusters of different colors are separated by vacancies [5]. In addition to this, the number of vacancies is controlled by a crystal field which plays the role of a chemical potential. The incorporation of an external magnetic field favors one spin type over the other one, giving rise to dissimilar entry fluxes for each agent type.

A physical analog of the model comparing segregation between clusters with two liquids was characterized in [6], an approach from the statistics physics to the segregation phenomena was carried out in [7], a relation with a 1-spin model where agents can enter or leave the city (open city) was established in [8]. From the social and economic fields, a housing market and different tolerance values for black and white people were introduced in the simulation [9]. Another interesting contribution was able to predict the relocation of higher-status households in suburban zones by using a cellular-automata [10].

In recent years focus has shifted to complex agent-based models which include ethical issues: the influence of altruistic agents which greatly affects the final state of the system improving the overall happiness [11]. The influence of altruist and fair agents is evaluated in [12], the behavior of agents with adaptive tolerance to their neighborhood in [13] and open city simulations, in which agents can leave or enter the city, were used to model gentrification processes in [5].

Nevertheless, these models consider the classical rectangular lattice whose applications to urban realities are limited in most cases. The GIS technology solved this issue by mapping real cities and measuring segregation in them.

2.2. Segregation (GIS)

Despite being a reality in urban environments, the measurement of segregation is a complex problem that emerges as a special spatial arrangement across multiple dimensions of culture, religion, economy, race, and others [14]. The clustering of these groups in parts of the city gives rise to the creation of ghettos. Five dimensions of segregation: evenness, exposure-isolation, concentration, centralization and clustering were proposed in [15]. It seems that these variables overlap, simplifying the measure of segregation to one index in most cases. One of the most used estimators is the dissimilarity index [16], which compares how evenly one population sub-group is spread out geographically compared to another population sub-group. However, this index does not take advantage of all the spatial information [14]. Therefore, segregation estimators taking advantage of the clustered nature of the phenomena were developed. Although several local indicators of spatial association (LISA) statistics were proposed [17], the local Moran´s indices [18] remain one of the most used.

Nevertheless, all the measures previously cited rely on data obtained from groups of individuals. Hence, these measures are exposed to the modifiable areal unit problem (MAUP) [22,23]. This issue is created by the modification of the boundaries in geographical units which are often demarcated artificially, i.e., they are not natural divisions. Therefore new measures relying on personal patterns which characterize people from the same neighborhood were proposed [19]. Nowadays, segregation is usually considered a multiscale phenomenon, and microdata is used when available [20,21,24].

Although an approach based on the minimization of the MAUP requires an individual level analysis, it also makes us dependent on the microdata level which is not always available or can be affected by variations such as the one provoked by the COVID-19 pandemic.

2.3. Related Works

Works particularly related to our contribution are those using the Schelling model on networks. The major role played by mild tolerance preferences in the segregation phenomena was compared for lattice and networks in [25]. This work also explained that polarization mechanisms occur not only in regular spatial networks but also in more general social networks. The performance of some segregation indices is particularized for several network types in [26]. Besides, the dynamics of the model are also characterized, thus finding that the system evolves toward steady states in which a maximum level of segregation is reached. In contrast to the previous works, where similar results are found despite the inherent differences between networks, the importance of cliques is underlined in [27]. Cliques are complete subgraphs inside another graph. They can be understood as clusters that reinforce segregation effects. Hence, suggestions against real city situations that can generate structures similar to cliques are put forward.

None of the previous network models have a direct connection with an urban structure, in contrast to the GIS-related works. In [28], several examples of the interaction between GIS and agent-based models are presented. One of them is specially related to this work, given that is a segregation model in which each census tract represents an agent in the Schelling model. A segregation model loosely coupled with GIS information is proposed in [29]. In this contribution, several agents can occupy the same polygon. In addition to this, the significance of barriers in segregation is evaluated. Basically, borders, such as rivers and highways, tend to amplify this urban reality. This finding is in good agreement with the significance of cliques in networks previously discussed [27].

The models we have briefly explained in this section assume a perspective of monetary equality for the agents, i.e., no economic gap between the groups. In addition to this, no housing price is associated with city areas. Therefore, they lead to a portrayal focused on clusterization where no identification of the economically handicapped group is possible. To put it in other words, ghettos can not be located by adopting these frameworks.

3. Methods and Model

3.1. Network

In this subsection, we explain the process of building our network. We consider a small zone in the north of Washington D.C. for pedagogical purposes. First, we need access to this census tract data. Their boundaries divide the region into different polygons, as can be seen in Figure 2a. It is worth noting that information on the different regions, associated with geographical, economic, or racial aspects could be linked to each node.

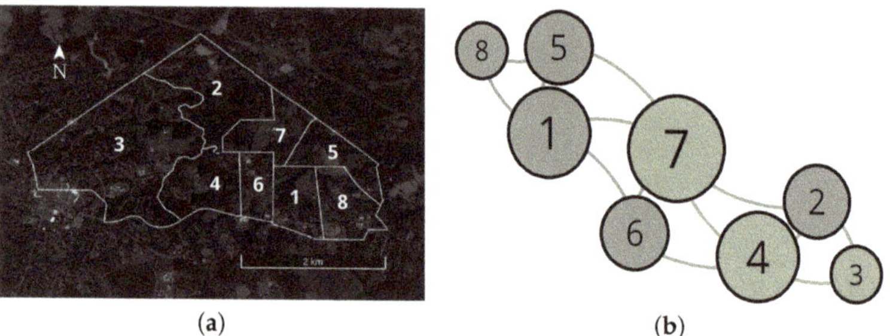

(a)   (b)

Figure 2. (**a**) Our region is divided in different polygons by census tracts. Each polygon is numbered to facilitate its identification. (**b**) Each node corresponds to a polygon with the same identification number as the figure on the left. Edges link overlapping polygons from the previous illustration. Node sizes are related to their degree.

Our network is defined by the nodes that represent each polygon from the map, and edges are established between neighbor polygons, as it is shown in Figure 2b. As a means to obtain these relations, QGIS software is applied [30]. Then, this data is mapped into a network by using the networkx package [31].

3.2. Dissatisfaction Index

Once our network is established we focus our attention on the agents. As in the Schelling model, we consider two kinds of agents defined by its color: red and blue. Each agent occupies a node, while some nodes may remain empty.

The fraction of different neighbors for the agent in the node i can be written as $f_d(i) = N_d/(N_s + N_d)$, where N_s and N_d are, respectively, the number of similar (s) and dissimilar (d) agents in the neighborhood, i.e., the linked nodes. In contrast to lattice models, where the numbers of neighbors remain fixed, the number of neighbors can vary from one node to another in our network.

The state of the agent in the node i is measured via the dissatisfaction index, $I_{dis}(i)$ as:

$$I_{dis}(i) = f_d(i) - T + D(i) \pm H. \tag{1}$$

Here, the first two terms are related to social preferences. T is the tolerance, the maximum fraction of different agents that an agent can withstand while remaining happy. This value acts as a threshold. These two terms arise from the Schelling model: if $f_d(i) \leq T$, the fraction of different neighbors would be under or be equal to the tolerance value, hence the agent would be happy, being $I_{dis}(i) \leq 0$. With only these two terms, our model is equivalent to the Schelling one. Nonetheless, we have extended our work considering economic contributions. Mean housing price in the node is denoted as $D(i)$ [5]. Lower $D(i)$ values imply cheaper living places, thus producing a higher level of happiness. Finally, we have portrayed the economic inequality between groups as H, understood as half the economic gap between them. This term favors red agents, $-H$, while blue agents are monetarily handicapped, $+H$. As can be inferred from the previous explanation, lower values of I_{dis} yield a higher level of happiness. To put in in another words, an agent is happy when $I_{dis} \leq 0$ and unhappy if $I_{dis} > 0$.

We study an open city framework where agents may abandon the city if they are unhappy, and enter otherwise. Therefore, a large value of H increases the number of red agents in the network while reducing the blue population. In other words, the interplay between $D(i) \pm H$ can generate zones depleted of blue population, which has the bias of the economic gap, $+H$. This term may produce positive values of $I_{dis}(i)$, pointing out the unhappiness of the agent, who can be relocated to a better location if it is available or leave the city. The interpretation is straightforward, the segregated group has no economical access to some city zones, while the red group, which is happier due to the financial advantage, $-H$, resides in them.

As a means to understand the final state reached, we examine the contribution of each term from Equation (1). The first two terms on the right-hand side give rise to the clustering effect: agents of the same kind minimize these terms by grouping them. If there are no dissimilar agents in the neighborhood, $f_d(i) = 0$, thus the only effective term is $-T$. An alternative reading of the economic terms is unalike housing prices for red and blue agents: on one hand, living places for the red group can be considered to be priced as $D(i) - H$, while for the economically handicapped blue agents become $D(i) + H$. In other words, prices for blue agents are increased by a factor of $2H$ taking as reference the red group. Consequently, the blue group may only settle in the most affordable living places.

It should be noted that the economic terms can balance the contribution from the neighborhood preferences, as Equation (1) points out. For example, agents who are surrounded by neighbors of the same kind may be happy even if the contribution of the housing price and the financial gap create some discomfort, resembling ghettos' reality.

3.3. System Dynamics

We start from a random initial configuration with equal proportions of red and blue agents and a small percentage of vacancies, 5%. Nonetheless, the final number of vacancies in the network is controlled by the interplay between $D(i) + H$, so this initial percentage does not vary from our final results. As we consider an open city model, agents can enter or leave the city depending on their dissatisfaction level. It must be noted that I_{dis} are calculated by means of Equation (1).

At each iteration, we choose an internal or external exchange with equal probabilities. On one hand, if the change is internal, two nodes i and j are randomly selected. However, i must be occupied with an agent, while the node j is a vacancy. If the exchange verifies $I_{dis}(j) \leq I_{dis}(i)$ the agent is relocated into the node j, and now the node i becomes a vacancy. Otherwise, the relocation is rejected. On the other hand, if the exchange is external, a random node is selected. If the node i is a vacancy, the node is occupied by an agent coming from outside, if the satisfaction condition is fulfilled, i.e., $I_{dis}(i) \leq 0$. The color of this agent is randomly selected (50/50 chance). If the node i has an agent, the agent leaves the system if $I_{dis}(i) > 0$. This iteration is repeated until the system reaches an equilibrium state where all changes are discarded.

In contrast to the Schelling model, happy agents can relocate inside the city if the new place implies a higher level of happiness that the current one. It is also worth mentioning that all the unhappy agents have abandoned the city at the end of the simulation.

4. Results

4.1. Segregation in Washington D.C.

Aiming to study a real case of segregation, we focus our attention on Washington D.C. (USA). This city is located on the eastern shore of the Potomac River, bordering the states of Maryland and Virginia, as we can see in Figure 3a. The establishment of black ghettos in this city has a historic origin. During the 1960s the city became majority Black and this population was concentrated in the south of Anacostia River, which is depicted in Figure 3b. Wards 7 and 8 suffered from disinvestment, and public housing was located in these zones [32]. Consequently, we expect a large concentration of ghettos in this area, which has been segregated for more than 50 years. Although there is no official classification of ghettos, deprived communities such as Anacostia or Deanwood can be currently found in this zone [33].

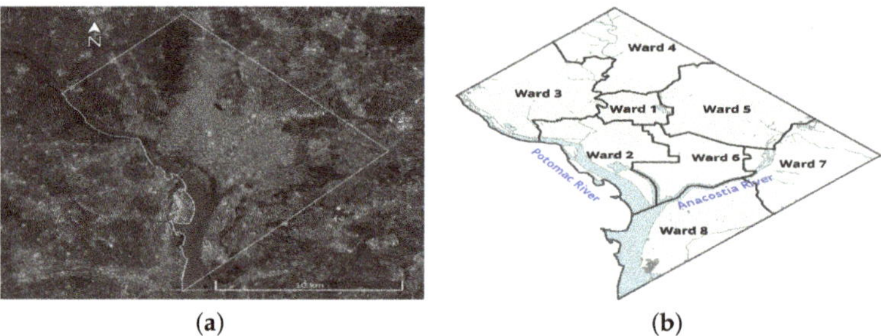

(a)   (b)

Figure 3. (**a**) Orthoimage of Washington D.C. (**b**) Division of Washington D.C. by wards. River names are represented in blue.

4.2. Model

Now, we apply the method from Section 3.1 to Washington D.C. (USA) considering its division by census tracts. Census tracts are small, relatively permanent statistical subdivisions of a county or statistically equivalent entities with a population size between 1200 and 8000 people [34]. Data for each tract can be accessed via the *tidycensus* R package [35]. After the process, we obtain a network with 179 nodes and 535 edges. Additional information on each census tract may be introduced into each node.

We choose the parameters values for Equation (1): T, $D(i)$ and $H(i)$. T is a measure of tolerance. In a classic Schelling model, where each agent has 8 neighbors, a tolerance of $T = 0.25$ implies that the agent is happy if only two or fewer individuals are different. As it can be seen in Figure 5b, there are tracts where the white population represents a really high or a very low part of the population. Therefore, we choose $T = 0.25$, a low value of tolerance that may explain these extreme racial concentrations. Even though no direct measure of the housing market can be obtained from the database, we can infer it from the median house income Figure 5a. This variable seems to decrease from the top to the bottom. Thus, it would be possible to define a vertical gradient for $D(i)$. Extreme cases are 0, assigned to the census tract with the highest latitude, and -1, for the lowest one. The rest of the values lie in the range $[-1, 0]$ and are proportional to each centroid latitude. This housing market distribution underlines that the most expensive living places are in the north while the most affordable regions are in the south. Then, the information is mapped into the network assigning a $D(i)$ value for each node, as can be seen in Figure 4a. Finally, $H(i) = 0.25$ implies a large financial gap as it can be deduced from the scale of Figure 5a.

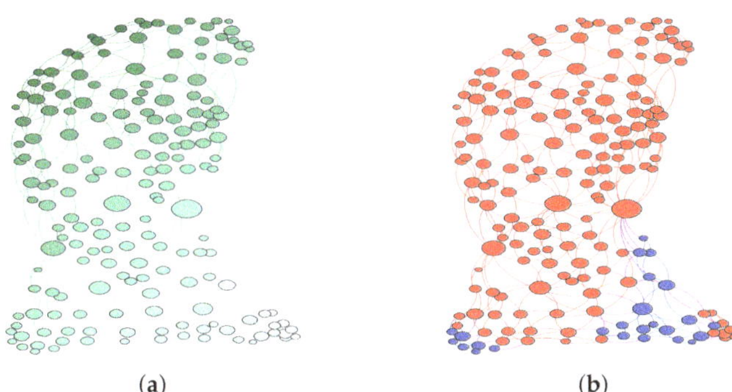

(a)   (b)

Figure 4. (**a**) Network representation of the housing price $D(i)$ for each node. Darker tones are related to greater values and vice versa. All values are in the range $[-1, 0]$. (**b**) Snapshot of the system's final state for one run. Red and blue agents are represented by their respective colors.

As it is illustrated in Figure 4b, blue nodes that represent ghettos tend to coincide with the most affordable areas of the city, depicted as clearer nodes in Figure 4a. The social interpretation is straightforward, segregation also occurs in the economic aspect and people from ghettos do not have the monetary resources to relocate to another place inside the city.

4.3. Spatial Analysis and Machine Learning Results

Although segregation is a multiscale phenomenon, we have selected data from the American Community Survey [35] for two of its main expressions: racial and economic. A measure of the economic level is the median house income, which is depicted in Figure 5a. As can be seen in the figure, the upper part exhibits brighter colors, associated with higher incomes, than the lower ones. Thus, economically handicapped people will be located near the bottom as a consequence of their financial situation, given that the housing market is less expensive in these zones. To study the racial distribution, we calculate the white people's fraction, defined from now on as WPF, from the quotient $n_w / (n_w + n_b)$ where n denotes the population in the census tract corresponding to white (w) and black (b) races. This coefficient is illustrated in Figure 5b and follows a similar color scheme to the median house income map from Figure 5a. Brighter tones are in the upper part of the town, especially concentrated in the west, where the fraction of white people is close to one. In contrast, places for people with financial issues are occupied by a high percentage of black people. These locations can be found in the southeast part of the city and correspond to wards 7 and 8, previously depicted in Figure 3b.

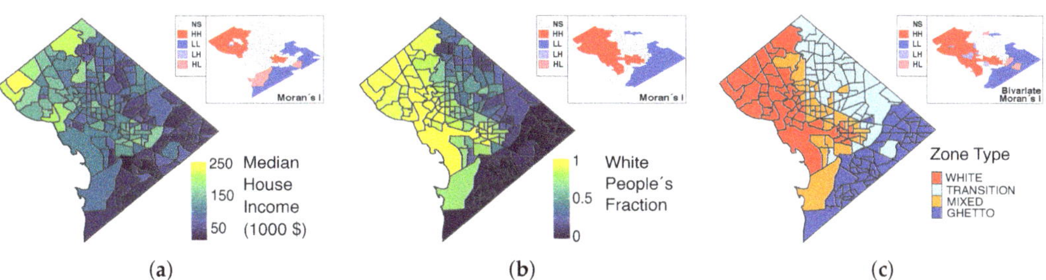

(a)   (b)   (c)

Figure 5. (**a**) Median house income distribution for Washington D.C. (**b**) White people's fraction for our region of interest. (**c**) Zone types found by the EM clusterization algorithm. The insets in (**a**) and (**b**) represent the local Moran's I. The inset from (**c**) illustrates the local Moran's bivariate index choosing as variables the median house income and the white people's fraction. In these insets, NS means Not Significant, while H and L denote High and Low values, respectively. Spatial Analysis software Geoda was used for the insets [36].

Economical segregation is estimated by the local Moran's I in each census tract, as can be seen in the inset of Figure 5a. This index is understood as how the value of a variable in a tract relates to the same value in its surroundings [18]. Thus, the red color in the inset highlights the clustering of rich zones, mainly located in the northwest zone, whereas blue regions in the southeast indicate the grouping of tracts with a handicapped economy. The same analysis for the WPF is graphed in the inset of Figure 5b. Results from both insets Figure 5a,b exhibit similar patterns. Hence, to test the spatial correlation between economy and race, we use the bivariate local moran's I [37], which is depicted in the inset of Figure 5c. This image exhibits a similar pattern to the other insets pointing out the strong correlation between both variables.

In order to identify which census tracts are ghettos, we use the Expectation-Maximization (EM) clusterization algorithm [38]. The EM method is a generalization of the maximum likelihood estimation to the incomplete data case. In particular, the EM algorithm tries to obtain the parameters that maximize the logarithmic probability of the observed data. This method is implemented in Weka, a free ML software [39]. We call this method from the R environment [40] by using Rweka [41], an interface that allows us to run it. Data supplied to this algorithm are the median house income, the WPF, and the latitude of the centroids for each census tract. These data are normalized from 0 to 1.

The EM method finds four clusters that we have defined as zone types, as can be observed in Figure 5c. The zone types labeled as *white* and *ghettos* describe regions with opposite situations: high WPF with abundant financial resources and ghettos populated by black people with economical issues. Between them, the *mixed* zone describes a region where the WPF and the median house income have increased a little taking as reference *ghetto* zones. The *transition* zone is similar to the white zone but the high WPF and income have declined, thus, it can be considered a transition layer from *white* to *mixed* or *ghetto* areas. Once some tracts are identified as ghettos, we study which kind of segregation is mainly responsible for these areas. Expressed in another way, we try to know if a simple classification for these zone types exists. As a mean, we make use of the J48 algorithm from Rweka [41], which is the code name assigned to the ID3 tree classifier [42]. This algorithm begins with the original dataset as the root node. On each iteration of the algorithm, it selects the attribute with the largest information gain value, thus creating a new node. Their result is 176 instances correctly classified out of 179. The only variable that the classifier retains is the WPF, standing out its importance. Ghettos are identified as tracts where the value of this variable is under 0.098.

4.4. Comparison

As can be seen in the previous sections, we have obtained the ghettos' location in two ways: a network model and ML methods. First, we compare these results from a qualitative point of view, as it is depicted in Figure 6. Nodes from the network are identified with census tracts, thus allowing us to map our results into the real city. Classification procedures from the EM algorithm found four types of zones. These sectors are labeled as ghettos and non-ghetto census tracts. It must be noted the resemblance between the model and the ML method, given that both of them locate ghettos in the south region of the city.

We have also evaluated the accuracy of the model, taking the outcome of the ML method as correct. The accuracy is defined as the quotient R/NT where R is the number of census tracts correctly identified and NT is the number of total census tracts, being categorized into the binary classification previously explained. We have simulated 1000 runs finding a mean accuracy of 80% with a standard deviation of 7%.

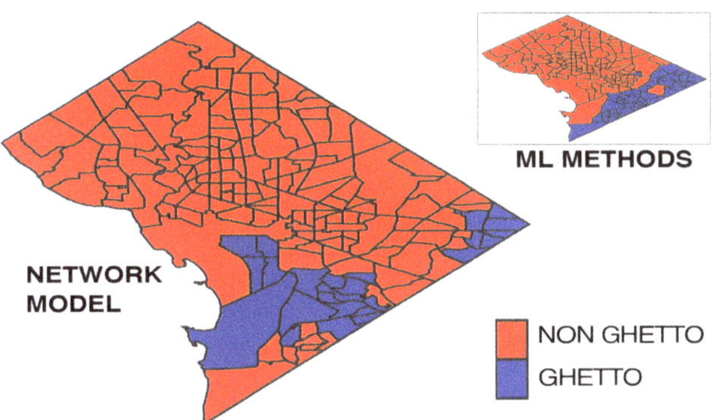

Figure 6. Mapping from the network model snapshot (Figure 4b) to the real city. In order to compare these results with the ones from ML, illustrated in Figure 5c, the inset exhibit the same zone types.

As a means to evaluate the obtained results from this novel approach, it is also interesting to compare our contribution with related works from Section 2. In [43], Washington D.C. is studied at various levels: 27 zip codes, 188 census tracts, and 433 block groups. Although the main aim of this paper was to study the effects of the modifiable areal unit problem (MAUP), and the data correspond to twenty years ago, we can observe that the proportion of the non-white population is high in the southern part of the city, showing good agreement with our results. In [24] they applied the Schelling model at the resolution of individual buildings and families to study the ethnic segregation of an area of Tel Aviv. The paper demonstrated good qualitative agreement between the Schelling model and real urban segregation for a period of 40 years. Economic terms were not considered for this study, highlighting the importance of the ethnic group as a basis for segregation. Recently, considering the multiple scales at which segregation occurs, a multilevel index of dissimilarity was defined in [20]. This index was applied to study the residential segregation of various ethnic groups in England and Wales. The results were consistent with a process whereby minority groups had spread out into more mixed neighborhoods. However, the defined index is a general measure of segregation, a general value that can hide spatial heterogeneity. In [21] two indices are used: the local spatial dissimilarity and the exposure index. The region of interest was a neighborhood in Naples (Italy). This contribution highlighted the major role played by some policy decisions to reinforce segregation, as it happened in Washington D.C. Another interesting contribution comes from the transportation geography field [19]. The method they put forward decomposed the social interaction potential into interactions within and between social groups. Therefore, they related this potential to segregation by a different race. Data were analyzed at the census tracts level, while a matrix disaggregation technique allow them to obtain the transport fluxes by race. The study allows the identification of hotspots of segregation and integration for the case study of Detroit. In our work, the segregation hotspots are located via spatial analysis.

5. Discussion

In order to understand the segregation in Washington D.C., we begin the discussion with spatial analysis. Moran's indexes from the insets in Figure 5a,b pinpoint a clustering of white and rich people in the northwest, while the southeast is strongly segregated, showing opposite characteristics. Besides, the bivariate index suggests a strong spatial correlation between the median house income and the white people's fraction, as it is depicted in the inset of Figure 5c.

Another issue inherent to SA is the MAUP. Basically, as census tracts or other divisions of the regions can change their boundaries through time, data would have an inherent error. For the Washington D.C. case, the problem was analyzed by [43]. The evaluation of the non-white fraction to the level of entire D.C, tracts, block groups, and ZIP codes gives values of 0.692, 0.724, 0.718, and 0.561, in the previous order. All the estimations are near except the one for the zip code. Therefore, the ZIP code has a large error, being the other measures similar. Thus as we choose the census tracts divisions and the error of our Schelling extended model is larger, the MAUP does not seem an issue in our case. In addition to this, census tract divisions of wards 7 and 8 have a river as a boundary with the rest of the city, as can be observed in Figure 3b. Hence, this natural boundary which is used for the census tracts has not been modified, thus reducing the MAUP even further.

Once the spatial analysis was carried out, we studied the data related to the median house income, WPF, and latitude from the ML perspective. The EM clusterer found four zone types, as it is shown in Figure 5c. These results were in good agreement with our SA, and they were taken as the correct classification of the census tracts. In addition to this, a tree classifier ID3 found that the main variable in the previous classification was the WPF, meaning that race plays a key role in the zone type. This suggests that the main motive of segregation is racial, even nowadays. However, this racial segregation is strongly linked with economic inequalities, as the SA analysis demonstrated.

Finally, the model results are compared with those from EM clustering reduced to the binary classification of ghettos and non-ghettos, as it is depicted in Figure 6. Good agreement is found between our model and the ML predictions, obtaining an accuracy of $80 \pm 7\%$. One possible explanation is that the interplay between $D \pm H$ limits the zones where blue clusters can be found to the south of the city. Therefore, it portrays the real situation where ghettos are on the southern shore of the Anacostia river.

6. Conclusions

The extended Schelling model in networks provides a new framework for the study and understanding of ghettos' establishment and their location. Starting with some basic data information from the GIS system, such as the neighborhood relationships between census tracts and a vertical gradient in the house pricing, our network, and its properties are defined. Then, an extended Schelling model, including the economic terms model, runs on it, allowing us to identify which census tracts can be classified as ghettos. It must be noted that this approximation is useful when microdata or reliable data are not available for the zone. As an example, we must mention that due to the impact of the COVID-19 pandemic, the Census Bureau changed the 2020 American Community Survey release into a series of experimental estimates, instead of the standard one [34].

The strong segregation is a consequence of the policies adopted during the 1960s, when the black population was concentrated in the south of Anacostia River (see Figure 3b), as was discussed in Section 4.1. Then, the disinvestment in the zone caused the population to fall into the poverty trap. This term alludes to self-reinforcing mechanisms that cause poverty to persist unless there is outside intervention [44]. In fact, as we discussed in Section 3.2, we can consider different prices for the same spot depending on group membership. This fact resembles redlining practices which can be summarized as an increase in the interest rate or even credit denial of a loan due to cultural or racial bias [45]. This procedure gave rise to an even further concentration of the deprived black population in the ghetto zones. In addition to this, the river acts as a boundary between this part and the rest of the city which leads to an increase in the zone isolation [29]. To sum up, the actual setting is strongly influenced by these past practices and the economic gap could act as a *de facto* redlining procedure, not allowing the relocation of economically handicapped people into better locations.

Nevertheless, the model has some limitations: other cities with more complicated structures can create a complex housing market difficult to define. For instance, cities in Europe tend to be radially structured, i.e., they have expanded from the center towards the outskirts where ghettos are mainly located.

A way to enhance this work is to include three parameters in the dissatisfaction index. One is linked to both segregation terms, and the others are associated with the housing market and half the financial gap. In this case, we could maximize the final accuracy of the model by using an optimization procedure over the parameters previously included.

Author Contributions: Conceptualization, D.O. and E.K.; methodology, D.O.; software, D.O.; validation, D.O and E.K.; data curation, D.O.; writing—original draft preparation, D.O.; writing—review and editing, E.K.; visualization, D.O.; supervision, E.K.; project administration, E.K. All authors have read and agreed to the published version of the manuscript.

Funding: This research was funded by the Spanish Government through grants PGC2018-094763-B-I00 and PID2019-105182GB-I00.

Institutional Review Board Statement: Not applicable.

Informed Consent Statement: Not applicable.

Data Availability Statement: Data were obtained from the American Community Survey (2014–2019) via tidycensus [35]. Data from the median house income, and white and black populations correspond to $B19013\_001$, $C02003\_003$, and $C02003\_004$ variables, respectively. White people refer to persons whose race has a European origin, while black denotes African American people exclusively. Orthophoto images from Washington D.C. in Figures 2a and 3a are taken from Google Earth.

Acknowledgments: We acknowledge financial support from the Spanish Government through grants PGC2018-094763-B-I00 and PID2019-105182GB-I00.

Conflicts of Interest: The authors declare no conflict of interest.

Abbreviations

The following abbreviations are used in this manuscript:

BEG	Blume-Emery-Griffiths
EM	Expectation-Maximization
GIS	Geographic Information Systems
ML	Machine Learning
MAUP	Modifiable Areal Unit Problem
USA	United States of America
SA	Spatial Analysis
WPF	White People's Fraction

References

1. Merriam Webster Dictionary. Available online: https://www.merriam-webster.com/ (accessed on 22 July 2022).
2. Encyclopedia Britannica. Available online: https://www.britannica.com/topic/segregation-sociology (accessed on 22 July 2022).
3. Schelling, T.C. Dynamic models of segregation. *J. Math. Sociol.* **1971**, *2*, 143–186. [CrossRef]
4. Blume, M.; Emery, M.J.; Griffiths, R.B. Ising Model for the λ Transition and Phase Separation in He^3 and He^4 Mixtures. *Phys. Rev. A* **1971**, *4*, 1071–1077 [CrossRef]
5. Ortega, D.; Rodríguez-Laguna, J.; Korutcheva, E. Avalanches in an extended Schelling model: An explanation of urban gentrification. *Phys. A Stat. Mech. Appl.* **2021**, *573*, 125943. [CrossRef]
6. Vinković, D.; Kirman, A. A physical analogue of the Schelling model. *Proc. Natl. Acad. Sci. USA* **2006**, *103*, 19261–19265. [CrossRef]
7. Dall'Asta, L.; Castellano, C.; Marsili, M. Statistical physics of the Schelling model of segregation. *J. Stat. Mech. Theory Exp.* **2008**, *7*, L07002. [CrossRef]
8. Gauvin, L.; Nadal, J.-P.; Vannimenus, J. Schelling segregation in an open city: A kinetically constrained Blume-Emery-Griffiths spin-1 system. *Phys. Rev. E* **2010**, *81*, 066120. [CrossRef]
9. Zhang, J. A dynamic model of residential segregation. *J. Math. Sociol.* **2004**, *280*, 147–170. [CrossRef]
10. Fossett, M. Ethnic Preferences, Social Distance Dynamics, and Residential Segregation: Theoretical Explorations Using Simulation Analysis. *J. Math. Sociol.* **2006**, *30*, 185–273. [CrossRef]
11. Jensen, P.; Matreux, T.; Cambe, J.; Larralde, H.; Bertin, E. Giant catalytic effect of altruists in Schelling's segregation model. *Phys. Rev. Lett.* **2018**, *120*, 208301. [CrossRef]
12. Flaig, J.; Houy, N. Altruism and fairness in Schelling's segregation model. *Phys. A Stat. Mech. Appl.* **2019**, *527*, 121298. [CrossRef]
13. Urselmans, L.; Phelps, S. A Schelling model with adaptive tolerance. *PLoS ONE* **2018**, *3*, e0193950. [CrossRef] [PubMed]
14. Yao, J.; Wong, D.; Bailey, N.; Minton, J. Segregation Measures: A Methodological Review. *Tijdschr. Econ. Soc. Geogr.* **2018**, *110*, 235–250. [CrossRef]

15. Massey, D.S.; Denton, N.A. The Dimensions of Residential Segregation. *Soc. Forces* **1988**, *67*, 281–315. [CrossRef]
16. Duncan, O.D.; Duncan, B. A methodological analysis of segregation indexes. *Am. Sociol. Rev.* **1955**, *20*, 210–217. [CrossRef]
17. O'sullivan, D.; Wong, D.W. A Surface based approach to measuring spatial segregation. *Geogr. Anal.* **2007**, *39*, 147–168. [CrossRef]
18. Anselin, L. Local Indicators of Spatial Association—LISA. *Geogr. Anal.* **1995**, *27*, 93–115. [CrossRef]
19. Farber, S.; O'kelly, M.; Miler, H.J.; Neutens, T. Measuring Segregation Using Patterns of Daily Travel Behavior: A Social Interaction Based Model of Exposure. *J. Transp. Geogr.* **2015**, *49*, 26–38. [CrossRef]
20. Harris, R. Measuring the scales of segregation: Looking at the residential separation of White British and other schoolchildren in England using a multilevel index of dissimilarity. *Trans. Inst. Br. Geogr.* **2017**, *42*, 432–444. [CrossRef]
21. Reitano, M.; Cerreta, M.; Poli, G. Evaluating Socio-Spatial Exclusion: Local Spatial Indices of Segregation and Isolation in Naples (Italy). In Proceedings of the ICCSA2020, Cagliari, Italy, 1–4 July 2020.
22. Jelinski, D.E.; Wu, J. The modifiable areal unit problem and implications for landscape ecology. *Landsc. Ecol.* **1996**, *11*, 129–140. [CrossRef]
23. Fotheringham, A.S.; Wong, D.W.S. The modifiable areal unit problem in multivariate statistical analysis. *Environ. Plan A* **1991**, *23*, 1025–1044. [CrossRef]
24. Benenson, I.; Hatna, E.; Or, E. From Schelling to Spatially Explicit Modeling of Urban Ethnic and Economic Residential Dynamics. *Sociol. Methods Res.* **2009**, *37*, 463–497. [CrossRef]
25. Fagiolo, G.; Valente, M.; Vriend, N.J. Segregation in networks. *J. Econ. Behav. Organ.* **2007**, *64*, 316–336. [CrossRef]
26. Cortez, V.; Rica, S. Dynamics of the Schelling Social Segregation Model in Networks. *Procedia Comput. Sci.* **2015**, *61*, 60–65. [CrossRef]
27. Banos, A. Network effects in Schelling's model of segregation: New evidences from agent-based simulation. *Environ. Plann. B Plann. Des.* **2010**, *38*, 393–405 [CrossRef]
28. Crooks, A.; Malleson, M.; Manley, E.; Heppenstall, A. *Agent-Based Modelling & Geographical Information Systems. A Practical Primer*; SAGE: London, UK, 2019.
29. Crooks, A.T. Constructing and implementing an agent-based model of residential segregation through vector GIS. *Int. J. Geogr. Inf. Sci.* **2009**, *24*, 661–675 [CrossRef]
30. QGIS Development Team (2022). QGIS Geographic Information System. Open Source Geospatial Foundation Project. Available online: http://qgis.osgeo.org (accessed on 15 May 2022).
31. Hagberg, A.; Swart, P.; Chult, D. Exploring Network Structure, Dynamics, and Function Using NetworkX. In Proceedings of the SciPy2008, Pasadena, CA, USA, 19–24 August 2008.
32. Asch, C.M.; Musgrove, G.D. *Chocolate City: A History of Race and Democracy in the Nation's Capital*; University of North Carolina Press: Chapel Hill, NC, USA, 2017.
33. Area Vibes. Available online: https://www.areavibes.com/washington-dc/most-dangerous-neighborhoods/ (accessed on 24 July 2022).
34. United States Census Bureau. Available online: https://www.census.gov (accessed on 20 June 2022).
35. Walker, K.; Herman, M. Tidycensus: Load US Census Boundary and Attribute Data as 'tidyverse' and 'sf'-Ready Data Frames R Package Version 1.2.1. 2022. Available online: https://CRAN.R-project.org/package=tidycensus (accessed on 20 April 2022).
36. Anselin, L.; Ibnu, S.; Youngihn, K. GeoDa: An Introduction to Spatial Data Analysis. *Geogr. Anal.* **2006**, *38*, 5–22. [CrossRef]
37. Wang, Z.; Liu, Y.; Zhang, Y.; Liu, Y.; Wang, B.; Zhang, G. Spatially Varying Relationships between Land Subsidence and Urbanization: A Case Study in Wuhan, China. *Remote Sens.* **2022**, *14*, 291. [CrossRef]
38. Do, C.B.; Batzoglou, S. What is the expectation maximization algorithm? *Nat. Biotechnol.* **2008**, *26*, 897–899. [CrossRef]
39. Hall, M.; Frank, E.; Holmes, G.; Pfahringer, B.; Reutemann, P.; Witten, I.H. The WEKA Data Mining Software: An Update. *SIGKDD Explor.* **2009**, *11*, 10–18. [CrossRef]
40. R Core Team. *R: A Language and Environment for Statistical Computing*; R Foundation for Statistical Computing: Vienna, Austria, 2013. Available online: http://www.R-project.org/ (accessed on 20 May 2022).
41. Hornik, K.; Buchta, C.; Zeileis, A. Open-Source Machine Learning: R Meets Weka. *Comput. Stat.* **2009**, *24*, 225–232. [CrossRef]
42. Quinlan, J.R. Induction of Decision Trees. *Mach. Learn.* **1986**, *11*, 81–106. [CrossRef]
43. Wong, D. *WorldMinds: Geographical Perspective on 100 Problems*; Kluwer Publishers: Dordrecht, The Netherlands, 2004; pp. 571–575.
44. Azariadis, C.; Stachurski, J. *Handbook of Economic Growth*; Elsevier: Amsterdam, The Netherlands, 2005; Chapter 5: Poverty Traps.
45. Rothstein, R. *The Color of Law*; Liveright Publishing Corporation: New York, NY, USA, 2018.

Article

Hybrid Deep Learning Algorithm for Forecasting SARS-CoV-2 Daily Infections and Death Cases

Fehaid Alqahtani [1], Mostafa Abotaleb [2,*], Ammar Kadi [3,*], Tatiana Makarovskikh [2], Irina Potoroko [3], Khder Alakkari [4,*] and Amr Badr [5]

[1] Department of Computer Science, King Fahad Naval Academy, Al Jubail 35512, Saudi Arabia
[2] Department of System Programming, South Ural State University, 454080 Chelyabinsk, Russia
[3] Department of Food and Biotechnology, South Ural State University, 454080 Chelyabinsk, Russia
[4] Department of Statistics and Programming, Faculty of Economics, University of Tishreen, Tartous P.O. Box 2230, Syria
[5] Faculty of Science, School of Science and Technology, University of New England, Armidale, NSW 2350, Australia
* Correspondence: abotalebmostafa@bk.ru (M.A.); ammarka89@gmail.com (A.K.); khderalakkari1990@gmail.com (K.A.)

Abstract: The prediction of new cases of infection is crucial for authorities to get ready for early handling of the virus spread. Methodology Analysis and forecasting of epidemic patterns in new SARS-CoV-2 positive patients are presented in this research using a hybrid deep learning algorithm. The hybrid deep learning method is employed for improving the parameters of long short-term memory (LSTM). To evaluate the effectiveness of the proposed methodology, a dataset was collected based on the recorded cases in the Russian Federation and Chelyabinsk region between 22 January 2020 and 23 August 2022. In addition, five regression models were included in the conducted experiments to show the effectiveness and superiority of the proposed approach. The achieved results show that the proposed approach could reduce the mean square error (RMSE), relative root mean square error (RRMSE), mean absolute error (MAE), coefficient of determination (R Square), coefficient of correlation (R), and mean bias error (MBE) when compared with the five base models. The achieved results confirm the effectiveness, superiority, and significance of the proposed approach in predicting the infection cases of SARS-CoV-2.

Keywords: hybrid deep learning; time series; LSTM; Stacked LSTM; CNN-LSTMs; BDLSTM; CNN; GRU; modeling; SARS-CoV-2

MSC: 35-00; 35-01; 35-02; 35-03; 35-04; 35-06; 35-11

Citation: Alqahtani, F.; Abotaleb, M.; Kadi, A.; Makarovskikh, T.; Potoroko, I.; Alakkari, K.; Badr, A. Hybrid Deep Learning Algorithm for Forecasting SARS-CoV-2 Daily Infections and Death Cases. *Axioms* **2022**, *11*, 620. https://doi.org/10.3390/axioms11110620

Academic Editor: Oscar Humberto Montiel Ross

Received: 26 September 2022
Accepted: 4 November 2022
Published: 7 November 2022

Publisher's Note: MDPI stays neutral with regard to jurisdictional claims in published maps and institutional affiliations.

1. Introduction

The outbreak of the coronavirus infection known as SARS-CoV-2 was reported in Wuhan city, China, in December 2019 SARS-CoV-2, and it spread to more than 200 countries in less than a year [1]. The world health organization (WHO) called it COVID-19, which stands for "Coronavirus Disease 2019," which is the second version of the previously known severe acute respiratory syndrome SARS (SARS-COV) and identified in short as SARS-CoV-2 [2]. There have been regular restrictions to avoid the infection spreading in all countries, including Russia. In almost all of the countries currently being impacted by the SARS-CoV-2 pandemic, the rate at which patients are becoming infected with and succumbing to the disease is alarmingly high [3]. The treatment of patients who required intensive care was one of the most influential factors in determining the death and case rates associated with (SARS-CoV-2). A significant challenge for healthcare systems all over the world is posed by the administration of SARS-CoV-2 treatment to patients who require acute or critical respiratory care [4]. Artificial intelligence and machine learning, two non-clinical computer-aided rapid fixes, are needed to battle (SARS-CoV-2) and halt its global expansion [5]. Intelligent healthcare is increasingly relying on AI, in particular machine learning algorithms [6]. More and more, these technologies are referred to be the brains of intelligent healthcare services [7]. Deep learning, a kind of machine learning in artificial intelligence, comprises networks that can learn from unstructured or unlabeled data without supervision [8]. SARS-CoV-2 is just one of the numerous applications that have heavily incorporated deep learning [9]. These solutions are also required in order to prevent the disease from becoming more widespread. Techniques for making predictions regarding the future are based on the evaluation of the past [10]. People are under the impression that nothing will be the same as it was before as a result of the widespread coronavirus pandemic, which has numerous global implications. The three most significant things being explored at the moment are figuring out the causes, implementing preventative measures, and attempting to develop an effective cure [11]. In Russia, there are more than 20 million confirmed cases and 386 thousand death cases as on September 2022 [12]. Continued research is being conducted on related diseases, as well as public health policies and containment mechanisms. Quarantine procedures vary from nation to nation, but their overall goal is the same: to slow or stop the spread of infectious diseases in order to keep hospitals operational and able to meet the rising demand for medical care [13]. If the number of patients diagnosed with SARS-CoV-2 continues to rise, it is possible that healthcare facilities will be unable to meet the needs of their patients and provide the services they require. This is the worst-case scenario that can be anticipated. It is crucial that the nations' health capabilities be used properly and that the demand for the supplies needed for medical infrastructure is predictable when infection rates are also taken into consideration [14]. This is because it is important that both the health capacities of the countries and the infection rates be taken into account. In this regard, it is recommended that public health strategies be developed and implemented [15]. As a consequence, deep learning (DL) models are considered precise tools that may aid in the development of prediction models [16]. The recurrent neural network (RNN) and the long short-term memory (LSTM) are the ones that are being explored in the (SARS-CoV-2) forecasting process because they utilize temporal data, despite the fact that several neural networks (NNs) have been reported in the past [17]. Deep learning networks, such as RNN and LSTM, were utilized in this investigation. These networks were selected because, by analyzing time series data, they were able to provide an accurate forecast of what would occur in the future [18]. An SIR model is a type of epidemiological model that estimates the total number of people in a closed community that could potentially become infected with an infectious illness over a period of time. This category of models gets its name from the fact that they use coupled equations to relate the number of susceptible people to one another $S_{(t)}$, the number of people infected $I_{(t)}$, and the number of people who are recovered $R_{(t)}$, so the initial letters of the three terms that make up the SIR model were shortened to form the acronym (susceptible, infected, and recovered) [19]. The simulation of the SARS-CoV-2 in the Isfahan province of Iran from 14 February 2020 to 11 April 2020 was the subject of one of the first articles published. The authors of this study made a prognosis of the remaining infectious cases using three different scenarios. These scenarios ranged from one another in terms of the extent of social distancing required. In spite of the fact that it was able to estimate infectious cases in shorter time intervals, the developed SIR model was not successful in predicting the actual spread and pattern of the epidemic over a longer period of time. Surprisingly, the majority of the published SIR models that were constructed in order to predict SARS-CoV-2 for different communities all suffer from the same conformity. The SIR models are predicated on assumptions that do not appear to be correct in the circumstances surrounding the SARS-CoV-2 epidemic. Therefore, in order to foresee the pandemic, more complex modeling methodologies and extensive knowledge of the biological and epidemiological features of the disease are required [20]. In addition to more conventional methods, these two models demonstrated a significant amount of success in the forecasting of temporal data. In the first place, recurrent neural networks (RNNs) have been put to use for the processing of time series and sequential data [18]. These networks are also useful for modeling sequence data. RNNs are a type of artificial neural network that is derived from feed-forward networks and exhibit behavior that is analogous to that of the human brain [21]. To put it another way, RNNs have the ability to predict outcomes based on sequence data, whereas other algorithms do not. After that, LSTMs, which have complex gated memory units designed to handle the vanishing-gradient problems that limit the efficiency of simple RNNs, have been used [22]. The average predicted errors for SARS-CoV-2 infection cases using machine learning models are substantially equal to those using statistical models. Machine learning algorithms can be used to forecast long-term time series [23]. They compared (TS-system) and (DLM-system) LSTM-BI-LSTM-GRU faults. Ensembling models provided fewer mistakes than (DLM-system) models at the level of four countries, and hence the ensembling model outperformed (DLM-system) deep learning models [24].

In this research, we aim to forecast SARS-CoV-2 cases (infection—death) in Russia and Chelyabinsk; the period extends (80–20) Using Hybrid deep learning models, which are based on different assumptions about data estimation.

2. Related Work

Researchers have been focusing on x-ray image diagnosis of SARS-CoV-2 and, on the other hand, using time series models and artificial intelligence for the prediction of daily infection, recovery, and death cases for SARS-CoV-2. X-ray images for SARS-CoV-2 were diagnosed using neural networks. In [25], they created a system using five models and deep learning algorithms: Xception, VGG19, ResNet50, DenseNet121, and Inception for binary classification of X-ray images for SARS-CoV-2. In order to aid medical efforts and lessen the strain on medical professionals while dealing with SARS-CoV-2, they provided deep learning models and algorithms that have been developed and evaluated. Based on machine learning and deep learning approaches, a survey of recent works for misleading information detection (MLID) in the health sectors is presented [26]. Other research focused on a database called COVIDGR-1.0 has all severity levels, from normal with positive RT-PCR to mild, moderate, and severe. With an accuracy of 97.72%, 0.95%, 86.90%, 3.20%, 61.80%, and 5.49% in severe, moderate, and mild SARS-CoV-2 severity levels, the technique produced excellent and steady results [27]. The use of user-generated data is envisioned as a low-cost method to increase the accuracy of epidemic tolls in marginalized populations. Utilizing the potential of user-posted data on the web is what they suggested [28]. In addition to social media channels, bogus news about the SARS-CoV-2 epidemic may be automatically classified and located using deep neural networks. In this investigation, the CNN model performs better than the other deep neural networks, with the greatest accuracy of 94.2% [29]. A brand-new interactive visualization system illustrates and contrasts the SARS-CoV-2 pandemic's pace of spread over time in various nations. The method used by the system, called knee detection, splits the exponential spread into many linear components. It may be used to analyze and forecast upcoming pandemics [30]. In [31], they provided a technique for extracting implicit responses from huge Twitter collections. Tweets were cleaned up and turned into a vector format that could be used by various machine-learning methods. For both informational and non-informational classes, the Deep Neural Network (DNN) classifier had the maximum accuracy (95.2%) and F1 score (73.6%). Other research has developed a brand-new relation-driven collaborative learning strategy for segmenting SARS-CoV-2 CT lung infections. Extensive research demonstrates that using shared information from non-SARS-CoV-2 lesions may enhance current performance by up to 3.0% in the dice similarity coefficient [32]. A domain-specific Bi-directional Encoder Representations from Transformer (BERT) language model called COVID-Twitter BERT (CT-BERT) has been introduced in recent sentiment analysis research on SARS-CoV-2. CT-BERT does not always perform better at comprehending sentiments than BERT. In comparison to a broad language model, a domain-specific language model would perform better. An auxiliary technique using BERT was developed to address performance concerns with the single-sentence categorization of SARS-CoV-2-related tweets [33].

In our work, we built a hybrid deep learning algorithm as part of our research, as well as an application that makes use of this algorithm, with the goal of forecasting the number of daily SARS-CoV-2 infections and death in the Russian Federation and the Chelyabinsk region. Therefore, in our work, we will be using hybrid deep learning models for modeling and forecasting SARS-CoV-2 infection and daily death cases in Russia and Chelyabinsk. Chelyabinsk is located in the Ural Federal District in central Russia [34]. The most important contribution made by this study is the development of DL prediction models that, when applied to historical and recent data, are capable of producing the most accurate forecasts of confirmed positive (SARS-CoV-2) cases and cases in which (SARS-CoV-2) was determined to be the cause of death in Russia and Chelyabinsk [35].

3. Data and Materials

When preparing data, deep learning faces some issues with long sequences in the database [36]. For the first problem, training is time-consuming and demands a lot of memory. Second problem, back-propagating extended sequences, results in an incorrectly trained model. Prepare and preprocess data before importing it into neural networks. Normalization and standardization problems are two aspects of data preparation. We used data normalization, a scaling procedure, to set the mean and standard deviation to 0 and 1, respectively [37]. We used daily data on SARS-CoV-2 infection and death cases in the Russian Federation and Chelyabinsk region. The dataset was obtained from the official website of the World Health Organization between the dates of 22 January 2020 and 23 August 2022. The dataset is then further prepared in such a way that the first eighty percent of the datasets are used for training purposes while the remaining twenty percent of the datasets are used for testing purposes (the last 20% of this dataset approximates the last 6 months (last 190 days)). The training dataset was used to train and improve the models, and 20% of the training data was utilized to analyze if the models were overfitting or underfitting the data. The performance of the model is evaluated with the help of the test set. Ref [38] provides both the method and the daily SARS-CoV-2 infection and death case data. Both of these can be accessed from our source.

Figure 1 showed a visual depiction of SARS-CoV-2 infection cases (left panel) and death cases (right panel) in Russia and Chelyabinsk repeatedly (Figure 1A,C). Figure 1A shows that the maximum month for total infection cases in Russia is February 2021. Figure 1C shows the same situation for infection cases in Chelyabinsk that same month (February 2021 and 2022). It had close to 100 thousand infection cases in 2022 when the mutant omicron appeared. We also note an upward trend in the development of death cases in Russia and Chelyabinsk (Figure 1B,D), with the emergence of volatility in death cases during the period. Figure 1B shows that the maximum month for total death cases in Russia is February 2022; Figure 1D shows that the maximum total number of death cases in November, December, and February in Chelyabinsk exceeded 800 death cases in November 2021. Then we find a decrease in the death cases after this month as a result of precautionary measures taken by both regions. One of the clear patterns in the visual is a similar trend in cases and death in both Russia and Chelyabinsk, which shows the unification of anti-SARS-CoV-2 policies. Using a heatmap enables us to extract some features from the SARS-CoV-2 data.

Figure 2 presents the heatmap for total monthly infection and death cases. Figure 2A shows that the maximum month for total infection cases in Russia is February 2021, and the same month in 2022 had close to 5 million infection cases in 2022 when the mutant omicron appeared. Figure 2B shows that the maximum month for total death cases in Russia is February 2022, and the same situation occurred in February 2021 when the mutant delta appeared. Figure 2C shows the same situation for infection cases in Chelyabinsk that same month (February 2021 and 2022). It had close to 100 thousand infection cases in 2022 when the mutant omicron appeared. Figure 2D shows that the maximum total number of death cases in November, December, and February in Chelyabinsk exceeded 800 death cases in November 2021.

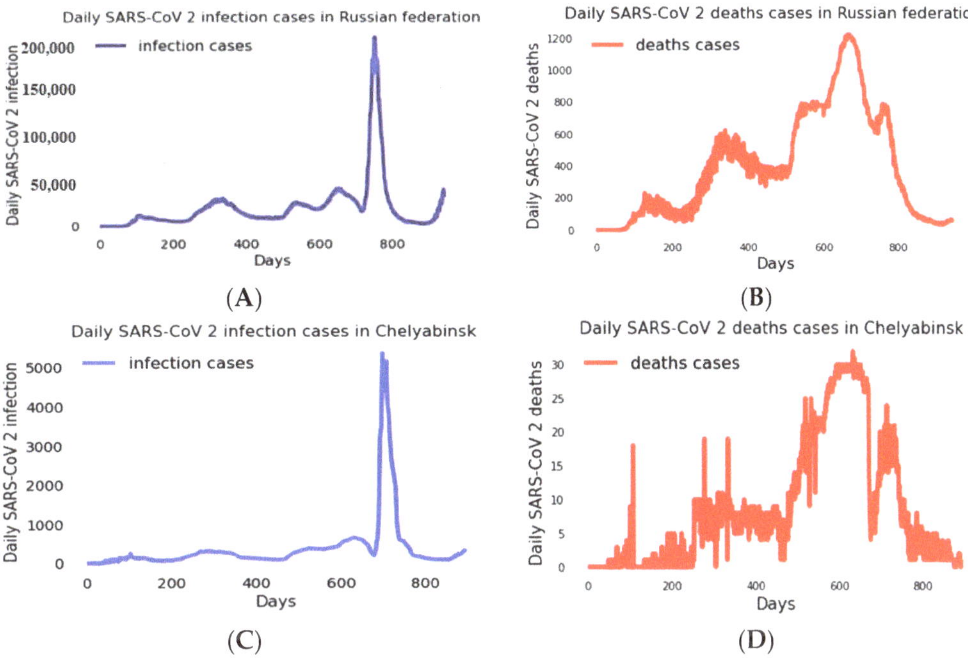

Figure 1. Daily infections and death cases SARS-CoV-2 in Russian Federation and Chelyabinsk. (**A**): Daily SARS-CoV-2 infection cases in Russian Federation. (**B**): Daily SARS-CoV-2 death cases in Russian Federation. (**C**): Daily SARS-CoV-2 infection cases in Chelyabinsk. (**D**): Daily SARS-CoV-2 death cases in Chelyabinsk.

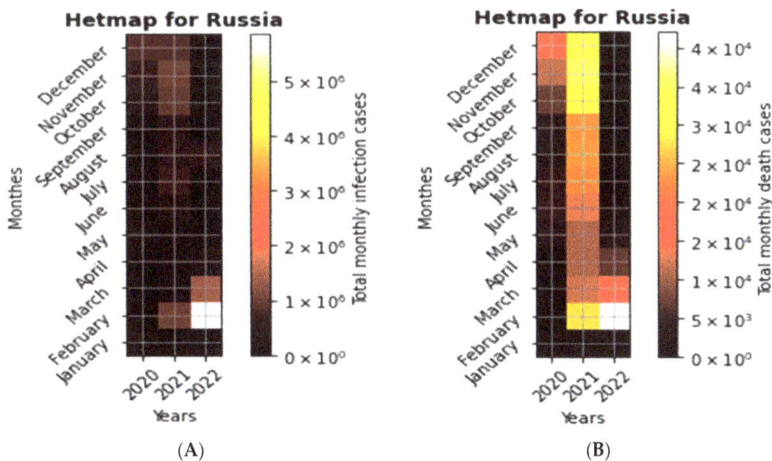

Figure 2. *Cont.*

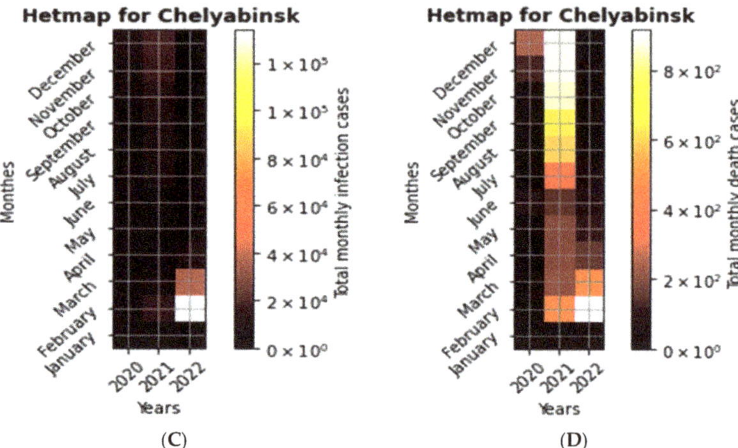

Figure 2. SARS-CoV-2 infection and death heatmap in Russian Federation and Chelyabinsk. for total monthly infection and death cases.

4. Proposed Framework Algorithm and Methodology

The mechanism that underlies our proposed approach for modeling and forecasting SARS-CoV-2 is depicted in Figure 3. The subsequent stages are carried out.

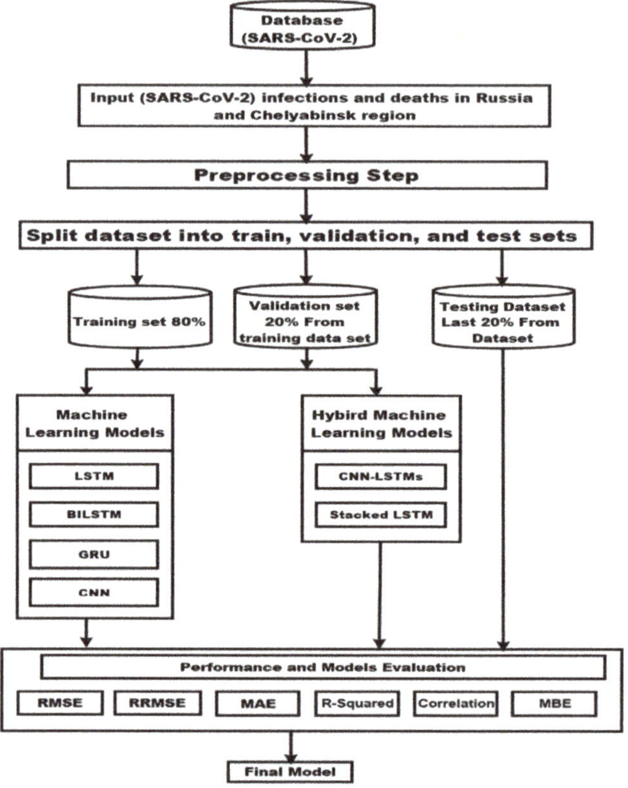

Figure 3. Proposed framework schematic schema.

4.1. Proposed Framework Algorithm

First step → Input time series data for daily infection and death cases into our algorithm. Then input parameters for the deep learning model (number of neural networks, number of epochs, Loss Function, and optimizer) start running the algorithm.

Second step → preprocessing step, training takes time and memory. Second, back-propagating extended sequences create a poorly trained model. Before importing data into neural networks, prep it. Normalization and standardization are data prep steps. Using data normalization, we set the mean and standard deviation to 0 and 1, respectively.

Third step → Separate the dataset into training, validation, and testing. SARS-CoV-2 infection and death cases. From 22 January 2020 to 23 August 2022, WHO website data was collected. We test our model using 20% of this dataset (the last 190 days). The dataset is divided such that 80% is used for training and 20% for testing. We utilized the training dataset to train and improve the models and 20% to test overfitting and underfitting. Test set is used to evaluate model performance.

Fourth step → Modeling In this stage, we execute our algorithm for LSTM, LSTMs (stacked LSTM), BDLSTM (Bidirectional LSTM), ConvLSTMs, and other forecasting models.

Fourth step → Performance and Models Evaluation

Fifth step → Forecasting using best models

4.2. Methodology

(A)   LSTM Model (long short-term memory model)

One of the first and most successful techniques for addressing vanishing gradients came in the form of long short-term memory (LSTM) due to [39].

The (long-term memory) part comes after simple recurrent neural networks have long-term memory in the form of weights. Weights change slowly during training, encoding general knowledge about the data. Moreover, the other part (short-term memory) is due to ephemeral activations, which go from each node to successive nodes. The LSTM model introduces an intermediate type of storage via the memory cell. A memory cell is a complex unit built from simpler nodes in a specific communication pattern with a new inclusion of multiplex nodes. A generalized LSTM unit consists of three gates (input, output, and a forget). The LSTM transition equations are given as follows [40].

Input gate: this gate makes the decision of whether or not the new information will be added to LSTM memory. This gate consists of two layers: (1) the sigmoid layer and (2) tanh layer. The first layer defines the values to be updated, and tanh layer creates a vector of new candidate values that will be added to LSTM memory. The output of these layers is calculated by:

$$i_t = \sigma\left(W^i x_t + U^i h_{t-1} + b^i\right) \tag{1}$$

$$u_t = \tan h(W^u x_t + U^u h_{t-1} + b^u) \tag{2}$$

where i_t: values updates, u_t: new candidate values, σ: sigmoid layer (or nonlinear function), x_t: represents a sequence of length t, b: is a constant bias, h: represents RNN memory at time step t. W and U are weight matrices.

Forget gate: the sigmoid function of this gate is used to decide what information to remove from LSTM memory. This decision is mainly made based on the value of h and x_t. The output of this gate is f, which is the value between 0 and 1, where 0 indicates completely eliminating the acquired value, and 1 indicates that the entire value is preserved. This output is calculated as:

$$f_t = \sigma\left(W^f x_t + U^f h_{t-1} + b^f\right) \tag{3}$$

where f_t: values updates, σ: sigmoid layer (or nonlinear function), x_t: represents a sequence of length t, b: is a constant bias, h: represents RNN memory at time step t. W and U are weight matrices.

Input gate: this gate first uses a sigmoid layer to decide which part of LSTM memory contributes to output. Next, it implements a nonlinear tanh function to set values between $-1, 1$. Finally, the result is multiplied by output of the sigmoid layer. The following equation represents the formulas for calculating output:

$$o_t = \sigma(W^o x_t + U^o h_{t-1} + b^o) \tag{4}$$

$$h_t = o_t \tanh_t c_{t-1} \tag{5}$$

where o_t: is an output gate, h_t: is represented as a value between $[1, -1]$.

Combining these two layers provides an update to LSTM where the current value is forgotten using forget layer by doubling the old value c_{t-1} followed by adding candidate value $i_t u_t$, The following equation represents its mathematical equation:

$$c_t = i_t u_t + f_t c_{t-1} \tag{6}$$

where c_t: is a memory cell. f_t are the results of forget gate, which is a value between 0 and 1 where 0 indicates completely rid-of value; 1 implies completely preserved value. The hypothetical combination between these units is illustrated in Figure 4

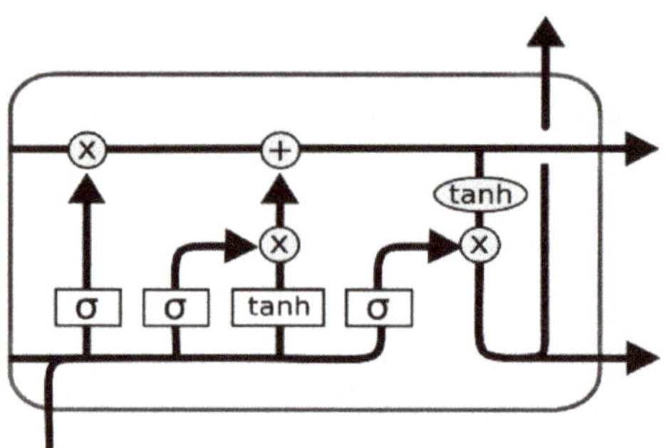

Figure 4. Long-short-term memory layer.

(B)   Stacked LSTM (Stacked long-short-term memory model)

Stacked LSTM model is an extension of LSTM model as it consists of multiple hidden layers where each layer contains multiple memory cells. It was introduced by [41]. They found that the depth of network was more important than the number of memory cells in a given layer to model skill layer for modeling the skill.

A stacked LSTM architecture can be defined as an LSTM model comprised of multiple LSTM layers. It provides a sequence output rather than a single value output to LSTM layer below. Specifically, one output per input time step rather than one output time step for all input time steps. This is illustrated in Figure 5.

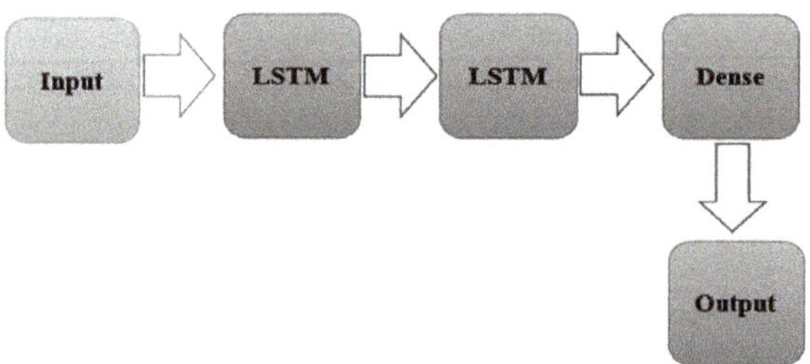

Figure 5. A stacked LSTM architecture.

(C)   Bi LSTM model (Bidirectional long-short-term memory model)

Bi LSTM model put two independent RNNs together. This architecture allows network to obtain back-and-forth information about the sequence at each time step [42].

Using Bi LSTM will run inputs in two ways, one from past to future and one from future to past; where this approach differs from unidirectional is that in LSTM running backward, you keep information from the future and using the two hidden states together are able at any time to hold the information from the past and future. Calculating the output y at time t is illustrated in Figure 6.

$$y_t = \sigma\big(W_y[h_t^{\rightarrow}, h_t^{\leftarrow}] + b_y\big) \tag{7}$$

where σ is nonlinear function, W_y: are weight matrices that are used in deep learning models, b_y: is a constant bias. h_t: are hidden states.

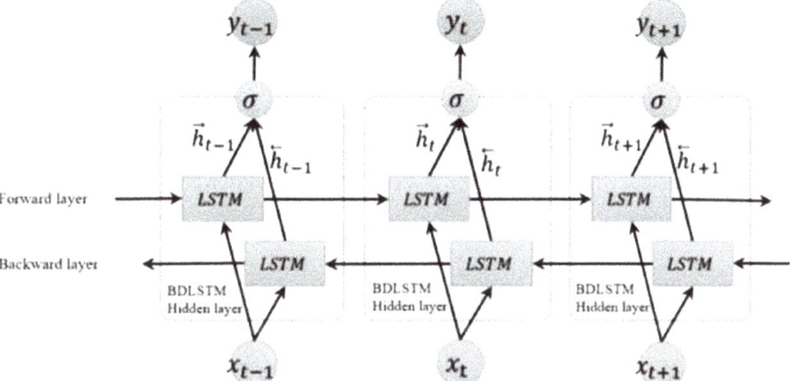

Figure 6. Bidirectional long-short-term memory layer with both forward and backward LSTM layers.

Is illustrated in Figure 5:

Figure 6 shows us how Bi LSTM model works, as it shows information sent from past and future time series (green color), from inputs x_t, which are collected in hidden layers h_t and extract features through nonlinear function σ to predict moment y_t.

(D)   GRU model (Gated Recurrent Unit model)

Gated Recurrent Unit (GRU) is an advanced and more improved version of LSTM. It is also a type of recurrent neural network. It uses less hyper parameters because of reset gate and update gate in contrast to three gates of LSTM. Update gate and reset gate are basically vectors and are used to decide which information should be passed to the output [43]. The reset gate controls how much of the previous state we need to remember. From there, update gate will allow us to control whether the new state is a copy of old state. Two gate outputs are given by two fully connected layers with sigmoid activation function; Figure 7 shows the inputs for both reset and update gates in GRU. Mathematically, output is calculated as follows:

$$r_t = \sigma(W^r x_t + U^r h_{t-1} + b^r) \tag{8}$$

$$z_t = \sigma(W^z x_t + U^z h_{t-1} + b^z) \tag{9}$$

where r_t: is reset gate, z_t: is update gate, σ: sigmoid activation function, W and U are weight parameters, h_{t-1}: the hidden state of the previous time step, b: is a constant bias. Next, we combine the reset gate with the regular refresh mechanism; it is given mathematically according to following equation:

$$i_t = \sigma\big(W^i x_t + U^i h_{t-1} + b^i\big) \tag{10}$$

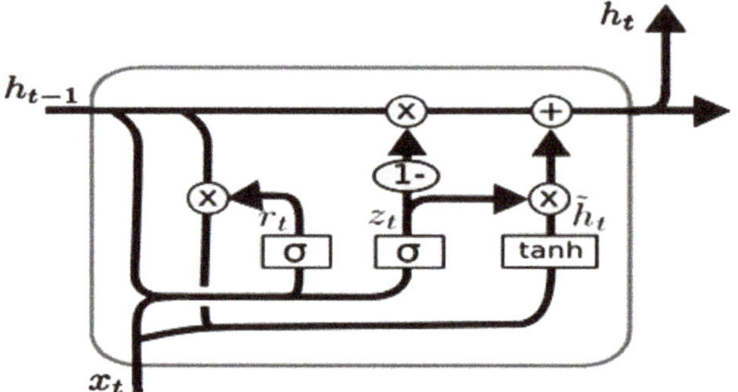

Figure 7. Gated Recurrent Unit (GRU) layer.

Which leads to the next candidate hidden state:

$$a_t = \tan h\left(wx_t + r_t U^i h_{t-1} + b^h\right) \tag{11}$$

where: a_t: candidate hidden state, $\tan h$: activation function, w and U are weight parametres, r_t: is reset gate, h_{t-1}: the hidden state of the previous time step, b: is a constant bias. Finally, we need to incorporate the effect of update gate. This determines how closely new hidden state is with old state versus how similar it is to new candidate state. Update gate can be used for this propose, simply by taking element-wise convex combinations of h_t and h_{t-1}. This leads to final update equation for GRU:

$$h_t = z_t h_{t-1} + (1 - z_t)a_t \tag{12}$$

where z_t: update gate, r_t: reset gate, a_t: activation function, h_t: hidden state output gate. The following Figure 7 illustrates this model:

(E) Conv and CNN-LSTM Model

The convolutional neural network consists of two convolutional layers; this allows for spatial advantage extraction. Where one-dimensional convolution operation is performed over the flow of data x_t^s at each time step t., a one-dimensional convolution kernel filter is used to acquire the local perceptual domain by a sliding filter [44]. The process of convolution kernel filter can be expressed as follows:

$$Y_t^s = \sigma(W_s * x_t^s + b_s) \tag{13}$$

where Y_t^s: output of convolutional layer, W_s: weights of the filter, x_t^s: input traffic flow at time t, σ: activation functions.

CNN-LSTM Model is combination of Conv and LSTM; the input of CNN-LSTM is a spatial-temporal traffic flow matrix x_t^s, as follows [2]:

$$x_t^s = \begin{bmatrix} x_{t-n}^s \\ x_{t-(n-1)}^s \\ \vdots \\ x_t^s \end{bmatrix} \begin{bmatrix} f_{t-n}^1 & f_{t-(n-1)}^1 & \cdots & f_t^1 \\ f_{t-n}^2 & f_{t-(n-1)}^2 & \cdots & f_t^2 \\ \vdots & \vdots & \ddots & \vdots \\ f_{t-n}^m & f_{t-(n-1)}^m & \cdots & f_t^m \end{bmatrix} \tag{14}$$

where $x_t^s = f_t^1 \ldots f_t^m$: denotes the traffic flow of the prediction region at time t, which represents the historical traffic flow of the POI to be predicted and its neighbors. As shown in Figure 8:

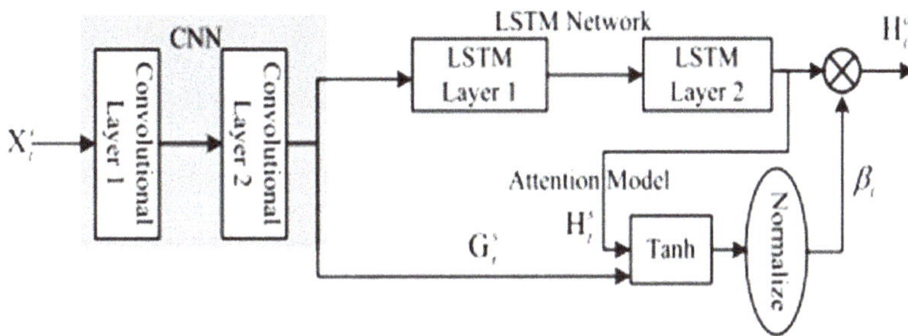

Figure 8. CNN-LSTMs Model is combination of Conv and LSTM.

Figure 8 shows us how CNN-LSTM model works; this is performed by adding CNN layer on the front end (left panel) followed by LSTM layers with a dense layer on output (right panel). CNN model works to extract features, and LSTM model works to interpret over time steps.

(F) Adam Optimization Algorithm

Stochastic gradient descent is extended by Adam optimization in order to update network weights in a more efficient manner. The method of adaptive moment estimation is used in stochastic optimization. This makes it possible for the rate of learning to adjust over the course of time, which is a vital concept to grasp, given that Adam also demonstrates this phenomenon. Adam is the result of combining the two variables (Momentum and RMSprop) as shown in Algorithms 1, which presents a method in greater detail and also Pseudo-code 1.

Adam proposed algorithm for stochastic optimization and for a slightly more efficient order of computation. g_t^2 indicates the elementwise square $g_t \odot g_t$. Good default settings for the tested machine learning problems are $\alpha = 0.001$, $\beta_1 = 0.9$, $\beta_2 = 0.999$, and $\epsilon = 10^{-8}$. All operations on vectors are element-wise. With β_1^t and β_2^t we denote β_1 and β_2 to the power t [19].

Algorithms 1: Adam algorithm for stochastic optimization [19].

Require: a : Stepsize
Require: $\beta_1, \beta_2 \in [0,1)$: Exponential decay rates for the moment estimates
Require: $f(\theta)$: Stochastic objective function with parameters θ
Require: θ_0 : Initial parameter vector $m_0 \leftarrow$
   0(Initialize 1st moment vector) $v_0 \leftarrow$
   0(Initialize 2nd moment vector) $t \leftarrow$
   0(Initialize timestep)
while θ not converged **do**
   $t + t_1$
   $g_t \leftarrow \nabla_\theta f_t(\theta_{t-1})$ (Get gradients w.r.t. stochastic objective at timestep t)
   $mt \leftarrow \beta_1 \cdot m_{t-1} + (1 - \beta_1) \cdot g_t$ (Update biased first moment estimate)
   $vt \leftarrow \beta_2 \cdot v_{t-1} + (1 - \beta_2) \cdot g_t^2$ (Update biased second raw moment estimate)
   $\hat{m}_t \leftarrow m_t / (1 - \beta_1^t)$ (Compute bias-corrected first moment estimate)
   $\hat{v}_t \leftarrow v_t / (1 - \beta_2^t)$ Compute bias-corrected second raw moment estimate)
   $\theta_t \leftarrow \theta_{t-1} - a \cdot \hat{m}_t / (\sqrt{\hat{v}_t} + \epsilon$ (Update parameters)
   end while
   return θ_t (Resulting parameters

Adaptive Moment Estimation (Adam)
Pseudo-code: Adam algorithm for stochastic optimization
Note:
We have two separate beta coefficients → one for each optimization part. We implement bias correction for each gradient
On iteration t:
Compute dW, db for current mini-batch
# #Momentum
v_dW = beta1 * v_dW + (1 − beta1) dW
v_db = beta1 * v_db + (1 − beta1) db
v_dW_corrected = v_dw/(1 − beta1 ** t)
v_db_corrected = v_db/(1 − beta1 ** t)
# #RMSprop
s_dW = beta * v_dW + (1 − beta2) (dW ** 2)
s_db = beta * v_db + (1 − beta2) (db ** 2)
s_dW_corrected = s_dw/(1 − beta2 ** t)
s_db_corrected = s_db/(1 − beta2 ** t)
# #Combine
W = W − alpha * (v_dW_corrected/(sqrt(s_dW_corrected) + epsilon))
b = b − alpha * (v_db_corrected/(sqrt(s_db_corrected) + epsilon))
Coefficients
alpha: the learning rate. 0.001.
beta1: momentum weight. Default to 0.9.
beta2: RMSprop weight. Default to 0.999.
epsilon: Divide by Zero failsave. Default to 10 ** −8.

(G) Performance indicators

To compare the prediction performance of the three models used:
Calculating root mean square error (RMSE) between the estimated data and actual data:

$$\text{RMSE} = \sqrt{\frac{\sum_{t=1}^{n}(\hat{y}_t - y_t)^2}{n}} \tag{15}$$

where $\hat{y}_t$: the forecast value, y_t: the actual value, n: number of fitted observed.
Calculating relative root mean square error (RRMSE):

$$\text{RRMSE} = \sqrt{\frac{\frac{1}{n}\sum_{t=1}^{n}(\hat{y}_t - y_t)^2}{\sum_{t=1}^{n}(\hat{y}_t)^2}} \tag{16}$$

Calculating mean absolute error (MAE):

$$\text{MAE} = \frac{1}{n}\sum_{t=1}^{n}|y_t - \hat{y}_t| \tag{17}$$

Calculating mean bias error (MBE):

$$\text{MBE} = \frac{\sum_{t=1}^{n}(y_t - \hat{y}_t)}{n} \tag{18}$$

Calculating Coefficient of correlation (R):

$$R = \frac{Cov(y_t, \hat{y}_t)}{\sqrt{V(y_t)\,V(\hat{y}_t)}} \tag{19}$$

Calculating Coefficient of determination (R Square):

$$R^2 = 1 - \frac{\sum_{t=1}^{n}(\hat{y}_t - \overline{y}_t)^2}{\sum_{t=1}^{n}(y_t - \overline{y}_t)^2} \tag{20}$$

The model that has the least values of (RMSE—RRMSE—MAE—MBE) and greater values of (R–R-Square) is the best model.

5. Results

To prove the effectiveness and superiority of the proposed approach, several experiments were conducted to predict SARS-CoV-2. Firstly, a set of baseline experiments were conducted using six base models, including LSTM, BDLSTM, GRU, LSTMs, and CONVLSTMs. The results of these models were compared to the achieved results using the Bi-LSTM, LSTM, CNN, and CNN-LSTMs algorithm for daily infection and death for *SARS-CoV-2* in Russia and Chelyabinsk, respectively. Table 1 presents the results of the testing for each of the base models along with the proposed approach based on the adopted evaluation criteria.

Table 1. Comparison of six methods evaluation testing 20% SARS-CoV-2 daily infection and death cases in Russian federation and Chelyabinsk.

Model	RMSE	RRMSE	MAE	R^2	r	MBE
(SARS-CoV-2)Infection Cases in Russia						
LSTM	9126.42	0.40	3653.27	0.93	1.00	3023.27
Stacked LSTM	35,612.77	1.56	12,646.76	−0.03	0.26	−10,796.24
BDLSTM	**2611.48**	**0.11**	**1417.74**	**0.99**	**1.00**	**−59.11**
GRU	13,105.75	0.57	4223.04	0.86	0.97	−3299.01
Conv	3397.80	0.33	1936.18	0.86	0.96	−1277.09
CNN-LSTMs	2583.41	0.25	1717.80	0.92	0.98	−1315.08
(SARS-CoV-2)Death Cases in Russia						
LSTM	**24.46**	**0.12**	**20.19**	**0.99**	**1.00**	**13.85**
Stacked LSTM	32.29	0.15	27.62	0.98	1.00	22.80
BDLSTM	24.98	0.12	20.97	0.99	1.00	16.61
GRU	27.07	0.13	23.33	0.99	1.00	19.77
Conv	88.80	0.70	46.65	0.37	0.99	39.03
CNN-LSTMs	58.11	0.46	37.69	0.73	0.99	16.52
(SARS-CoV-2)Infection Cases in Chelyabinsk region						
LSTM	160.23	0.43	59.46	0.91	1.00	57.78
Stacked LSTM	583.25	1.55	188.00	0.14	0.03	−177.87
BDLSTM	64.47	0.17	25.46	0.99	1.00	21.97
GRU	64.98	0.17	25.38	0.99	1.00	20.51
Conv	**24.69**	**0.13**	**14.36**	**0.96**	**0.98**	**3.86**
CNN-LSTMs	122.46	0.65	86.77	−0.02	0	−19.01
SARS-CoV-2Death Cases in Chelyabinsk region						
LSTM	1.84	0.35	1.44	0.88	0.94	0.22
Stacked LSTM	1.91	0.37	1.46	0.87	0.94	0.15
BDLSTM	2.03	0.39	1.63	0.85	0.94	0.68
GRU	1.79	0.35	1.39	0.89	0.94	−0.03
Conv	2.83	0.90	2.19	−0.44	0.75	1.87
CNN-LSTMs	**1.60**	**0.51**	**1.29**	**0.54**	**0.78**	**0.63**

As presented in the table, the proposed approach could achieve the best values over all the evaluation criteria, which confirms the superiority of the proposed approach. The achieved RMSE on the test set using the proposed approach **BDLSTM** for infection cases of **SARS-CoV-2 in Russia** is (2611.48). In addition, RRMSE, MAE, R^2, r, and MBE of the test set using the proposed approach **BDLSTM** is (0.11), (1417.74), (0.99), (1), and (−59.11). These values prove the effectiveness of the proposed approach. The achieved RMSE on the test set using the proposed approach **LSTM** for death cases of **SARS-CoV-2 in Russia** is (24.46). In addition, RRMSE, MAE, R^2, r, and MBE of the test set using the proposed approach **LSTM** is (0.12), (20.19), (0.99), (1), and (13.85). These values prove the effectiveness of the proposed approach. The achieved RMSE on the test set using the proposed approach **Conv** for infection cases of **SARS-CoV-2 in the Chelyabinsk region** is (24.69). In addition, RRMSE, MAE, R^2, r, and MBE of the test set using the proposed approach **Conv** are (0.13), (14.36), (0.96), (0.98), and (3.86). These values prove the effectiveness of the proposed approach. The achieved RMSE on the test set using the proposed approach **CNN-LSTMs** for death cases of **SARS-CoV-2 in the Chelyabinsk region** is (1.60). In addition, RRMSE, MAE, R^2, r, and MBE of the test set using the proposed approach **CNN-LSTMs** are (0.51), (1.29), (0.54), (0.78), and (0.63). These values prove the effectiveness of the proposed approach.

Table 2 shows us the large difference between the maximum and minimum values of all variables and thus affects the shape of the distribution. Thus, the estimators here (Mean, Median, Mode, and SD) are useless because they are breakdown points. We notice from the table that the largest difference is for the variable number of infections in Russia, from 0 to 202,211 cases, which leads to a kurtosis that gives a pointed top of the distribution as its value is much greater than three and a greater value for standard error (more difficulty in predicting), with the distribution skewed towards the right as the frequency of values greater than the average is greater for this variable. as the injury variable in Russia took 700 days to move from the lowest value to the largest value. The same thing happened for infection Chelyabinsk, with less difference between max and min values leading to less S.D. As for death cases, we notice a negative kurtosis, which indicates less volatility for both variables and, therefore, a smaller S.D than infection cases with a slight Skewness due to the convergence of the values from the arithmetic mean, and therefore, the cases of death are less developed than the cases of injury with the preventive measures that have been taken in these areas.

Table 2. Descriptive statistics of SARS-CoV-2.

	Mean	S.E	Median	Mode	S.D	Kurtosis	Skewness	Mini	Max
Infection in Russia	20,002.25	940.40	11,409	0	28,908.88	18.08	4.015	0	202,211
Death in Russia	397.79	10.94	354	0	336.374	−0.50	0.70	0	1222
Infection Chelyabinsk	383.25	25.07	180	0	750.20	21.87	4.58	0	5354
Death Chelyabinsk	8.76	0.31	6	0	9.35	−0.11	1.06	0	32

The table shows us that the best model for predicting SARS-CoV-2 infection cases in Russia is (BDLSTM) because it has the least values of (RMSE—RRMSE—MAE—MBE) and, therefore, the least difference between the real and estimated values using the model. We also note that the model is able to explain the volatility in a variable through the high value of the coefficient of determination (R Square = 99%); there is a perfect linear correlation between the estimated and actual values. As before, we note that the best model for SARS-CoV-2 death cases in Russia is (LSTM), and for SARS-CoV-2 infection cases in the Chelyabinsk region is (CONV), and for SARS-CoV-2 death cases in the Chelyabinsk region is (CNN-LSTMs). As these models achieve convergence between the actual and estimated values of the training and test data, noting their ability to capture extreme values (Maximum and Minimum value). This is illustrated by the following figures:

Figure 9 shows us the convergence of data on actual daily infection of SARS-CoV-2 in Russia with estimated using the BDLSTM model (training–testing), so we notice a great convergence between the actual and estimated data and the ability of the model to clarify volatility in infection of SARS-CoV-2 and capture structural points, and thus this model can be used to predict in daily infection of SARS-CoV-2 in Russia.

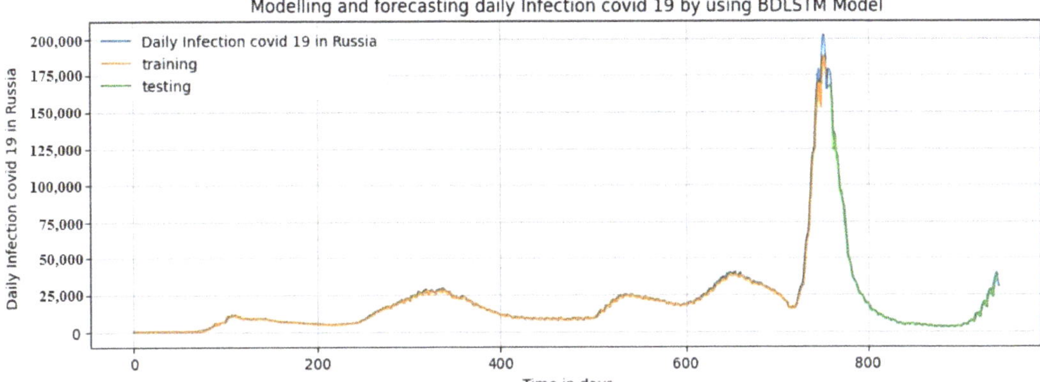

Figure 9. Comparison of the forecasting SARS-CoV2 infection cases and the real infection cases for BDLSTM.

Figure 10 shows us the convergence of data on actual daily death SARS-CoV-2 in Russia with estimated using the LSTM model (training–testing), so we notice a great convergence between the actual and estimated data and the ability of model to clarity volatility in death SARS-CoV-2 and capture trends change and thus this model can be used to predict in daily death SARS-CoV-2 in Russia.

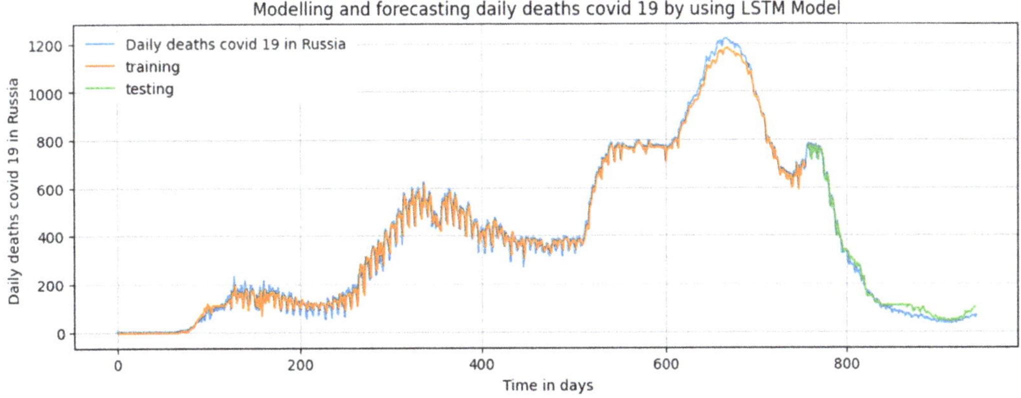

Figure 10. Comparison of the forecasting SARS-CoV-2 death cases and the real infection cases for LSTM.

Figure 11 shows us the convergence of data on actual daily infection SARS-CoV-2 in the Chelyabinsk region estimated using the CNN model (training–testing), so we notice a great convergence between the actual and estimated data and the ability of the model to clarify volatility in SARS-CoV-2 infection and capture structural points, and thus this model can be used to predict in daily SARS-CoV-2 infection in the Chelyabinsk region.

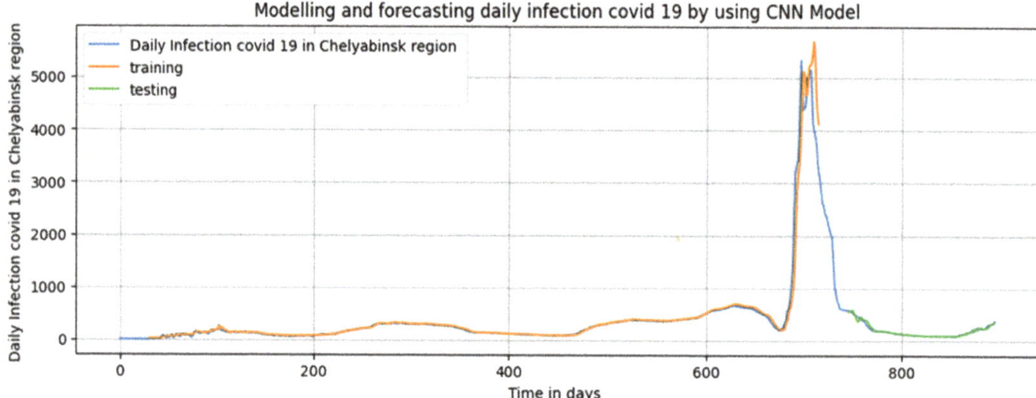

Figure 11. Comparison of the forecasting SARS-CoV-2 infection cases and the real infection cases for CNN.

Figure 12 shows us the convergence of data on actual daily death SARS-CoV-2 in the Chelyabinsk region with estimated using the CNN-LSTMs model (training–testing), so we notice a great convergence between the actual and estimated data and the ability of the model to clarify volatility in death SARS-CoV-2 and capture structural points, and thus this model can be used to predict in daily death SARS-CoV-2 in the Chelyabinsk region. The hyper-parameters for deep learning models are shown in Table 3.

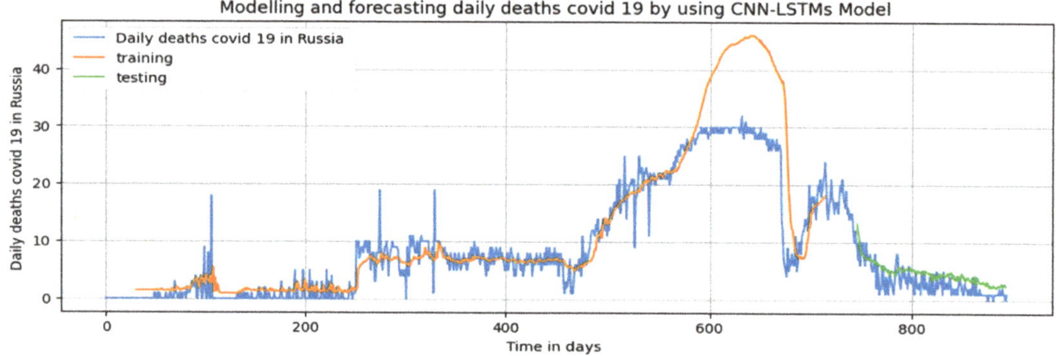

Figure 12. Comparison of the forecasting SARS-CoV-2 death cases and the real infection cases for CNN-LSTMs.

Table 3. Hyper-parameter setting for models.

Parameter	Infection in	Death	Infection	Death
Area	Russia	Russia	Chelyabinsk	Chelyabinsk
Model	BDLSTM	LSTM	Conv	ConvLSTMs
Activation function	Relu	Relu	Relu	Relu
Number of hidden units in LSTM layer	200	200	200	200
LSTM layer activation function	Relu	Relu	Relu	Relu
Timestep	2	2	2	10
Batch size	1	1	1	1
Optimizer	Adam	Adam	Adam	Adam
Learning rate	0.001	0.001	0.001	0.001
Loss function	MSE	MSE	MSE	MSE
Epochs	200	200	200	200

6. Conclusions and Future Research

In this study, a hybrid deep learning model's algorithm was used to improve the performance of a standard LSTM network in the analysis and forecasting of SARS-CoV-2 infections and death cases in the Russian Federation and the Chelyabinsk region. This was accomplished by using a combination of traditional LSTM networks and hybrid deep learning models. In order to demonstrate that the strategy being offered is effective, a dataset is gathered for the purposes of analysis and prediction. The suggested method was evaluated by applying it to datasets obtained from an official data source that was representative of the Russian Federation and the Chelyabinsk region. The utilization of these six key performance indicators allows for the performance of the suggested methodology to be evaluated and analyzed. In addition, the performance of the suggested method is evaluated and compared to that of the other five prediction models in order to demonstrate that the proposed method is superior. The compiled data provided unmistakable evidence that the strategy being recommended (Hybrid Deep-Learning models) are not only successful but also significantly more advantageous and important. On the other hand, it serves as a reference for the health sector in Russia, in particular, as well as the World Health Organization (WHO), as well as, more generally, for the health sectors in other nations. As for future research directions, it is planned to enable medium- and long-term forecasting of time series in weakly structured situations, to develop mechanisms for correcting long-term forecasts, to force a set of forecasting models to account for forecasting quality in previous periods, and to consider the possibility of employing nonlinear forecasting models for weakly structured data. All of these, along with the use of additional criteria for the verification of the best models, can be used to expand and enhance the algorithm discussed in this study and create a new package in Python for modeling and forecasting not only SARS-CoV-2 data but any univariate-dimensional time series data.

Author Contributions: Methodology, M.A.; software, M.A. and T.M.; validation, M.A., A.K. and I.P.; formal analysis, F.A.; investigation, A.K.; resources, M.A.; data curation, M.A.; writing—original draft preparation, K.A.; writing—review and editing, F.A.; visualization, A.B.; supervision, M.A.; project administration, M.A.; funding acquisition, M.A. All authors have read and agreed to the published version of the manuscript.

Funding: This research received no external funding.

Institutional Review Board Statement: Not applicable.

Informed Consent Statement: Not applicable.

Data Availability Statement: Not applicable.

Acknowledgments: The research was supported by RSF grant 22-26-00079.

Conflicts of Interest: The authors declare no conflict of interest.

References

1. Chakraborty, I.; Maity, P. COVID-19 outbreak: Migration, effects on society, global environment and prevention. *Sci. Total Environ.* **2020**, *728*, 138882. [CrossRef] [PubMed]
2. House, C.; Naseefa, N.; Palissery, S.; Sebastian, H. Corona viruses: A review on SARS, MERS and COVID-19. *Microbiol. Insights* **2021**, *14*, 11786361211002481.
3. Sachs, J.; Schmidt-Traub, G.; Kroll, C.; Lafortune, G.; Fuller, G.; Woelm, F. *Sustainable Development Report 2020: The Sustainable Development Goals and COVID-19 Includes the SDG Index and Dashboards*; Cambridge University Press: Cambridge, UK, 2021.
4. Shekerdemian, L.S.; Mahmood, N.R.; Wolfe, K.K.; Riggs, B.J.; Ross, C.E.; McKiernan, C.A.; Heidemann, S.M.; Kleinman, L.C.; Sen, A.I.; Hall, M.W.; et al. Characteristics and Outcomes of Children with Coronavirus Disease 2019 (COVID-19) Infection Admitted to US and Canadian Pediatric Intensive Care Units. *JAMA Pediatr.* **2020**, *174*, 868–873. [CrossRef]
5. Li, J.-P.O.; Liu, H.; Ting, D.S.J.; Jeon, S.; Chan, R.V.P.; Kim, J.E.; Sim, D.A.; Thomas, P.B.M.; Lin, H.; Chen, Y.; et al. Digital technology, tele-medicine and artificial intelligence in ophthalmology: A global perspective. *Prog. Retin. Eye Res.* **2021**, *82*, 100900. [CrossRef]
6. Bohr, A.; Memarzadeh, K. The rise of artificial intelligence in healthcare applications. *Artif. Intell. Healthc.* **2020**, 25–60. [CrossRef]

7.   El-Sherif, D.M.; Abouzid, M.; Elzarif, M.T.; Ahmed, A.A.; Albakri, A.; Alshehri, M.M. Telehealth and Artificial Intelligence Insights into Healthcare during the COVID-19 Pandemic. *Healthcare* **2022**, *10*, 385. [CrossRef] [PubMed]

8.   Hossain, M.S.; Muhammad, G. Emotion recognition using deep learning approach from audio–visual emotional big data. *Inf. Fusion* **2019**, *49*, 69–78. [CrossRef]

9.   Abu Adnan Abir, S.M.; Islam, S.N.; Anwar, A.; Mahmood, A.N.; Than Oo, A.M. Building resilience against COVID-19 pandemic using artificial intelligence, machine learning, and IoT: A survey of recent progress. *IoT* **2020**, *1*, 506–528. [CrossRef]

10. Pan, Y.; Zhang, L. Roles of artificial intelligence in construction engineering and management: A critical review and future trends. *Autom. Constr.* **2021**, *122*, 103517. [CrossRef]

11. Ahin, M. Impact of weather on COVID-19 pandemic in Turkey. *Sci. Total Environ.* **2020**, *728*, 138810.

12. World Health Organization. Available online: https://www.who.int/publications/m/item/weekly-epidemiological-update-on-covid-19 (accessed on 4 May 2022).

13. Dutta, A.; Fischer, H.W. The local governance of COVID-19: Disease prevention and social security in rural India. *World Dev.* **2021**, *138*, 105234. [CrossRef] [PubMed]

14. Hossain, F.; Clatty, A. Self-care strategies in response to nurses' moral injury during COVID-19 pandemic. *Nurs. Ethic-* **2021**, *28*, 23–32. [CrossRef] [PubMed]

15. Murhekar, M.V.; Bhatnagar, T.; Thangaraj, J.W.V.; Saravanakumar, V.; Kumar, M.S.; Selvaraju, S.; Vinod, A. SARS-CoV-2 seroprevalence among the general population and healthcare workers in India, December 2020–January 2021. *Int. J. Infect. Dis.* **2021**, *108*, 145–155. [CrossRef] [PubMed]

16. Alassafi, M.O.; Jarrah, M.; Alotaibi, R. Time series predicting of COVID-19 based on deep learning. *Neurocomputing* **2022**, *468*, 335–344. [CrossRef]

17. Tanrma, Ö.; Al-Dulaimi, A.; Harman, A.G.G. Estimating and Analyzing the Spread of COVID-19 in Turkey Using Long Short-Term Memory. In Proceedings of the 2021 5th International Symposium on Multidisciplinary Studies and Innovative Technologies (ISMSIT), Ankara, Turkey, 21–23 October 2021; pp. 17–26. [CrossRef]

18. Ishfaque, M.; Dai, Q.; Haq, N.U.; Jadoon, K.; Shahzad, S.M.; Janjuah, H.T. Use of Recurrent Neural Network with Long Short-Term Memory for Seepage Prediction at Tarbela Dam, KP, Pakistan. *Energies* **2022**, *15*, 3123. [CrossRef]

19. Mostafa Salaheldin Abdelsalam, A.; Makarovskikh, T. The research of mathematical models for forecasting COVID-19 cases. In Proceedings of the International Conference on Mathematical Optimization Theory and Operations Research, Irkutsk, Russia, 5–10 July 2021; Springer: Cham, Switzerland.

20. Mohamadou, Y.; Halidou, A.; Kapen, P.T. A review of mathematical modeling, artificial intelligence and datasets used in the study, prediction and management of COVID-19. *Appl. Intell.* **2020**, *50*, 3913–3925. [CrossRef]

21. Duan, D.; Wu, X.; Si, S. Novel interpretable mechanism of neural networks based on network decoupling method. *Front. Eng. Manag.* **2021**, *8*, 572–581. [CrossRef]

22. Agarwal, A.; Mishra, A.; Sharma, P.; Jain, S.; Ranjan, S.; Manchanda, R. Using LSTMfor the Prediction of Disruption in ADITYA Tokamak. *arXiv* **2020**, *preprint*. arXiv:2007.06230.

23. Abotaleb, M.S.; Makarovskikh, T. Analysis of Neural Network and Statistical Models Used for Forecasting of a Disease Infection Cases. In Proceedings of the 2021 International Conference on Information Technology and Nanotechnology (ITNT), Samara, Russia, 20–24 September 2021; pp. 1–7. [CrossRef]

24. Makarovskikh, T.; Abotaleb, M. Comparison between Two Systems for Forecasting COVID-19 Infected Cases. In *Computer Science Protecting Human Society Against Epidemics. ANTICOVID 2021. IFIP Advances in Information and Communication Technology*; Byrski, A., Czachórski, T., Gelenbe, E., Grochla, K., Murayama, Y., Eds.; Springer: Cham, Switzerland, 2021; Volume 616. [CrossRef]

25. Makarovskikh, T.; Salah, A.; Badr, A.; Kadi, A.; Alkattan, H.; Abotaleb, M. Automatic classification Infectious disease X-ray images based on Deep learning Algorithms. In Proceedings of the 2022 VIII International Conference on Information Technology and Nanotechnology (ITNT), Samara, Russia, 23–27 May 2022; pp. 1–6. [CrossRef]

26. Darwish, O.; Tashtoush, Y.; Bashayreh, A.; Alomar, A.; Alkhaza'Leh, S.; Darweesh, D. A survey of uncover misleading and cyberbullying on social media for public health. *Clust. Comput.* **2022**, 1–27. [CrossRef]

27. Tabik, S.; Gomez-Rios, A.; Martin-Rodriguez, J.L.; Sevillano-Garcia, I.; Rey-Area, M.; Charte, D.; Guirado, E.; Suarez, J.L.; Luengo, J.; Valero-Gonzalez, M.A.; et al. COVIDGR Dataset and COVID-SDNet Methodology for Predicting COVID-19 Based on Chest X-Ray Images. *IEEE J. Biomed. Health Inform.* **2020**, *24*, 3595–3605. [CrossRef]

28. Aboubakr, H.A.; Amr, M. On improving toll accuracy for covid-like epidemics in underserved communities using user-generated data. In *1st ACM SIGSPATIAL International Workshop on Modeling and Understanding the Spread of COVID-19*; ACM: New York, NY, USA, 2020.

29. Tashtoush, Y.; Alrababah, B.; Darwish, O.; Maabreh, M.; Alsaedi, N. A Deep Learning Framework for Detection of COVID-19 Fake News on Social Media Platforms. *Data* **2022**, *7*, 65. [CrossRef]

30. Biswas, P.; Saluja, K.S.; Arjun, S.; Murthy, L.; Prabhakar, G.; Sharma, V.K.; Dv, J.S. COVID-19 Data Visualization through Automatic Phase Detection. *Digit. Gov. Res. Pract.* **2020**, *1*, 1–8. [CrossRef]

31. Karajeh, O.; Darweesh, D.; Darwish, O.; Abu-El-Rub, N.; Alsinglawi, B.; Alsaedi, N. A classifier to detect informational vs. non-informational heart attack tweets. *Future Internet* **2021**, *13*, 19. [CrossRef]

32. Zhang, Y.; Liao, Q.; Yuan, L.; Zhu, H.; Xing, J.; Zhang, J. Exploiting Shared Knowledge from Non-COVID Lesions for Annotation-Efficient COVID-19 CT Lung Infection Segmentation. *IEEE J. Biomed. Health Inform.* **2021**, *25*, 4152–4162. [CrossRef]

33. Lin, H.Y.; Moh, T.-S. Sentiment Analysis on COVID Tweets Using COVID-Twitter-BERT with Auxiliary Sentence Approach. In *2021 ACM Southeast Conference, Virtual, 15–17 April 2021*; ACM: New York, NY, USA, 2021. [CrossRef]
34. Salamatov, A.A.; Davankov, A.Y.; Malygin, N.V. Socio-economic consequences of the COVID-19 Pandemic: The quality of Life of the Population in the Chelyabinsk Region in comparison with the Ural Federal District and Russia. *Bull. Chelyabinsk State Univ.* **2022**, 72–80. [CrossRef]
35. Karasu, S.; Altan, A. Crude oil time series prediction model based on LSTM network with chaotic Henry gas solubility optimization. *Energy* **2022**, *242*, 122964. [CrossRef]
36. Khan, S.D.; AlArabi, L.; Basalamah, S. Toward Smart Lockdown: A Novel Approach for COVID-19 Hotspots Prediction Using a Deep Hybrid Neural Network. *Computers* **2020**, *9*, 99. [CrossRef]
37. Ali, P.J.M.; Faraj, R.H.; Koya, E. Data normalization and standardization: A technical report. *Mach Learn. Technol. Rep.* **2014**, *1*, 1–6.
38. Abotaleb, M. Hybrid Deep Learning Algorithm. Available online: https://github.com/abotalebmostafa11/Hybrid-deep-learning-Algorithm (accessed on 26 September 2022).
39. Hochreiter, H.; Schmidhuber, J. Long short term memory. *Neural Comput.* **1997**, *9*, 1735–1780. [CrossRef]
40. Huynh, H.; Dang, L.; Duong, D. A new model for stock price movements prediction using deep neural network. In proceedings of the eighth international symposium on information and communication technology. *ACM* **2017**, 57–62. [CrossRef]
41. Graves, A.; Mohamed, A.; Hinton, G. Speech recognition with deep recurrent neural networks. In Proceedings of the 2013 IEEE International Conference on Acoustics, Speech and Signal Processing, Vancouver, BC, Canada, 26–31 May 2013; pp. 6645–6649. [CrossRef]
42. Graves, A.; Fernandez, S.; Schmidhuber, J. Multi-dimensional recurrent neural networks. In Proceedings of the 2007 International Conference on Artificial Neural Networks, Porto, Portugal, 9–13 September 2007.
43. Gulli, A.; Pal, S. *Deep Learning with Keras*; Packt Publishing Ltd.: Birmingham, UK, 2017.
44. Sutskever, I.; Vinyals, O.; Le, Q.V. Sequence to sequence learning with neural networks. In *Advances in Neural Information Processing Systems*; Curran Associates, Inc.: Red Hook, NY, USA, 2014; pp. 3104–3112.

Article

Score-Guided Generative Adversarial Networks

Minhyeok Lee [1] and Junhee Seok [2,*]

[1]   School of Electrical and Electronics Engineering, Chung-Ang University, Seoul 06974, Republic of Korea
[2]   School of Electrical Engineering, Korea University, Seoul 02841, Republic of Korea
*   Correspondence: jseok14@korea.ac.kr

Abstract: We propose a generative adversarial network (GAN) that introduces an evaluator module using pretrained networks. The proposed model, called a score-guided GAN (ScoreGAN), is trained using an evaluation metric for GANs, i.e., the Inception score, as a rough guide for the training of the generator. Using another pretrained network instead of the Inception network, ScoreGAN circumvents overfitting of the Inception network such that the generated samples do not correspond to adversarial examples of the Inception network. In addition, evaluation metrics are employed only in an auxiliary role to prevent overfitting. When evaluated using the CIFAR-10 dataset, ScoreGAN achieved an Inception score of 10.36 ± 0.15, which corresponds to state-of-the-art performance. To generalize the effectiveness of ScoreGAN, the model was evaluated further using another dataset, CIFAR-100. ScoreGAN outperformed other existing methods, achieving a Fréchet Inception distance (FID) of 13.98.

Keywords: generative adversarial network; image generation; image synthesis; GAN; generative model; Inception score; scoreGAN

MSC: 68T45

Citation: Lee, M.; Seok, J. Score-Guided Generative Adversarial Networks. *Axioms* **2022**, *11*, 701. https://doi.org/10.3390/axioms 11120701

Academic Editor: Joao Paulo Carvalho

Received: 3 November 2022
Accepted: 3 December 2022
Published: 7 December 2022

Publisher's Note: MDPI stays neutral with regard to jurisdictional claims in published maps and institutional affiliations.

1. Introduction

A recent advancement in artificial intelligence is the implementation of deep learning algorithms to generate synthetic samples [1–3]. These types of neural networks are able to learn how to map inputs to outputs after being trained on large datasets. In the past few years, researchers have used deep learning algorithms to create synthetic samples in various domains such as music, images, and speech [4–6]. One important application of synthetic sample generation is in the field of data augmentation [3,7]. Data augmentation is a technique used in machine learning to increase the size of the training datasets. Synthetic samples can be used to create new data points that are similar to existing data points, but may have different labels or attributes. This can help improve the performance of machine learning algorithms by providing them with more data to train on.

Due to their innovative training algorithm and superb performance in image generation tasks, generative adversarial networks (GANs) have been widely studied in recent years [8–12]. GANs generally employ two artificial neural network (ANN) modules, called a generator and a discriminator, which are trained with an adversarial process to detect and deceive each other. Specifically, the discriminator aims at detecting synthetic samples that are produced by the generator; meanwhile, the generator is trained by errors that are obtained from the discriminator. By such a competitive learning process, the generator can produce fine synthetic samples of which features are incredibly similar to those of actual samples [13,14].

However, the performance evaluation of GAN models is a challenging task since the quality and diversity of generated samples should be assessed from the human perspective [15,16]; furthermore, unbiased evaluations are also difficult because each person can have different views on the quality and diversity of samples. Therefore, several studies have introduced quantitative metrics to evaluate GAN models in a measurable manner [16,17].

The Inception score is one of the most representative metrics to evaluate GAN models for image generation [16]. A conventional pretrained ANN model for image classification, called the Inception network [18], is employed to assess both the quality and diversity of the generated samples, by measuring entropies of inter- and intra-samples in terms of estimated probabilities for each class. The Fréchet Inception distance (FID) is another metric to measure GAN performance, in which the distance between feature distributions of real samples and generated samples are calculated [17].

From the adoption of the evaluation metrics, the following questions then arise: Can the evaluation metrics be used as targets for the training of GAN modelssince the metrics reasonably represent the quality and diversity of samples? By backpropagating gradients of the score or distance, is it possible to maximize or minimize them? Such an approach seems feasible since the metrics are generally differentiable; therefore, the gradients can be computed and backpropagated.

However, simply backpropagating the gradients and training with the metrics correspond to learning adversarial examples in general [19,20]. Since the complexity of ANN models is significantly high, we can easily make a sample be incorrectly predicted, by adding minimal noises into the sample; this noisy sample is called the adversarial example [20]. Therefore, in short, a fine quality and rich diversity of samples can have a high Inception score, while the reverse is not always true.

Barratt and Sharma [21] studied this problem and found that directly maximizing the score does not guarantee that the generator produces fine samples. They trained a GAN model to maximize the Inception score; then, the trained model produced image samples with a very high Inception score. While the Inception score of real samples in the CIFAR-10 dataset is around 10.0, the produced images achieved an Inception score of 900.15 [21]. However, the produced images were entirely different from the real images in the CIFAR-10 dataset; instead, they looked like noises.

In this paper, to address such a problem and utilize the evaluation metric as a training method, we propose a score-guided GAN (ScoreGAN) that employs an evaluator ANN module using pretrained networks with the evaluation metrics. While the aforementioned problems exist in ordinary GANs, ScoreGAN solves the problems through two approaches as follows.

First, ScoreGAN uses the evaluation metric as an auxiliary target, while the target function of ordinary GANs is mainly used. Using the evaluation metric as the only target causes overfitting of the network used for the metric, instead of learning meaningful information from the network, as shown in related studies [21]. Thus, the evaluation metric is employed as the auxiliary target in ScoreGAN.

Second, in order to backpropagate gradients and train the generator in ScoreGAN, we employ a different pretrained model called MobileNet [22]. This prevents the generator from overfitting on the Inception network. To the best of our knowledge, employing a pretrained MobileNet with an additional score function for the training of the generator has not been explored thus far. Additionally, this approach allows us to validate that the generator has actually learned features, rather than simply memorizing details from the Inception network. In this process, we can assess whether ScoreGAN is able to achieve a high Inception score without using the Inception network, which can prove the effectiveness of ScoreGAN.

The main contributions of this paper are as follows:

- The score-guided GAN (ScoreGAN) that uses the evaluation metric as an additional target is proposed.
- The proposed ScoreGAN circumvents the overfitting problem by using MobileNet as an evaluator.
- Evaluated by the Inception score and cross-validated through the FID, ScoreGAN demonstrates state-of-the-art performance on the CIFAR-10 dataset and CIFAR-100

dataset, where its Inception score in the CIFAR-10 is 10.36 ± 0.15, and the FID in the CIFAR-100 is 13.98.

2. Background

Generative models aim to learn sample distributions and produce realistic samples. For instance, generative models can be trained with an image dataset; then, a successfully trained generative model produces realistic, but synthetic images for which the features are extremely similar to the original images in the training set. The GAN is one the representative generative models, which uses deep learning architectures and an algorithm with game theory. In recent years, diffusion models have been employed as generative models and demonstrated superior performances [2,23,24]. In Section 2.1, we discuss a variant of the GAN called the controllable GAN, which is the baseline of the proposed model. Additionally, two metrics to assess the produced images by the generative models are presented in Sections 2.2 and 2.3.

2.1. Controllable Generative Adversarial Networks

The conventional GAN model consists of two ANN modules, i.e., the generator and the discriminator. The two modules are trained by playing a game to deceive or detect each other [15,25]. The game to train a GAN can be represented as follows:

$$\hat{\boldsymbol{\theta}}_D = \underset{\boldsymbol{\theta}_D}{\arg\min} \big\{ L_D(1, D(X; \boldsymbol{\theta}_D)) + L_D(0, D(G(Z; \hat{\boldsymbol{\theta}}_G); \boldsymbol{\theta}_D)) \big\}, \tag{1}$$

$$\hat{\boldsymbol{\theta}}_G = \underset{\boldsymbol{\theta}_G}{\arg\min} \{ L_D(1, D(G(Z; \boldsymbol{\theta}_G); \boldsymbol{\theta}_D)) \}, \tag{2}$$

where G and D denote the generator and the discriminator, respectively, X is a training sample, Z represents a noise vector, $\boldsymbol{\theta}$ is a set of weights of an ANN model, and L_D indicates a loss function for the discriminator.

However, the ordinary GAN can hardly produce the desired samples since each feature in a dataset is randomly mapped into each variable of the input noise vector. Therefore, it is hard to discover which noise variable corresponds to which feature. To overcome this problem, conditional variants of GAN that introduce conditional input variables have been studied [26–28].

Controllable GAN (ControlGAN) [29] is one of the conditional variants of GANs that uses an independent classifier and the data augmentation techniques to train the classifier. While a conventional model, called auxiliary classifier GAN (ACGAN) [28], has an overfitting issue on the classification loss and a trade-off for using the data augmentation technique [29], ControlGAN breaks the trade-off through introducing the independent classifier, as well as the data augmentation technique. The training of ControlGAN is performed as follows:

$$\hat{\boldsymbol{\theta}}_D = \underset{\boldsymbol{\theta}_D}{\arg\min} \big\{ L_D(1, D(X; \boldsymbol{\theta}_D)) + L_D(0, D(G(Z, \mathcal{L}; \hat{\boldsymbol{\theta}}_G); \boldsymbol{\theta}_D)) \big\}, \tag{3}$$

$$\hat{\boldsymbol{\theta}}_G = \underset{\boldsymbol{\theta}_G}{\arg\min} \big\{ L_D(1, D(G(Z; \boldsymbol{\theta}_G); \boldsymbol{\theta}_D)) + \gamma_t \cdot L_C(\mathcal{L}, C(G(Z, \mathcal{L}; \boldsymbol{\theta}_G); \hat{\boldsymbol{\theta}}_C)) \big\}, \tag{4}$$

$$\hat{\boldsymbol{\theta}}_C = \underset{\boldsymbol{\theta}_C}{\arg\min} \{ L_C(\mathcal{L}, C(X; \boldsymbol{\theta}_C)) \}, \tag{5}$$

where C represents the independent classifier, $\mathcal{L}$ denotes the input labels, and γ_t is a learning parameter that modulates the training of the generator in terms of the classification loss.

2.2. The Inception Score

To assess the quality and diversity of the generated samples by GANs, the Inception score [16] is one of the most conventional evaluation metrics, which has been extensively employed in many studies [8,14,16,21,26,27,29]. For the quantitative evaluation of GANs, the Inception score introduces the Inception network, which was initially used for image classification [18]. The Inception network is pretrained to solve the image classification task over the ImageNet dataset [30], which contains more than one million images of 1000 different classes; then, the network learns the general features of various objects.

Through the pretrained Inception network, the quality and diversity of the generated samples can be obtained from two aspects [16,21]: First, the high quality of an image can be guaranteed if the image is firmly classified into a specific class. Second, a high entropy in the marginal probability of the generated samples indicates a rich diversity of the samples since such a condition signifies that the generated samples are different from each other.

Therefore, the entropies of the intra- and inter-samples are calculated over the generated samples; then, these two entropies compose the Inception score as follows:

$$IS\big(G(\cdot\,;\hat{\boldsymbol{\theta}}_G)\big) = \exp\left(\frac{1}{N}\sum KL\big(Pr(Y|\hat{X})||Pr(Y)\big)\right),\qquad(6)$$

where $\hat{X}$ denotes a generated sample, KL indicates the Kullback–Leibler (KL) divergence, namely the relative entropy, and N is the number of samples in a batch. Since a high KL divergence signifies a significant difference between the two probabilities, thus a higher Inception score indicates greater qualities and a wider variety of samples. Generally, ten sets, each of which contains 5000 generated samples, are used to calculate the Inception score [16,21].

2.3. The Fréchet Inception Distance

The FID is another metric to evaluate the generated samples in which the Inception network is employed as well [17]. Instead of the predicted probabilities, the FID introduces the feature distribution of the generated samples that can be represented as the outputs of the penultimate layer of the Inception network.

With the assumption that the feature distribution follows a multivariate normal distribution, the distance between the feature distributions of the real samples and generated samples is calculated as follows:

$$FID\big(X,\hat{X}\big) = \big\|\mu_X - \mu_{\hat{X}}\big\|_2^2 + Tr\left(\Sigma_X + \Sigma_{\hat{X}} - 2\cdot\sqrt{\Sigma_X\Sigma_{\hat{X}}}\right),\qquad(7)$$

where X and $\hat{X}$ are the data matrices of the real samples and generated samples, respectively, and Σ denotes the covariance matrix of a data matrix. In contrast to the Inception score, a lower FID indicates the similarity between the feature distributions since the FID measures a distance.

3. Methods

In this paper, we propose ScoreGAN, which uses an additional target, derived from the evaluation metrics in Section 2.2. The proposed ScoreGAN uses the Inception score as a target of the generator. However, directly targeting the Inception score leads to an overfitting issue; thus, in ScoreGAN, a pretrained MobileNet is used for the training. Then, the trained model is evaluated with the conventional Inception score and FID using the Inception network. This method is elaborated in Section 3.1. The training details of ScoreGAN are described in Section 3.2.

3.1. Score-Guided Generative Adversarial Network

The main idea of ScoreGAN is straightforward: For its training, the generator in ScoreGAN utilizes an additional loss that can be obtained from the evaluation metric for GANs. Since it has been verified that the evaluation metric strongly reflects the quality and diversity of the generated samples [8,16], it is expected that the performance of GAN models can be enhanced by optimizing the metrics.

Therefore, the architecture of ScoreGAN corresponds to ControlGAN with an additional evaluator; the evaluator is used to calculate the score, then gradients are backpropagated to train the generator. The other neural network structures are the same as those of ControlGAN.

However, due to the high complexity of GANs, it is not guaranteed that such an approach can work properly, as described in the previous section. Directly optimizing the Inception score can cause overfitting over the network that is used to compute the metric; then, the overfitted GANs produce noises instead of realistic samples even if the score of the generated noise is high [21].

In this paper, we circumvent this problem through two different approaches, i.e., employing the metric as an auxiliary cost instead of the main target of the generator and adopting another pretrained network as an evaluator module as a replacement of the Inception network.

3.1.1. The Auxiliary Costs Using the Evaluation Metrics

ScoreGAN mainly uses the ordinary GAN cost in which the adversarial training process is performed while the evaluation metric is utilized as an auxiliary cost. Therefore, the training of the generator in ScoreGAN is conducted by adding the cost of the evaluation metric to (4). Such a method using an auxiliary cost has been introduced in ACGAN [28]; then, the method has been widely studied in many recent works [27], including ControlGAN [29]. As a result of the recent works, it has been demonstrated that the auxiliary costs serve as a "rough guide" for a generator to be trained with additional information. The proposed technique using the evaluation metrics in this paper corresponds to a variant of such a method, where the metrics are used as rough guides to generate high-quality and a rich variety of samples. In short, the generator in ScoreGAN aims at maximizing a score in addition to the original cost, which can be represented as follows:

$$\hat{\theta}_G = \arg\min_{\theta_G}\left\{ \mathcal{L}_G - \delta \cdot IS(\hat{X}) \right\}, \tag{8}$$

where $\mathcal{L}_G$ denotes the regular cost for a generator, such as the optimization target in (4), δ is a parameter for the score, and IS is the score that can be obtained from the evaluator. Since (6) is differentiable with respect to G, θ_G can be optimized by the gradients in such a manner.

3.1.2. The Evaluator Module with MobileNet

To obtain the IS in (8), originally, the Inception network [18] is required as the evaluator in ScoreGAN since the metrics are calculated through the network. However, as described in the previous sections, directly optimizing the score leads to overfitting the network, thereby making the generator produce noises instead of fine samples. Furthermore, if the Inception network is used for the training, it is challenging to validate whether the generator actually learns features rather than memorizes the network, since the generator trained by the Inception network certainly achieves a high Inception score, regardless of the actual learning.

Therefore, ScoreGAN introduces another network, called MobileNet [22], as the evaluator module, in order to maximize the score. MobileNet [22,31,32] is a comparatively small classifier for mobile devices, which is trained with the ImageNet dataset as well. Due to its compact network size, enabling GANs to be trained, MobileNet is used in this study. The score is calculated over the feature distribution of MobileNet; then, the generator aims to maximize the score, as described in (8). For MobileNet, the pretrained model in the Keras library is used in this study.

Furthermore, to prevent overfitting on MobileNet, ScoreGAN uses a regularized score, which can be represented as follows:

$$RIS_{mobile}(\hat{X}) := \min\left\{ IS_{mobile}(X), IS_{mobile}(\hat{X}) \right\}, \tag{9}$$

where RIS represents the regularized score and IS_{mobile} denotes the score calculated by the same manner as (6) through MobileNet instead of the Inception network. Since a perfect GAN model can achieve a high score that is similar to the score of real data, thus, it is expected that the maximum value of the score that a GAN model can attain is the score of real data. Therefore, such an approach in (9) assists the GAN training by reducing the overfitting of the target network.

The evaluation, however, is performed with the Inception network, as well as the Inception score, instead of MobileNet and IS_{mobile}, which can generalize the performance of ScoreGAN. If ScoreGAN is trained to optimize MobileNet, the training ensures maximizing the score obtained with MobileNet, irrespective of the learning of actual features. Therefore, to validate the performance, the model must be evaluated with the original metric, the Inception score.

Furthermore, the model is further evaluated and cross-validated through the FID. Since the score and the FID measure different aspects of the generated samples, the maximization of the score does not guarantee obtaining a low FID. Instead, only if ScoreGAN produces realistic samples that are highly similar to real data in terms of feature distributions, the model can achieve a lower FID than the baseline. Therefore, by using the FID, we can properly cross-validate the model even if the score is used for the target.

3.2. Network Structures and Regularization

Since ScoreGAN employs the ControlGAN structure as the baseline and integrates an evaluator measuring the score with the baseline, ScoreGAN consists of four ANN modules, namely the generator, discriminator, classifier, and evaluator. In short, ScoreGAN additionally uses the evaluator, attached to the original ControlGAN framework. The structure of ScoreGAN is illustrated in Figure 1.

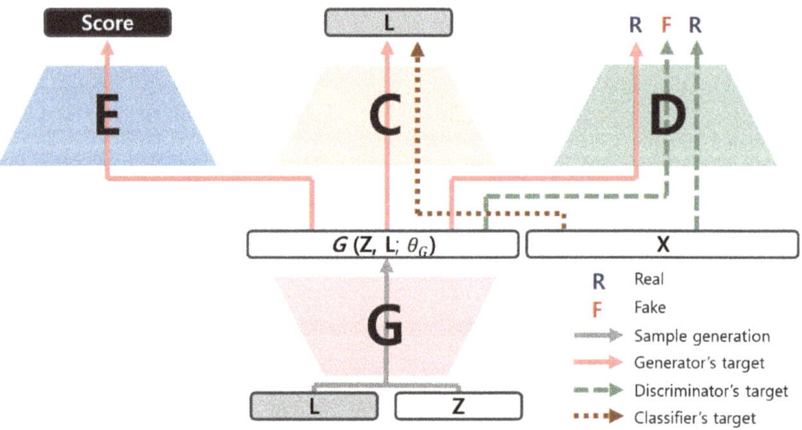

Figure 1. The structure of ScoreGAN. The training of each module is represented with arrows. E: evaluator; C: classifier; D: discriminator; G: generator.

As described in Figure 1 and (8), the generator is trained by targeting the three other ANN modules to maximize the score and minimize the losses, simultaneously. Meanwhile, the discriminator tries to distinguish between the real samples and generated samples. The classifier is trained only with the real samples in which the data augmentation is applied; then, the loss for the generator can be obtained with the trained classifier. The evaluator is a pretrained network and fixed during the training of the generator; thereby, the generator learns general features of various objects from the pretrained evaluator by maximizing the score of the evaluator.

Due to the vulnerable nature of the training of GANs, regularization methods for the ANN modules in GANs are essential [33,34]. Accordingly, ScoreGAN also uses the regularization methods that are widely employed in various GAN models for its training. Spectral normalization [35] and the hinge loss [36] that are commonly used in state-of-the-art GAN models are employed in ScoreGAN as well. The gradient penalty with a weight parameter of 10 is used [33]. Furthermore, according to recent studies that show the regularized discriminator requires intense training [8,35], multiple training iterations for the discriminator are applied; the discriminator is trained over five times per one training iteration of the generator. For the generator and the classifier, the conditional batch normalization (cBN) [37] and layer normalization (LN) [38] techniques are used, respectively.

For the neural network structures in ScoreGAN, we followed a typical architecture that is generally introduced in many other studies [27,39]. The detailed structures are shown in Table 1. Two time-scale update rule (TTUR) [17] is employed with learning rates of 4×10^{-4} and 2×10^{-4} for the discriminator and the generator, respectively. The learning rates halve after 50,000 iterations; then, the models are further trained with the halved learning rates for another 50,000 iterations. The Adam optimization method is used with the parameters of $\beta_1 = 0$ and $\beta_2 = 0.9$, which is the same setting as the other recent studies [29,35]. The maximum threshold for the training from the classifier was set to 0.1. The parameter δ in (8) that modulates the training from the evaluator was set to 0.5.

Table 1. Architecture of neural network modules. The values in the brackets indicate the number of convolutional filters or nodes of the layers. Each ResBlock is composed of two convolutional layers with pre-activation functions.

Generator	Discriminator	Classifier
$Z \in \mathbb{R}^{128}$	$Z \in \mathbb{R}^{32 \times 32 \times 3}$	$Z \in \mathbb{R}^{32 \times 32 \times 3}$
Dense $(4 \times 4 \times 256)$	ResBlock Downsample (256)	ResBlock $(32) \times 3$ ResBlock Downsample (32)
ResBlock Upsample (256)	ResBlock Downsample (256)	ResBlock $(64) \times 3$ ResBlock Downsample (64)
ResBlock Upsample (256)	ResBlock (256)	ResBlock $(128) \times 3$ ResBlock Downsample (128)
ResBlock Upsample (256)	ResBlock (256)	ResBlock $(128) \times 3$
cBN; ReLU; Conv (3); Tanh	ReLU; Global Pool; Dense (1)	LN; ReLU; Global Pool; Dense (10)

4. Results

In this section, we discuss the performance of ScoreGAN with respect to the Inception score, the FID, and the quality of the generated images. In the experiments, three images datasets called CIFAR-10, CIFAR-100, and LSUN were used. Three subsections in this section explain the performance results on each dataset. The characteristics of the datasets are described in Table 2.

Table 2. Datasets used in the experiments.

Name	Image Res.	No. of Samples	Descriptions
CIFAR-10	32×32	50,000	10 classes of small objects 5000 images per class
CIFAR-100	32×32	50,000	100 classes of small objects 500 images per class
LSUN	down-sampled to 128×128	around 10 million	10 classes of indoor and outdoor scenes around 120,000 to 3,000,000 per class

4.1. Image Generation with CIFAR-10 Dataset

The proposed ScoreGAN was evaluated over the CIFAR-10 dataset, which is conventionally employed as a standard dataset to assess the image generation performance of GAN models in many studies [26,27,29,35,39–42]. The training set of the CIFAR-10 dataset is composed of 50,000 images that are from 10 different classes. To train the models, we used a minibatch size of 64, and the generator was trained over 100,000 iterations. The other settings and the structure of ScoreGAN that was used to train the CIFAR-10 dataset are described in the previous section. Since the proposed ScoreGAN introduces an additional evaluator compared to ControlGAN, we used ControlGAN as the baseline; thereby, we can properly assess the effect of the additional evaluator.

To evaluate the image generation performance of the models, the Inception score and FID were employed. As described in the previous sections, since the Inception score is the average of the relative entropy between each prediction and the marginal predictions, a higher Inception score signifies better-quality and a rich diversity of the generated samples; conversely, a lower FID indicates that the feature distributions of the generated samples are similar to those of the real samples. Notice that, for ScoreGAN, the Inception score and FID are measured after the training iterations (100,000). It is expected that we can enhance the performance results if the models are repeatably measured during the training, and then, we selected the best model among the iterations, as conducted in several studies [8,39].

Table 3 shows the performance of GAN models in terms of the Inception score and FID. While the neural network architectures of the GAN are the same as ControlGAN, the proposed ScoreGAN demonstrates superior performance compared to ControlGAN, which verifies the effectiveness of the additional evaluator in ScoreGAN. The Inception score increased by 20.5%, from 8.60 to 10.36, which corresponds to state-of-the-art performance among the existing models thus far. The FID also decreased by 21.1% in ScoreGAN compared to ControlGAN in which the FID values of ScoreGAN and ControlGAN are 8.66 and 10.97, respectively. Random examples that are generated by ScoreGAN are shown in Figure 2.

Figure 2. Random examples of the generated images by ScoreGAN with the CIFAR-10 dataset. Each column represents each class in the CIFAR-10 dataset. All images have a 32 × 32 resolution.

Table 3. Performance of GAN models over the CIFAR-10 dataset. IS indicates the Inception score; FID indicates the Fréchet Inception distance. The best performances are highlighted in bold.

Methods	IS	FID
Real data	11.23 ± 0.20	-
ControlGAN [29]	8.61 ± 0.10	-
ControlGAN (w/Table 1; baseline)	8.60 ± 0.09	10.97
Conditional DCGAN [40]	6.58	-
AC-WGAN-GP [33]	8.42 ± 0.10	-
CAGAN [27]	8.61 ± 0.12	-
Splitting GAN [41]	8.87 ± 0.09	-
BigGAN [8]	9.22	14.73
MHingeGAN [39]	9.58 ± 0.09	**7.50**
ScoreGAN	$\mathbf{10.36 \pm 0.15}$	8.66

The results of this study appear to validate the effectiveness of both the additional evaluator and auxiliary score present in ScoreGAN. It can be said that the generator in ScoreGAN appears to properly learn general features through the pretrained evaluator and is then enforced to produce a variety of samples by maximizing the score. This is reflected not only in an increase in the Inception scores, but also in a decrease in the FID scores. Since the FID measures the similarity between feature distributions, it is less related to the objective of ScoreGAN. Therefore, this enhancement of the decreased FIDs could be evidence that ScoreGAN does not overfit on the Inception scores, and the proposed evaluator enhances the performance. Furthermore, since ScoreGAN does not use the Inception network as the evaluator and the score, it is difficult to regard the generated samples by ScoreGAN as adversarial examples of the Inception network, as shown in the examples in Figure 2, where the images are far from noises.

The detailed Inception score and FID over iterations are shown in Figure 3. As shown in the figures, the training of ControlGAN becomes slow after 30,000 iterations, while the proposed ScoreGAN continues its training. For example, the Inception score of ControlGAN at 35,000 iterations is 8.48, which is 98.6% of the final Inception score, while, at the same time, the Inception score of ScoreGAN is 9.34, which corresponds to 90.2% of its final score. The FID demonstrates similar results to those of the Inception score. In ControlGAN, the FID decreases by 10.7% from 50,000 to 100,000 iterations; in contrast, it declines by 26.9% in ScoreGAN. Such a result implies that the generator in ScoreGAN can be further trained by the proposed evaluator, although the training of the discriminator is saturated.

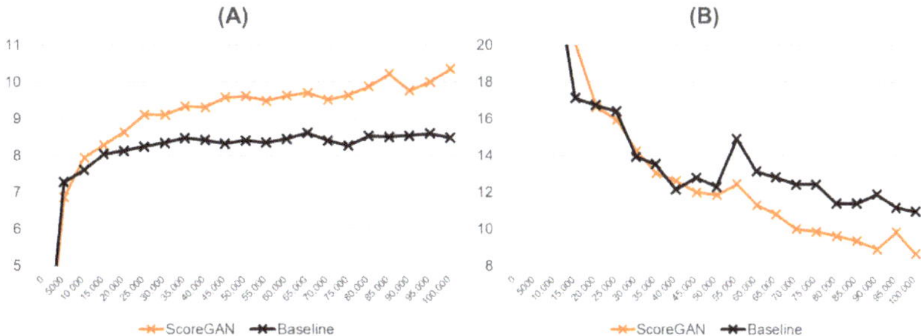

Figure 3. The performance of ScoreGAN in terms of the Inception score and Fréchet Inception distance over iterations. (**A**) The Inception scores; (**B**) the Fréchet Inception distance (FID). The baseline is ControlGAN with the same neural network architecture, identical to that of ScoreGAN.

4.2. Image Generation with CIFAR-100 Dataset

To generalize the effectiveness of ScoreGAN, the CIFAR-100 dataset was employed for the evaluation of the GAN models. The CIFAR-100 dataset is similar to the CIFAR-10 dataset, where each dataset contains 50,000 images of size 32×32 in the training set. The difference between the CIFAR-100 dataset and the CIFAR-10 dataset is that the CIFAR-100 dataset is composed of 100 different classes. Therefore, it is generally regarded that the training of the CIFAR-100 dataset is more challenging than that of the CIFAR-10 dataset. The architectures used in this experiment are shown in Appendix A.

Since existing methods in several recent studies have been evaluated over the CIFAR-100 dataset [43], we compared the performance between ScoreGAN and the existing methods. The performance in terms of the Inception score and FID is demonstrated in Table 4. The results show that ScoreGAN outperforms the other existing models. While the same neural network architectures are used in both methods, the performance of ScoreGAN is significantly superior to that of the baseline. For instance, the FID significantly declines from 18.42 to 13.98, which corresponds to a state-of-the-art result. Random examples of the generated images with ScoreGAN trained with CIFAR-100 are shown in Figure 4.

Figure 4. Random examples of the generated images by ScoreGAN with the CIFAR-100 dataset. Each column represents each class in the CIFAR-100 dataset. All images have a 32×32 resolution.

Table 4. Performance of the GAN models over the CIFAR-100 dataset. IS indicates the Inception score; FID indicates the Fréchet Inception distance. The best performances are highlighted in bold.

Methods	IS	FID
Real data	14.79 ± 0.18	-
ControlGAN (baseline)	9.32 ± 0.11	18.42
MSGAN [43]	-	19.74
SNGAN [42]	9.30 ± 0.08	15.6
MHingeGAN [39]	**14.36 ± 0.09**	17.30
ScoreGAN	13.11 ± 0.16	**13.98**

While the Inception score of ScoreGAN is slightly lower than that of MHingeGAN [39], such a disparity results from a difference in the assessment of the scores, in which, for MHingeGAN, the Inception score is continuously measured during the training iterations; then, the best score is selected among the training iterations. In contrast, the Inception score of ScoreGAN is computed only once after 100,000 iterations. Furthermore, in terms of the FID, ScoreGAN demonstrates superior results, compared to MHingeGAN. Furthermore, it is reported that the training of MHingeGAN over the CIFAR-100 dataset collapses before 100,000 iterations.

4.3. Image Generation with LSUN Dataset

For an additional experiment, ScoreGAN was applied to another dataset, called LSUN [44]. LSUN is a large-scale image dataset with 10 million images in 10 different scene categories, such as bedroom and kitchen. Furthermore, different from the CIFAR-10 and CIFAR100 datasets, LSUN is composed of high-resolution images; therefore, we evaluated ScoreGAN with LSUN to verify that the proposed framework can be performed with high-resolution images. In this experiment, ScoreGAN produces 128×128 resolution images.

The training process is the same as the previous experiments with the CIFAR datasets, while different training parameters were used; a learning rate of 5×10^{-5} was used for both the generator and discriminator, and the weights of the discriminator were updated two times for each update of the generator. Furthermore, the number of layers of the generator and discriminator was increased due to the resolution of the produced images. Since the resolution of the images is four times that of the CIFAR datasets, two additional residual modules were employed, which correspond to four additional convolutional layers for both the generator and discriminator.

Examples of the generated images by ScoreGAN are shown in Figure 5. The proposed model produced fine images for each category in the LSUN dataset. These results confirm that the proposed model can be applied to higher-resolution images than those in the CIFAR datasets, which demonstrates the generality of the performance of the proposed model. The result of the additional experiments signifies that the proposed model can be trained with various image datasets that have many image categories, such as CIFAR-100, as well as datasets with high-resolution images, such as LSUN.

Figure 5. Random examples of the generated images by ScoreGAN with the LSUN dataset. The images are a 128×128 resolution. Each column represents each class in the LSUN dataset, i.e., bedroom, bridge, church outdoor, classroom, conference room, dining room, kitchen, living room, restaurant, and tower.

5. Conclusions

In this paper, the proposed ScoreGAN introduces an evaluator module that can be integrated with conventional GAN models. While it is known that the regular use of the Inception score to train a generator corresponds to making noise-like adversarial examples of the Inception network, we circumvented this problem by using the score as an auxiliary target and employing MobileNet instead of the Inception network. The proposed ScoreGAN was evaluated over the CIFAR-10 dataset and CIFAR-100 dataset. As a result, ScoreGAN demonstrated an Inception score of 10.36, which is the best score among the existing models. Furthermore, evaluated over the CIFAR-100 dataset in terms of the FID, ScoreGAN outperformed the other models, where the FID was 13.98.

Although the proposed evaluator is integrated with the ControlGAN architecture and demonstrated fine performance, it needs to be further investigated whether the evaluator module properly performs when it is additionally used for other GAN models. Since the evaluator module can be employed along with various GANs, the performance can be enhanced by adopting other GAN models. Furthermore, in this paper, only the Inception score is introduced to train the generator while the other metric to assess GANs, i.e., the FID, can be used as a score. Such a possibility to use the FID as a score should be further studied as well for future work.

Author Contributions: Conceptualization, M.L. and J.S.; methodology, M.L.; software, M.L.; validation, M.L.; formal analysis, M.L.; investigation, M.L. and J.S.; writing—original draft preparation, M.L.; writing—review and editing, M.L. and J.S.; supervision, J.S. All authors have read and agreed to the published version of the manuscript.

Funding: This research was supported by the National Research Foundation of Korea (NRF) grants funded by the Korean government (MSIT) (Nos. 2022R1A2C2004003 and 2021R1F1A1050977).

Institutional Review Board Statement: Not applicable.

Informed Consent Statement: Not applicable.

Data Availability Statement: Not applicable.

Conflicts of Interest: The authors declare no conflict of interest.

Appendix A. Neural Network Architectures of ScoreGAN for the CIFAR-100 Dataset

Table A1. Architecture of the neural network modules for the training of the CIFAR-100 dataset. The values in the brackets indicate the number of convolutional filters or nodes of the layers. Each ResBlock is composed of two convolutional layers. The difference between the architecture for the CIFAR-10 dataset is at the classifier, in which 256 filters are used in the last three ResBlocks.

Generator	Discriminator	Classifier
$Z \in \mathbb{R}^{128}$	$Z \in \mathbb{R}^{32 \times 32 \times 3}$	$Z \in \mathbb{R}^{32 \times 32 \times 3}$
Dense $(4 \times 4 \times 256)$	ResBlock Downsample (256)	ResBlock $(32) \times 3$ ResBlock Downsample (32)
ResBlock Upsample (256)	ResBlock Downsample (256)	ResBlock $(64) \times 3$ ResBlock Downsample (64)
ResBlock Upsample (256)	ResBlock (256)	ResBlock $(128) \times 3$ ResBlock Downsample (128)
ResBlock Upsample (256)	ResBlock (256)	ResBlock $(256) \times 3$
cBN; ReLU; Conv (3); Tanh	ReLU; Global Pool; Dense (1)	LN; ReLU; Global Pool; Dense (100)

References

1. Ramesh, A.; Dhariwal, P.; Nichol, A.; Chu, C.; Chen, M. Hierarchical text-conditional image generation with clip latents. *arXiv* **2022**, arXiv:2204.06125.
2. Ho, J.; Jain, A.; Abbeel, P. Denoising diffusion probabilistic models. *Adv. Neural Inf. Process. Syst.* **2020**, *33*, 6840–6851.
3. Aggarwal, A.; Mittal, M.; Battineni, G. Generative adversarial network: An overview of theory and applications. *Int. J. Inf. Manag. Data Insights* **2021**, *1*, 100004. [CrossRef]
4. Kim, W.; Kim, S.; Lee, M.; Seok, J. Inverse design of nanophotonic devices using generative adversarial networks. *Eng. Appl. Artif. Intell.* **2022**, *115*, 105259. [CrossRef]
5. Park, M.; Lee, M.; Yu, S. HRGAN: A Generative Adversarial Network Producing Higher-Resolution Images than Training Sets. *Sensors* **2022**, *22*, 1435. [CrossRef]
6. Lee, M.; Tae, D.; Choi, J.H.; Jung, H.Y.; Seok, J. Improved recurrent generative adversarial networks with regularization techniques and a controllable framework. *Inf. Sci.* **2020**, *538*, 428–443.
7. Cai, Z.; Xiong, Z.; Xu, H.; Wang, P.; Li, W.; Pan, Y. Generative adversarial networks: A survey toward private and secure applications. *ACM Comput. Surv. (CSUR)* **2021**, *54*, 1–38. [CrossRef]
8. Brock, A.; Donahue, J.; Simonyan, K. Large scale GAN training for high fidelity natural image synthesis. In Proceedings of the International Conference on Learning Representations (ICLR), New Orleans, LA, USA, 6–9 May 2019.
9. Karras, T.; Laine, S.; Aila, T. A style-based generator architecture for generative adversarial networks. In Proceedings of the IEEE Conference on Computer Vision and Pattern Recognition (CVPR), Long Beach, CA, USA, 15–20 June 2019; pp. 4401–4410.
10. Zhu, J.Y.; Park, T.; Isola, P.; Efros, A.A. Unpaired image-to-image translation using cycle-consistent adversarial networks. In Proceedings of the IEEE International Conference on Computer Vision (ICCV), Venice, Italy, 22–29 October 2017; pp. 2223–2232.
11. Kim, S.W.; Zhou, Y.; Philion, J.; Torralba, A.; Fidler, S. Learning to simulate dynamic environments with GameGAN. In Proceedings of the IEEE Conference on Computer Vision and Pattern Recognition (CVPR), Seattle, WA, USA, 14–19 June 2020.
12. Lee, M.; Seok, J. Estimation with uncertainty via conditional generative adversarial networks. *Sensors* **2021**, *21*, 6194. [CrossRef]
13. Yi, X.; Walia, E.; Babyn, P. Generative adversarial network in medical imaging: A review. *Med. Image Anal.* **2019**, *58*, 101552.
14. Zhang, H.; Goodfellow, I.; Metaxas, D.; Odena, A. Self-attention generative adversarial networks. In Proceedings of the International Conference on Machine Learning (ICML), Long Beach, CA, USA, 9–15 June 2019; pp. 7354–7363.

15. Goodfellow, I.; Pouget-Abadie, J.; Mirza, M.; Xu, B.; Warde-Farley, D.; Ozair, S.; Courville, A.; Bengio, Y. Generative adversarial nets. In Proceedings of the Advances in Neural Information Processing Systems (NeurIPS), Montreal, QC, Canada, 8–13 December 2014; pp. 2672–2680.
16. Salimans, T.; Goodfellow, I.; Zaremba, W.; Cheung, V.; Radford, A.; Chen, X. Improved techniques for training GANs. In Proceedings of the Advances in Neural Information Processing Systems (NeurIPS), Barcelona, Spain, 5–10 December 2016; pp. 2234–2242.
17. Heusel, M.; Ramsauer, H.; Unterthiner, T.; Nessler, B.; Hochreiter, S. GANs trained by a two time-scale update rule converge to a local nash equilibrium. In Proceedings of the Advances in Neural Information Processing Systems (NeurIPS), Long Beach, CA, USA, 4–9 December 2017; pp. 6626–6637.
18. Szegedy, C.; Vanhoucke, V.; Ioffe, S.; Shlens, J.; Wojna, Z. Rethinking the Inception architecture for computer vision. In Proceedings of the IEEE Conference on Computer Vision and Pattern Recognition, Las Vegas, NV, USA, 27–30 June 2016; pp. 2818–2826.
19. Yuan, X.; He, P.; Zhu, Q.; Li, X. Adversarial examples: Attacks and defenses for deep learning. *IEEE Trans. Neural Netw. Learn. Syst.* **2019**, *30*, 2805–2824. [CrossRef]
20. Goodfellow, I.J.; Shlens, J.; Szegedy, C. Explaining and harnessing adversarial examples. In Proceedings of the International Conference on Learning Representations (ICLR), San Diego, CA, USA, 7–9 May 2015.
21. Barratt, S.; Sharma, R. A note on the inception score. *arXiv* **2018**, arXiv:1801.01973.
22. Sandler, M.; Howard, A.; Zhu, M.; Zhmoginov, A.; Chen, L.C. Mobilenetv2: Inverted residuals and linear bottlenecks. In Proceedings of the IEEE Conference on Computer Vision and Pattern Recognition (CVPR), Salt Lake City, UT, USA, 18–22 June 2018; pp. 4510–4520.
23. Zhu, Y.; Wu, Y.; Olszewski, K.; Ren, J.; Tulyakov, S.; Yan, Y. Discrete contrastive diffusion for cross-modal and conditional generation. *arXiv* **2022**, arXiv:2206.07771.
24. Song, J.; Meng, C.; Ermon, S. Denoising Diffusion Implicit Models. In Proceedings of the International Conference on Learning Representations, Addis Ababa, Ethiopia, 26–30 April 2020.
25. Creswell, A.; White, T.; Dumoulin, V.; Arulkumaran, K.; Sengupta, B.; Bharath, A.A. Generative adversarial networks: An overview. *IEEE Signal Process. Mag.* **2018**, *35*, 53–65. [CrossRef]
26. Miyato, T.; Koyama, M. cGANs with projection discriminator. In Proceedings of the International Conference on Learning Representations (ICLR), Vancouver, BC, Canada, 30 April–3 May 2018.
27. Ni, Y.; Song, D.; Zhang, X.; Wu, H.; Liao, L. CAGAN: Consistent adversarial training enhanced GANs. In Proceedings of the International Joint Conference on Artificial Intelligence (IJCAI), Stockholm, Sweden, 13–19 July 2018; pp. 2588–2594.
28. Odena, A.; Olah, C.; Shlens, J. Conditional image synthesis with auxiliary classifier GANs. In Proceedings of the International Conference on Machine Learning (ICML), Sydney, Australia, 6–11 August 2017; pp. 2642–2651.
29. Lee, M.; Seok, J. Controllable generative adversarial network. *IEEE Access* **2019**, *7*, 28158–28169. [CrossRef]
30. Russakovsky, O.; Deng, J.; Su, H.; Krause, J.; Satheesh, S.; Ma, S.; Huang, Z.; Karpathy, A.; Khosla, A.; Bernstein, M. Imagenet large scale visual recognition challenge. *Int. J. Comput. Vis.* **2015**, *115*, 211–252. [CrossRef]
31. Chen, H.Y.; Su, C.Y. An enhanced hybrid MobileNet. In Proceedings of the International Conference on Awareness Science and Technology (iCAST), Fukuoka, Japan, 19–21 September 2018; pp. 308–312.
32. Qin, Z.; Zhang, Z.; Chen, X.; Wang, C.; Peng, Y. FD-MobileNet: Improved MobileNet with a fast downsampling strategy. In Proceedings of the IEEE International Conference on Image Processing (ICIP), Athens, Greece, 7–10 October 2018; pp. 1363–1367.
33. Gulrajani, I.; Ahmed, F.; Arjovsky, M.; Dumoulin, V.; Courville, A.C. Improved training of wasserstein GANs. In Proceedings of the Advances in Neural Information Processing Systems (NeurIPS), Long Beach, CA, USA, 4–9 December 2017; pp. 5767–5777.
34. Lee, M.; Seok, J. Regularization methods for generative adversarial networks: An overview of recent studies. *arXiv* **2020**, arXiv:2005.09165.
35. Miyato, T.; Kataoka, T.; Koyama, M.; Yoshida, Y. Spectral normalization for generative adversarial networks. In Proceedings of the International Conference on Learning Representations (ICLR), Vancouver, BC, Canada, 30 April–3 May 2018.
36. Lim, J.H.; Ye, J.C. Geometric GAN. *arXiv* **2017**, arXiv:1705.02894.
37. Dumoulin, V.; Shlens, J.; Kudlur, M. A learned representation for artistic style. In Proceedings of the International Conference on Learning Representations (ICLR), Toulon, France, 24–26 April 2017.
38. Ba, J.L.; Kiros, J.R.; Hinton, G.E. Layer normalization. *arXiv* **2016**, arXiv:1607.06450.
39. Kavalerov, I.; Czaja, W.; Chellappa, R. cGANs with Multi-Hinge Loss. *arXiv* **2019**, arXiv:1912.04216.
40. Wang, D.; Liu, Q. Learning to draw samples: With application to amortized mle for generative adversarial learning. *arXiv* **2016**, arXiv:1611.01722.
41. Grinblat, G.L.; Uzal, L.C.; Granitto, P.M. Class-splitting generative adversarial networks. *arXiv* **2017**, arXiv:1709.07359.
42. Shmelkov, K.; Schmid, C.; Alahari, K. How good is my GAN? In Proceedings of the European Conference on Computer Vision (ECCV), Munich, Germany, 8–14 September 2018; pp. 213–229.
43. Tran, N.T.; Tran, V.H.; Nguyen, B.N.; Yang, L. Self-supervised GAN: Analysis and improvement with multi-class minimax game. In Proceedings of the Advances in Neural Information Processing Systems (NeurIPS), Vancouver, BC, Canada, 8–14 December 2019; pp. 13232–13243.
44. Yu, F.; Seff, A.; Zhang, Y.; Song, S.; Funkhouser, T.; Xiao, J. Lsun: Construction of a large-scale image dataset using deep learning with humans in the loop. *arXiv* **2015**, arXiv:1506.03365.

Article

Improved Method for Oriented Waste Detection

Weizhi Yang [1,*], Yi Xie [2] and Peng Gao [3]

[1] School of Information and Intelligent Engineering, Guangzhou Xinhua University, Dongguan 523133, China
[2] School of Computer Science and Engineering, Sun Yat-Sen University, Guangzhou 510006, China
[3] College of Electronic Engineering, South China Agricultural University, Guangzhou 510642, China
* Correspondence: weizhiyangq@163.com

Abstract: Waste detection is one of the main problems preventing the realization of automated waste classification, which is a basic function for robotic arms. In addition to object identification in general image analysis, a waste-sorting robotic arm not only needs to identify a target object but also needs to accurately judge its placement angle so that it can determine an appropriate angle for grasping. In order to solve the problem of low-accuracy image detection caused by irregular placement angles, in this work, we propose an improved oriented waste detection method based on YOLOv5. By optimizing the detection head of the YOLOv5 model, this method can generate an oriented detection box for a waste object that is placed at any angle. Based on the proposed scheme, we further improved three aspects of the performance of YOLOv5 in the detection of waste objects: the angular loss function was derived based on dynamic smoothing to enhance the model's angular prediction ability, the backbone network was optimized with enhanced shallow features and attention mechanisms, and the feature aggregation network was improved to enhance the effects of feature multi-scale fusion. The experimental results showed that the detection performance of the proposed method for waste targets was better than other deep learning methods. Its average accuracy and recall were 93.9% and 94.8%, respectively, which were 11.6% and 7.6% higher than those of the original network, respectively.

Keywords: waste classification; angle detection box; dynamic smoothing; YOLOv5

MSC: 68T20; 68T45; 68U10

Citation: Yang, W.; Xie, Y.; Gao, P. Improved Method for Oriented Waste Detection. *Axioms* **2023**, *12*, 18. https://doi.org/10.3390/axioms12010018

Academic Editor: Oscar Humberto Montiel Ross

Received: 14 November 2022
Revised: 11 December 2022
Accepted: 21 December 2022
Published: 24 December 2022

1. Introduction

Waste disposal is an important problem worldwide that must be addressed. Classifying waste and implementing differentiated treatments can help to improve resource recycling and promote environmental protection. However, many countries and regions still rely on manual waste classification. The main drawbacks of this are twofold. First, the health of operators can be seriously threatened by the large number of bacteria carried by waste [1]. Second, manual sorting is not only costly but also inefficient. Consequently, automated waste management and classification approaches have received extensive attention [2].

Using a robotic arm is a common method for replacing the manual mode with automated waste sorting [3]. In order to enable the robot arm to correctly classify and grasp the target object, each robot arm needs to have the functions of object recognition and placement angle judgment.

Wu et al. [4] proposed a plastic waste classification method based on FV-DCNN. They extracted classification features from original spectral images of plastic waste and constructed a deep CNN classification model. Their experiments showed that the model could recognize and classify five categories of polymers. Chen et al. [5] proposed a lightweight feature extraction network based on MobileNetv2 and used it to achieve image classification of waste. Their experiments showed that the average accuracy of the classification with their dataset was 94.6%. Liu et al. [6] proposed a lightweight neural network based on MobileNet that can reduce the cost of industrial processes. Kang et al. [7] proposed an automated waste classification system based on the ResNet-34 algorithm. The experimental results showed that the classification had high accuracy, and the classification speed of the system was as quick as 0.95 s.

These models can recognize categories of waste intelligently based on convolutional neural networks, but they do not generate location boxes for the waste targets. In addition, when there are multiple categories of waste in an image, such models cannot achieve effective identification. Therefore, they cannot be applied directly to tasks such as automated waste sorting.

Cui et al. [8] used the YOLOv3 algorithm to detect domestic waste, decoration waste and large waste on the street. Xu et al. [9] proposed a five-category waste detection model based on the YOLOv3 algorithm and achieved 90.5% average detection accuracy with a self-made dataset. The dataset included paper waste, plastic products, glass products, metal products and fabrics. Chen et al. [10] proposed a deep learning detection network for the classification of scattered waste regions and achieved good detection results. Majchrowska et al. [11] proposed deep learning-based waste detection method for natural and urban environments. Meng et al. [12] proposed a MobileNet-SSD with FPN for waste detection.

These methods can achieve image-based waste detection, but they do not provide the grasping angle information for the target object. For a target object placed at any angle, these methods only provide a horizontal identification box. Therefore, the robotic arm cannot determine the optimal grasp mode for the shape and placement angle of a waste object, which may easily lead to the object falling or to grabbing failure, especially in cases involving a large aspect ratio, as a small angle deviation can lead to a large deviation in the intersection over union (IoU).

In addition to object identification in general image analysis, a waste-sorting robotic arm not only needs to identify a target object but also needs to accurately judge its placement angle so that the robotic arm can determine the appropriate grasping angle. YOLOv5 has a strong feature extraction structure and feature aggregation network, allowing it to achieve higher detection recall and accuracy. It also provides a series of methods that can be used to achieve data enhancement. YOLOv5 is a good choice for many common identification and classification problems due to its fast detection speed, high detection accuracy and easy deployment, making it popular in many practical engineering applications. Li et al. [13] and Chen et al. [14] proposed improved algorithms for vegetable disease and plant disease detection based on YOLOv5. Their experiments showed that the detection rates reached 93.1% and 70%, respectively, which were better than other methods. Ling et al. [15] and Wang et al. [16] proposed gesture recognition and smoke detection models, respectively, based on YOLOv5. Gao et al. [17] proposed a beehive detection model based on YOLOv5. However, the original YOLOv5 does not provide the grasping angle information required for a target object. For a target object placed at any angle, it only provides a horizontal identification box, as shown in Figure 1a. Therefore, the robotic arm cannot determine the optimal grasp mode for the shape and placement angle of a target object, which may easily lead to the object falling or to grabbing failure, especially in cases involving a large aspect ratio.

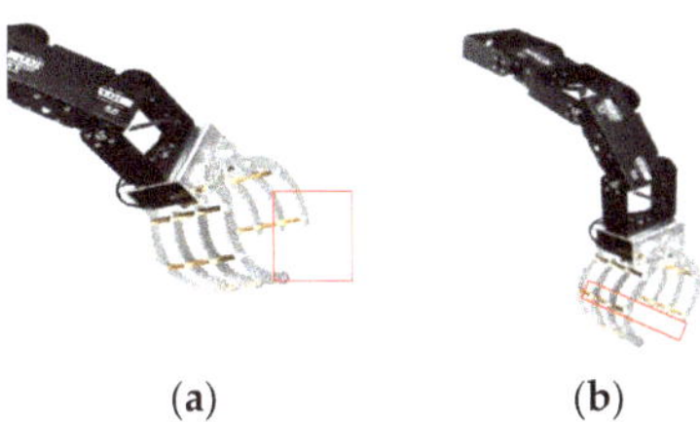

(a)   (b)

Figure 1. Grabbing with horizontal and oriented detection box. (**a**) Horizontal detection box; (**b**) oriented detection box.

In this work, we made two modifications to YOLOv5 to improve its suitability for automated waste sorting application scenarios. First, we added an angular prediction network in the detection head to provide grasping angle information for the waste object and developed a dynamic smoothing label for angle loss to enhance the angular prediction ability of the model. Second, we optimized the structure of the feature extraction and aggregation by enhancing the multi-scale feature fusion.

The contributions of this work are threefold:

(1) An optimized waste detection approach was designed based on YOLOv5 that provides higher detection accuracy for both general-sized waste and waste with a large aspect ratio;

(2) An angular prediction method is proposed for YOLOv5 that enables the rotation detection box to obtain the actual position of oriented waste;

(3) New optimization schemes are introduced for YOLOv5, including a loss function, feature extraction and aggregation.

2. Detection Method for Oriented Waste

2.1. Detection Scheme

As shown in Figure 2, the framework of the proposed waste detection scheme consists of five parts: the input layer, feature extraction backbone network, feature aggregation network, detection head and dynamic smoothing module. In this study, the backbone network mainly consisted of the focus module, the convolution module and an optimized HDBottleneckCSP module based on BottleneckCSP. The focus module reduces the number of computations and improves the speed in accordance with the slicing operation. The BottleneckCSP module is a convolution structure that demonstrates good performance in model learning. The backbone was used to extract the features from waste images and generate feature maps with three different sizes. The feature aggregation network converges and fuses multi-scale features generated from the backbone network to improve the representation learning ability for rotating waste angle features. The detection head generates the category, location and rotation angle for waste based on the multi-scale feature maps. Finally, the dynamic smoothing module partially densifies the "one-hot label encoding" of the angle labels for model training.

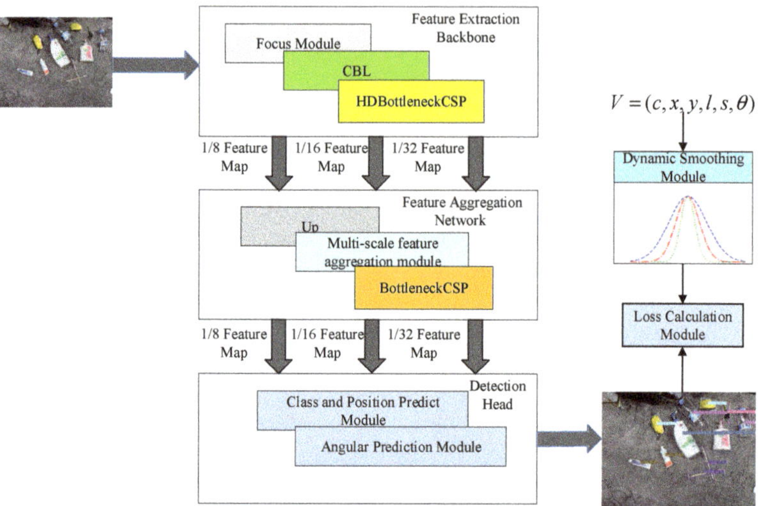

Figure 2. Rotation angle waste detection scheme.

2.2. Improvement of Detection Model

The original YOLOv5 model has the following limitations: (i) It can only generate a target detection box with a horizontal angle and not a rotation angle. (ii) The stack of bottleneck modules in BottleneckCSP is serial, which causes the middle-layer features to be lost. (iii) The feature aggregation network lacks end-to-end connection between the input and output feature maps.

To solve these problems, we optimized three aspects of YOLOv5: (i) We added an angular prediction network and loss function, as well as a dynamic angle smoothing algorithm for angular classification, to improve the angular prediction ability. (ii) We optimized the BottleneckCSP module of the backbone network to enhance the model's ability to extract the features of oriented waste. (iii) We optimized the feature aggregation network to improve the effect of multi-scale feature fusion.

2.2.1. Improvement of the Detection Head Network

The original YOLOv5 detector lacks a network structure for angular prediction and cannot provide the grasping angle information for waste objects. Therefore, the robotic arm cannot set the optimal grasp mode according to the placement angle of the waste, which easily leads to the object falling or to grabbing failure. Thus, we optimized the structure of the detection head.

Angular prediction can be realized as regression or classification. The regression mode produces a continuous prediction value for the angle but there is a periodic boundary problem, which leads to a sudden increase in the value of the loss function at the boundary of periodic changes, increasing the difficulty of learning [18]. For example, in the $180°$ long-side definition method, the defined label range is $(-90°, 90°)$. When the true angle of waste is $89°$ and the prediction is $-90°$, the error learned by the model is $179°$, but the actual error should be $1°$, which affects the learning of the model.

Therefore, we added convolution network branches in the detection head and defined the angle label with 180 categories obtained by rotating the long side of the target box clockwise around the center. The angle convolution network generates the angle prediction using information extracted from the multi-scale features obtained by the feature aggregation network.

In the detection head, the angle convolution network and the original network share the output of the feature aggregation network as the input feature graph. The output of the angle prediction network and the original network are merged as follows:

$$V = (\hat{c}, \hat{x}, \hat{y}, \hat{l}, \hat{s}, \hat{\theta})$$ (1)

where $\hat{c}$ is the predicted category of the waste, $\hat{x}$ and $\hat{y}$ are the predicted central coordinates of the object box, $\hat{l}$ and $\hat{s}$ are the predicted lengths of the longer side and shorter side of the object box and $\hat{\theta}$ is the predicted angle of oriented waste.

2.2.2. Angle Smoothing and Loss Function

The realization of angular prediction as classification can avoid the periodic boundary problem caused by regression, but there are still some limitations. The loss function of traditional category tasks is calculated as cross-entropy loss, and the form of the labels is "one-hot label encoding", as shown in Equations (2) and (3):

$$y_{ic} = \begin{cases} 1, c = \theta \\ 0, c \neq \theta \,\&\, c \in \{0, 1, \ldots, 179\} \end{cases}$$ (2)

$$L = -\frac{1}{N} \sum_i \sum_{c=0}^{179} y_{ic} \log(p_{ic})$$ (3)

where y_{ic} is the "one-hot label encoding" for the angle of sample i, θ is the angle of the oriented waste and p_{ic} is the prediction of the detection model.

Equation (8) shows that, for different incorrect predictions of the angle, the same loss value is obtained and the distance of the mistake cannot be quantified, which makes it difficult for model training to determine the angle of the oriented waste.

To solve this problem, we propose a dynamic smoothing label algorithm based on the circular smooth label (CSL) algorithm [18] to optimize the "one-hot label encoding" label of the angle.

The circular smooth label algorithm is shown in Equation (4):

$$\text{CSL}(x) = \begin{cases} g(x), \theta - r < x < \theta + r \,\&\, x \in \{0, 1, \ldots, 179\} \\ 0, \text{others} \end{cases}$$ (4)

where θ is the rotation angle value, r is the range of smoothness and $g(x)$ is the smoothing function. The angle label vector manifests as a "dense" distribution because $g(x)$ is within the range of smoothness.

The value of the smoothing function is shown in Equation (5):

$$0 < g(\theta - \varepsilon) = g(\theta + \varepsilon) \leq 1, |\varepsilon| \leq r$$ (5)

where, when $\varepsilon = 0$, the function has a maximum value of 1, and when $\varepsilon = r$, it is 0.

The CSL algorithm partially densifies the "one-hot label encoding". When the angular prediction of the model is in the range of smoothness, different loss values for different predicted degrees are obtained; thus, it can quantify the mistake in the angle category prediction. However, the performance of CSL is sensitive to the range of smoothness. If the range of smoothness is too small, the smoothing label will degenerate into "one-hot label encoding" and lose its effect, and it will be difficult to learn the information from the angle. If the range is too large, the deviation in the angle prediction will be large, which will lead to it missing the object, especially for waste with a large aspect ratio.

Therefore, we propose a dynamic smoothing function for the angle label to adjust the smoothing amplitude and range.

The dynamic smoothing function uses the dynamic Gaussian function to smooth the angle labels. It can be seen from Figure 3 that the smoothing amplitude and the range of the Gaussian function are controlled by the root mean square (RMS) value: the larger the RMS, the flatter the curve; the smaller the RMS, the steeper the curve and the smaller the smoothing range. Therefore, the RMS of the Gaussian function is gradually shrunk to achieve dynamic smoothing, as shown in Equation (6).

$$\text{DSM}(x) = \exp\left(-\frac{d^2(x,\theta)}{2 \times b^2}\right), x \in \{0, 1, \ldots, 179\} \tag{6}$$

We provide two efficient functions—linear annealing and cosine annealing—to adjust the RMS, as follows:

$$b = c + e \times \cos(\frac{0.5 \times \pi \times epoch}{epochs})$$

$$b = c - e \times \frac{epoch}{epochs}$$

where θ is the value of the rotation angle for the waste, which corresponds to the peak position of the function; x is the encoding range of the waste angle; b is the value of the RMS; and $d(x,\theta)$ is the circular distance between the encoding position and the angle values. For example, if θ is 179, $d(x,\theta)$ is 1 when x is 0; *epoch* and *epochs* represent the current number of training rounds and the maximum number of rounds of the model, respectively, and c and e are hyper-parameters.

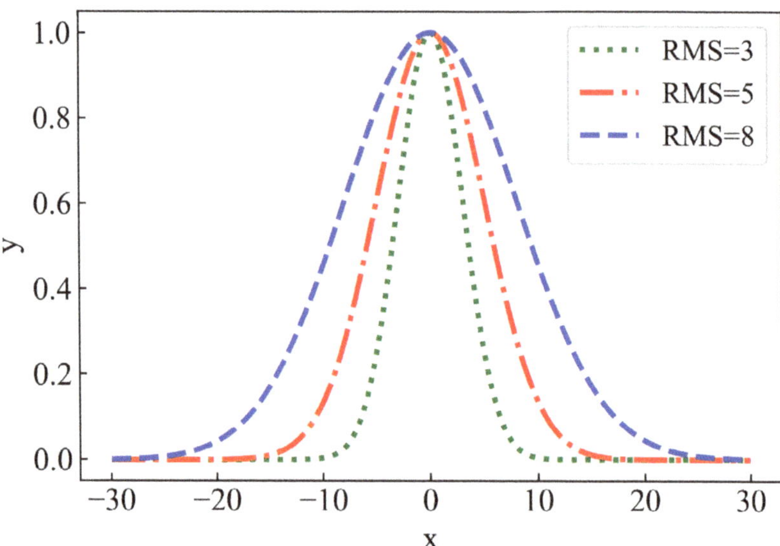

Figure 3. Gaussian function curves with different RMS.

It can be seen from Equation (6) that the DSM densifies the angle label according to the distance between the encoding position and the angle value dynamically. In the early stage of model training, b obtained large values because of the small epoch. At this time, the range of smoothing was large, and the model's learning of angles was reflected in the window area. When the smoothing range was more "loose", the model came closer to the neighborhood area of the optimal point; thus, it reduced the difficulty of angle learning and improved the recall rate in image waste detection. The range of angle smoothing decreased with the increase in the *epoch* value. The objective of the model was changed from the optimal region to the learning of the optimal point so that the deviation in the angular prediction would be smaller. The higher accuracy of the angle prediction improved the recall rate for the oriented waste, especially in cases with a large aspect ratio.

The angular loss of waste was calculated using the cross-entropy loss function based on the dynamic smoothing algorithm:

$$loss(a) = -\sum_{i=0}^{s^2} I_{ij}^{obj} \sum_{t=0}^{179} \{\hat{p}_i(t) \log[p_i(t)] + [1 - \hat{p}_i(t)] \log[1 - p_i(t)]\} \tag{7}$$

where $\hat{p}_i(t) = \text{DSM}(t)$. $p_i(t)$ is the prediction of the angle and s^2 is the quantification of the subdomain of the picture, and the model provides the prediction of the target for each subdomain. I_{ij}^{obj} is 0 or 1, which indicates whether there is a target. When the prediction is close to the true value, the cross-entropy has a smaller value.

In addition, the GIoU loss function [19] was used to calculate the regression loss of the detection boundary box. In Figure 4, A and B are the real box and the prediction box of the detection target, respectively. C is the smallest rectangle surrounding A and B. The green area is $|C| - |A \cup B|$.

The specific calculation is shown in Equations (8)–(10). GIoU not only pays attention to the overlap of the real box and the prediction box but also to the non-overlapping area, which allows it to solve the problem of the gradient not being calculated caused by A and B not intersecting.

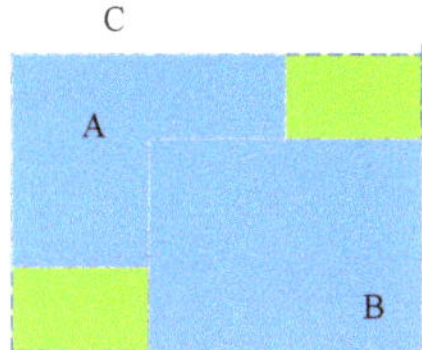

Figure 4. Illustration of GIoU.

$$\text{IoU}(A, B) = \frac{|A \cap B|}{|A \cup B|} \tag{8}$$

$$\text{GIoU}(A, B) = \text{IoU}(A, B) - \frac{|C| - |A \cup B|}{|C|} \tag{9}$$

$$loss(r) = 1 - \text{GIoU}(A, B) \tag{10}$$

In the equations, A and B are the real box and the prediction box of the detection target, respectively. C is the smallest rectangle surrounding A and B. The confidence loss function and category loss function are as shown by Equations (11) and (12):

$$loss(o) = -\sum_{i=0}^{s^2} \sum_{j=0}^{B} I_{ij}^{obj}[\hat{c}_i \log(c_i) + (1 - \hat{c}_i) \log(1 - c_i)] \\ -l_{noobj} \sum_{i=0}^{s^2} \sum_{j=0}^{B} I_{ij}^{noobj}[\hat{c}_i \log(c_i) + (1 - \hat{c}_i) \log(1 - c_i)] \tag{11}$$

$$loss(c) = -\sum_{i=0}^{s^2} I_{ij}^{obj} \sum_{c \in class} \{\hat{p}_i(c) \log[p_i(c)] + [1 - \hat{p}_i(c)] \log[1 - p_i(c)]\} \tag{12}$$

where I_{ij}^{obj} and I_{ij}^{noobj} indicate whether the prediction box j of the grid i is the target box, and λ_{noobj} indicates the weight coefficients.

The overall loss function of the improved model is a weighted combination of the above loss functions, as shown in Equation (13):

$$Loss = loss(r) + loss(o) + loss(c) + loss(a) \tag{13}$$

2.2.3. Improvement of Feature Extraction Backbone Network

The feature extraction backbone network was used to extract the features of the waste in the image. Due to the addition of angular prediction in the detection of oriented waste, there is a higher demand on the feature extraction to realize effective recognition, especially in cases involving a large aspect ratio due to a narrow area.

BottleneckCSP is the main module in the backbone of YOLOv5. The BottleneckCSP module is stacked using a bottleneck architecture. As shown in Figure 5a, the stacking of the bottleneck modules is serial. With the deepening of the network, the feature abstraction capability is gradually enhanced, but shallow features are generally lost [20]. Shallow features have lower semantics and can be more detailed due to the fewer convolution operations. Utilizing multi-level features in CNNs through skip connections has been found to be effective for various vision tasks [21–23]. The bypassing paths are presumed to be the key factor for easing the training of deep networks. Concatenating feature maps learned by different layers can increase the variation in the input of subsequent layers and improve efficiency [24,25]. In addition, attention mechanisms, which are methods used to assign different weights to different features according to their importance, have been found to be effective for the recognition of an image [26,27]. The coordinate attention mechanism (CA) [28] is one such mechanism that shows good performance. Therefore, as shown in Equation (14), we concentrated and merged the middle features of BottleneckCSP and added the CA module to enhance the feature extraction capability. The attention mechanism is optional in the module at different levels.

$$Z^{out} = g\left(Z^c_{h \times w \times (c \times t)}\right) \tag{14}$$

where

$$\begin{cases} Z^1 = f_1(x) \\ Z^t = f_t\left(Z^{t-1}\right) \\ Z^c = \left[Z^1_{h \times w \times (c \times t)}, Z^2_{h \times w \times (c \times t)}, \cdots, Z^t_{h \times w \times (c \times t)}\right] \end{cases}$$

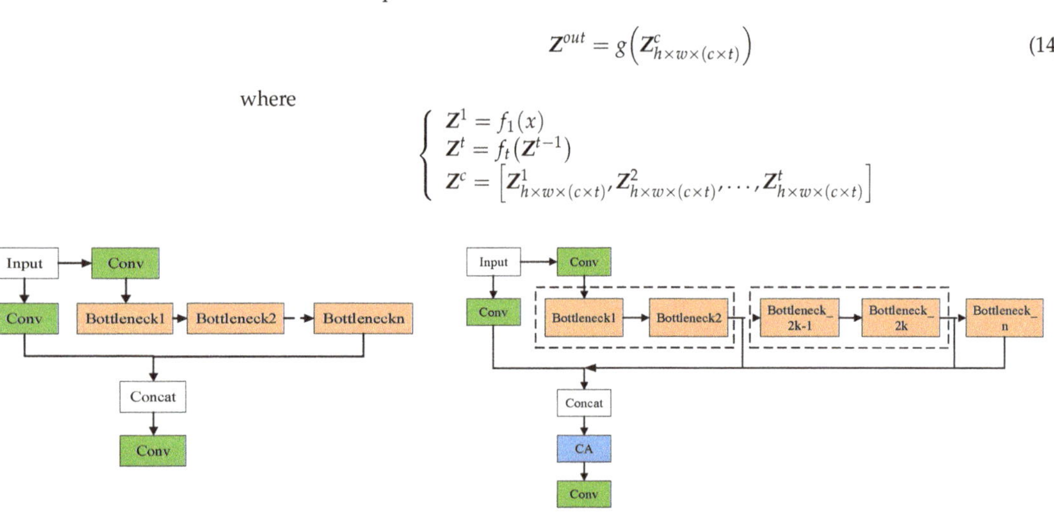

Figure 5. Comparison of BottleneckCSP before and after improvement. (**a**) BottleneckCSP module. (**b**) HDBottleneckCSP module.

Z is the feature map, x is the input of the BottleneckCSP module, f is the function mapping of the bottleneck module, and g represents the CA attention operation.

Due to the "residual block" connection in the bottleneck architecture, excessive feature merging between bottlenecks leads to feature redundancy, which is not suitable for model training, and the increased number of parameters means that more resources are consumed. Therefore, the characteristic layers were connected using "interlayer merging", as shown in Figure 5b. The optimized module was named HDBottleneckCSP.

The CA module structure in HDBottleneckCSP is shown in Figure 6. The input feature maps are coded along the horizontal and vertical coordinates to obtain the global field and to encode position information, respectively, which helps the network to detect the locations of targets more accurately.

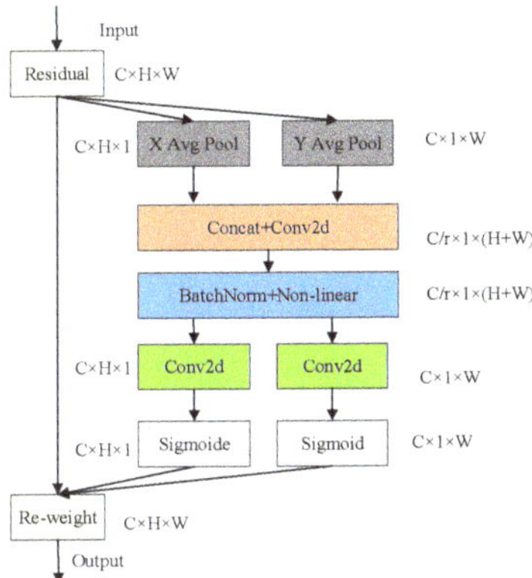

Figure 6. CA attention module.

As shown in Equation (15), the CA module generates vertical and horizontal feature maps for the input feature map and then transforms them through a 1×1 convolution. The generated $A \in \mathbb{R}^{C/r \times (H+W)}$ is the intermediate feature map for the spatial information in the horizontal and numerical directions, r is the down sampling scale and F_1 represents the convolution operation.

$$A = \delta\left(F_1\left(\left[\mathbf{Z}^h, \mathbf{Z}^w\right]\right)\right) \tag{15}$$

where A is divided into $A^h \in \mathbb{R}^{C/r \times H}$ and $A^w \in \mathbb{R}^{C/r \times W}$ in the spatial dimension. As shown in Equations (16) and (17), it is transformed into the same number of channels as the input feature map through the convolution operation, while g^h and g^w are used as the attention weight and participate in the feature map operation. The output result of the CA module is shown in Equation (18).

$$g^h = \delta\left(F_h\left(\left[A^h\right]\right)\right) \tag{16}$$

$$g^w = \delta\left(F_w\left(\left[A^w\right]\right)\right) \tag{17}$$

$$y_c(i, j) = x_c(i, j) \times g_c^h(i) \times g_c^w(j) \tag{18}$$

The optimized feature extraction backbone network structure is shown in Figure 7. It extracts features through the convolution module and the HDBottleneckCSP module and generates feature maps with three sizes by downsampling (1/8, 1/16 and 1/32).

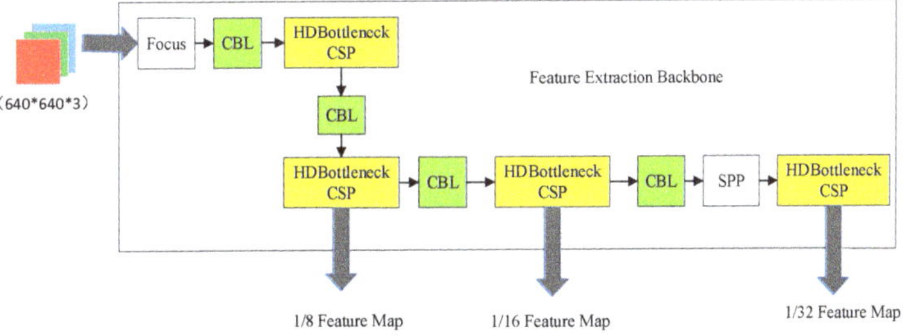

Figure 7. Backbone network structure for feature extraction.

2.2.4. Improvement of Feature Aggregation Network

The YOLOv5 feature aggregation network consists of feature pyramid networks [29] (FPNs) and path aggregation networks [30] (PANets). The structure of a PANet is shown in Figure 8a. The PANet aggregates features along two paths: top-down and bottom-up. However, the aggregated features are deep features with high semantics, and the shallow features with high resolution are not fused. In order to make use of the input features more effectively, we used P2P-PANet to replace the PANet based on BiFPN [31], as shown in Figure 8b.

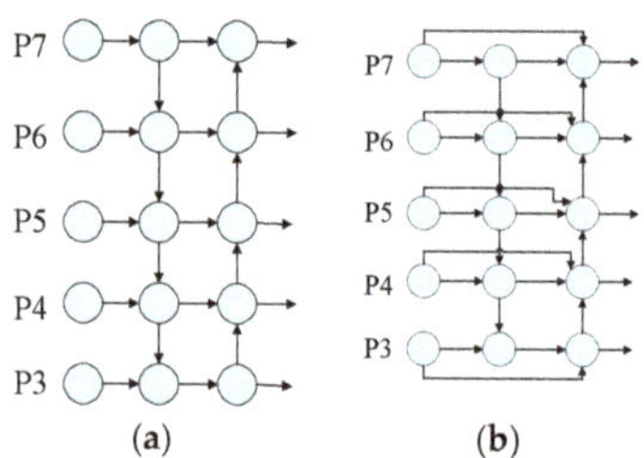

Figure 8. Network structures of PANet and P2P-PANet. (**a**) PANet network structure. (**b**) P2P-PANet network structure.

Compared to PANet, P2P-PANet adds end-to-end connection for the input-feature and output-feature maps, which establishes a "point-to-point" horizontal connection path from the low level to the high level, and it can realize the fusion of high-resolution and complex semantic features in an image without adding much cost. Through the extraction and induction of semantic information for the high-resolution and low-resolution feature maps, the angular feature information of rotating waste is further enhanced, and the detection ability of the model is improved.

The method for oriented waste detection after all the optimizations was named YOLOv5m-DSM and is shown in Figure 9. When a picture is input into the model, YOLOv5m-DSM extracts features using the backbone and generates downsampling feature maps with three different sizes for the detection of waste. The feature aggregation network undertakes feature aggregation and fusion to enhance the model's ability to learn features. The detection head generates the prediction information for waste targets based on the multi-scale features. In the model's training stage, the label of the training set is smoothed using the dynamic smoothing module, and the loss in the prediction, including class, angle and position, is calculated using the loss calculation module for iterative learning.

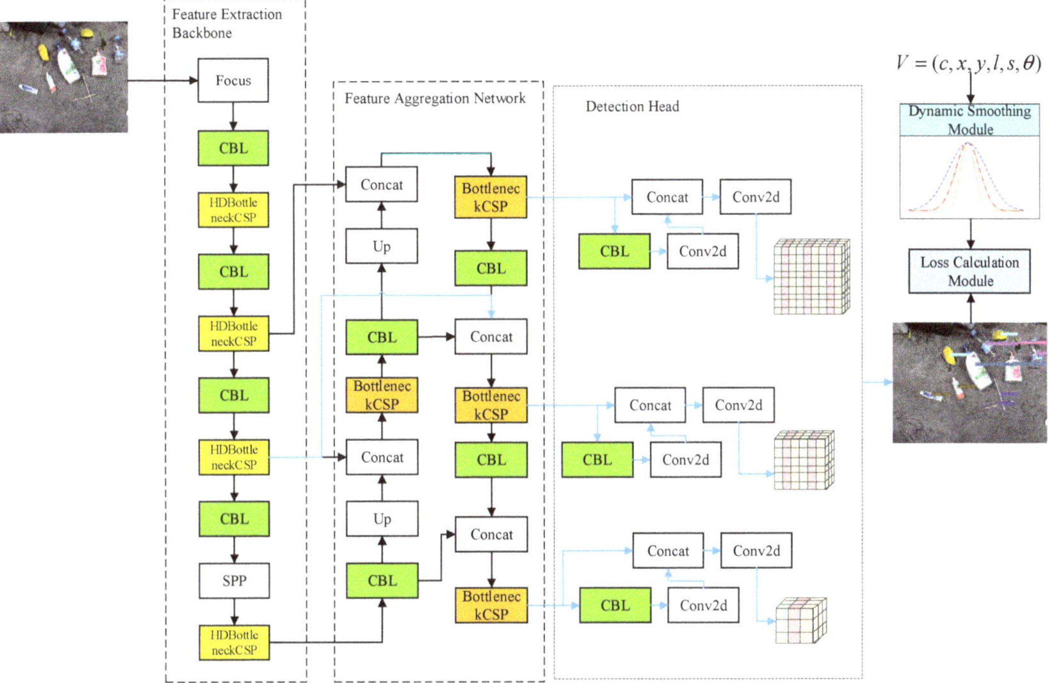

Figure 9. Method diagram for YOLOv5m-DSM.

3. Experimental Results and Analysis

3.1. Datasets

The dataset for the experiment contained eleven kinds of domestic waste, including a cotton swab, a stick, paper, a plastic bottle, a tube, vegetables, peels, a shower gel bottle, a coat hanger, clothes pegs and an eggshell. The vector of the label contained the category, the center x coordinate of the target box, the center y coordinate of the target box, the long-side value, the short-side value and the angle value. The angle was the angle between the long side of the target frame and the horizontal axis in the clockwise direction, with a range of $(0°, 180°)$.

3.2. Evaluation Index

In order to evaluate the performance of YOLOv5m-DSM, it was compared and analyzed using the recall (R), mean average precision (mAP) and other indicators. The recall is as follows:

$$R = \frac{TP}{TP + FN} \times 100\%$$ (19)

TP represents a "true positive" sample, and FN represents a "false negative" sample. The mean average precision formula is shown in Equation (20).

$$mAP = \frac{1}{m} \sum_{i=1}^{m} AP_i$$ (20)

The mean average precision refers to the average precision (AP) for each category of samples, which is calculated from the recall rate and precision (P) as follows:

$$R = \frac{TP}{TP + FN} \times 100\%$$ (21)

$$AP = \int_0^1 P(R)\,dR \tag{22}$$

3.3. Experimental Results and Analysis

In order to better show the advantages of the method described in this paper, YOLOv5-DSM was compared with mainstream horizontal box detection methods and rotating box detection methods in the experiments.

Table 1 shows a comparison of the detection effects for YOLOv5m-DSM and horizontal rectangular box detection methods, such as SSD-OBB, YOLOv3-OBB, YOLOv5s-OBB and YOLOv5m-OBB, which are angle classification network structures commonly added in detection heads based on their original models [32,33].

Table 1. Comparison between YOLOv5m-DSM and horizontal frame detection methods.

Method	Recall/%	mAP/%	AP of "Large Aspect Ratio Category"/%				
			Cotton Swab	Stick	Plastic Bottle	Shower Gel Bottle	Tube
SSD-OBB	82.1	74.9	42.6	41.9	76.2	71.6	67.8
YOLOv3-OBB	82.6	75.8	41.6	40.5	79.8	75.5	68.1
YOLOv5s-OBB	84.7	77.5	43.4	40.7	85.8	76.5	71.5
YOLOv5m-OBB	87.2	82.3	52.1	57.1	83.8	79.6	72.1
YOLOv5m-DSM (Cos)	94.5	93.3	70.1	80.7	100	100	99.9
YOLOv5m-DSM (Linear)	94.8	93.9	78.7	81.0	100	100	100

Table 1 shows that, compared with SSD-OBB, YOLOv3-OBB and YOLOv5s-OBB, the recall rate and average precision of YOLOv5m were better. Compared with the original network, YOLOv5m-DSM showed improvements of 7.6% and 11.6% in the recall rate and the average precision, respectively. This proves that the modified waste detection algorithm has obvious improvements. Furthermore, YOLOv5m-DSM showed a good detection effect for oriented waste with a large aspect ratio, demonstrating an obvious improvement over the original model. The good performance of DSM (Cos) and DSM (Linear) proves that the dynamic smoothing label was effective and strong.

The detection effects of YOLOv5m, YOLOv5m-OBB and YOLOv5m-DSM are shown in Figure 10. It can be seen from Figure 10a that the YOLOv5m network only generated a horizontal detection box. It did not provide the grasping angle information for the waste object. Therefore, the robotic arm could not set the optimal grasp mode according to the inherent shape and placement angle of the target object, which could easily lead to the object falling and to grabbing failure, especially in cases involving a large aspect ratio.

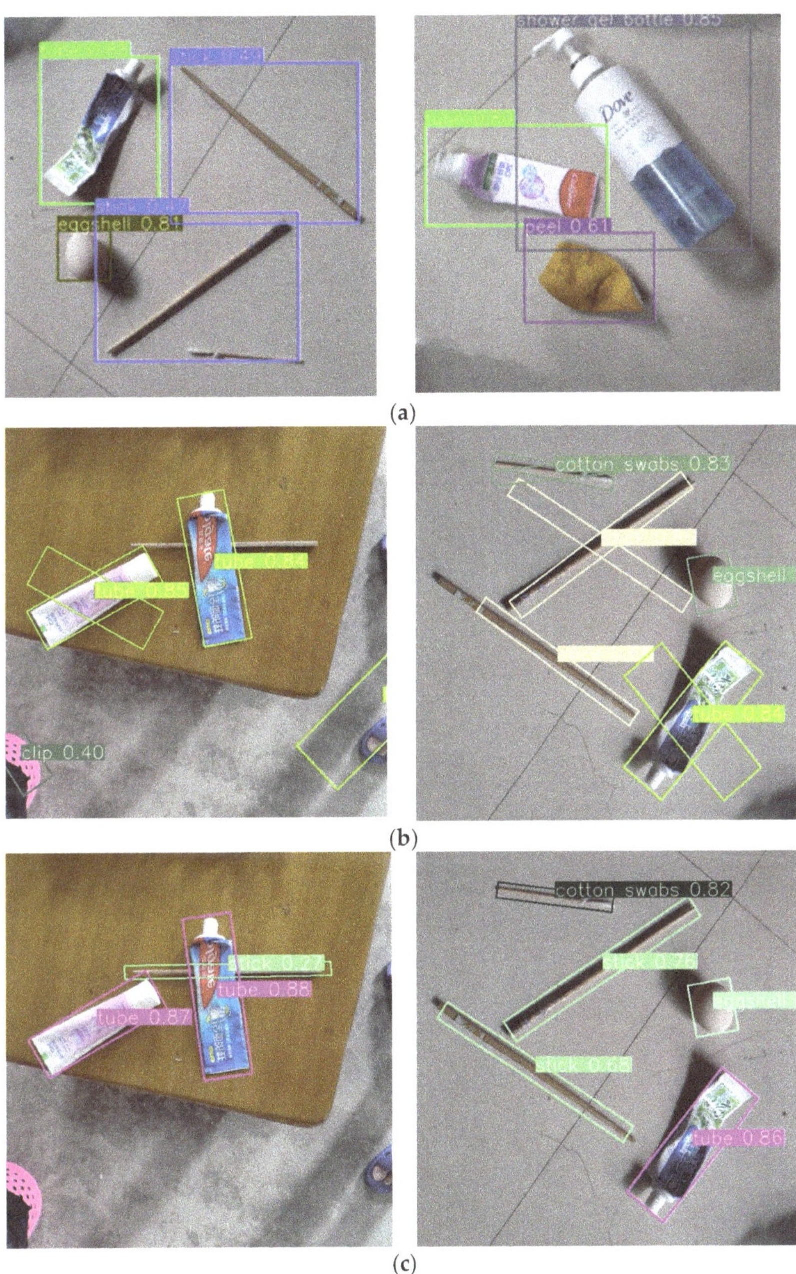

Figure 10. Comparison of detection effects of the methods. (**a**) Detection using YOLOv5m. (**b**) Detection using YOLOv5m-OBB. (**c**) Detection using YOLOv5m-DSM.

Figure 10b shows that, when the angular classification network was added to the detection head, YOLOv5m-OBB could generate a waste object detection box at any angle, but the angle of the generated detection box was not accurate enough, especially in cases involving a large aspect ratio. Due to the large aspect ratio, a slight deviation in the prediction box resulted in a smaller IoU for the prediction box and the true box, which resulted in difficulties in the model training. Therefore, a large aspect ratio makes effective learning difficult.

Figure 10c shows the detection results for YOLOv5m-DSM. It can be seen that YOLOv5m-DSM could generate a waste object detection box at any angle and could detect objects involving a large aspect ratio accurately. It can be seen that, with the optimization of the feature extraction backbone network and feature aggregation network, and after the optimization of the loss function through the dynamic smoothing algorithm, YOLOv5m-DSM had better precision and performance in the detection of oriented waste.

Table 2 shows a comparison of the detection effects of YOLOv5m-DSM and the mainstream rotating rectangular box detection methods.

Table 2. Comparison of YOLOv5m-DSM and rotating frame detection methods.

Method	Recall/%	mAP/%	Params (M)	GFLOPs	FPS
RoI Trans	88.6	87.3	55.4	265.4	5.8
Gliding-Vertex	92.4	89.6	41.4	224.8	7.6
R3Det	91.4	90.1	48.0	250.5	7.0
S2A-Net	94.3	93.1	38.9	153.8	8.3
YOLOv5m-DSM (Cos)	94.5	93.3	23.7	76.5	15.5
YOLOv5m-DSM (Linear)	94.8	93.9	23.7	76.5	15.5

It can be seen that, when compared to RoI Trans [34], the average recall rate and average precision of detection increased by 6.2% and 6.6%, respectively. Compared to Gliding-Vertex [35], they increased by 2.4% and 4.3% respectively. Compared to R3Det [36], they increased by 3.4% and 3.8%, respectively. The recall rate and average precision of the YOLOv5m-DSM model were also better than those of S2A-Net [37], and our method had fewer parameters and a detection rate twice as high. In addition, we extended the flops counter tool to calculate the floating point operations (FLOPs) in the methods, and the computation load of YOLOv5m-DSM was lower than the comparison algorithm, making the model more suitable for deployment and application in embedded devices.

3.4. Network Model Ablation Experiment

The network model ablation comparison experiment was used to evaluate the optimization effects of each improvement scheme. The experimental comparison results are shown in Table 3. Optimization 1 involved using the dynamic smoothing algorithm to densify the angle label conversion and calculate the loss function (1a is linear annealing and 1b is the cosine annealing angle). Optimization 2 involved the improvement of the feature extraction backbone network based on the proposed HDBottleneckCSP module. Optimization 3 involved an improvement of the feature aggregation network of the YOLOv5 based P2P-PANet structure.

Table 3. Ablation experiment.

Method	Angle	op 1a	op 1b	op 2	op 3	Recall/%	mAP/%
YOLOV5m	×	×	×	×	×	-	-
YOLOV5m-OBB	√	×	×	×	×	87.2	82.3
Optimization model 1	√	√	×	×	×	92.5	90.5
Optimization model 2	√	√	×	√	×	93.6	91.7
Optimization model 3	√	√	×	×	√	93.4	91.5
YOLOv5m-DSM (Cos)	√	×	√	√	√	94.5	93.3
YOLOv5m-DSM (Linear)	√	√	×	√	√	94.8	93.9

It can be seen from Table 3 that, after adding the linear dynamic smoothing algorithm and the corresponding loss function, the recall rate and mean average precision of optimization model 1 increased by 5.3% and 8.2%, respectively. After adding the linear dynamic smoothing algorithm and HDBottleneckCSP module, these values increased by 6.4% and 9.4% in optimization model 2, respectively. After adding the linear dynamic smoothing algorithm and P2P-PANet module, they increased by 6.2% and 9.2% in optimization model 3, respectively. For the YOLOV5m-DSM (Linear) model, the detection recall rate and average precision of the model increased by 7.6% and 11.6%, respectively, with the above optimization methods.

In order to analyze the effects of replacing the original module structure with the HD-BottleneckCSP structure and P2P-PANet network on the image waste detection algorithm more clearly, as well as the reasons for these effects, the intermediate characteristic graphs of YOLOv5 and YOLOv5m-DSM were extracted for comparison, as shown in Figure 10.

Figure 11a,b show an input image and label image, and Figure 11c–f and Figure 11g–j show the 1/8, 1/16 and 1/32 down sampling feature maps of YOLOv5 and YOLOv5m-DSM in the backbone network. It can be seen from Figure 11c,d,g,h that shallower feature information was extracted from the model after using the HDBottleneckCSP network, and the edge information and feature details of the waste were obtained more clearly. As can be seen from Figure 11e,i, two network structures obtained high-level semantic features through multi-layer convolution operations. Finally, from the comparison of Figure 11f,j, we can see that the YOLOv5m-DSM network generated a clearer edge for the target object, which led to an improvement in the recall and accuracy of the waste detection.

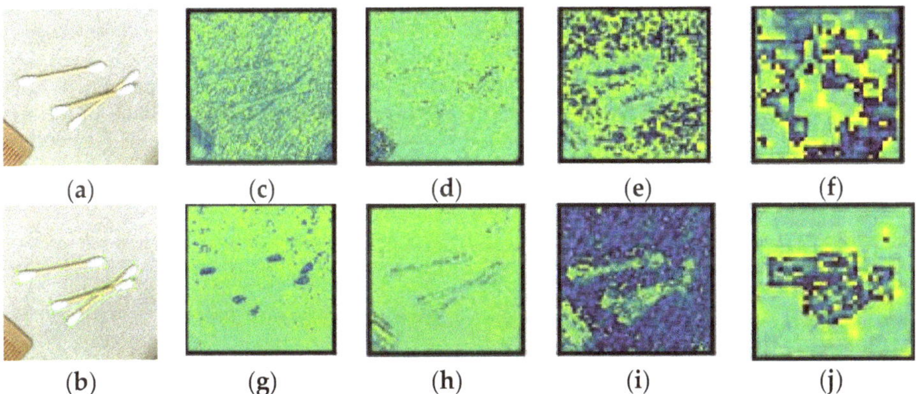

Figure 11. Intermediate characteristic diagrams of YOLOv5 and YOLOv5m-DSM. (**a**) Input image. (**b**) Label image. (**c–f**) Down sampling feature maps of YOLOv5. (**g–j**) Down sampling feature maps of YOLOv5m-DSM.

Table 4 shows a comparison of the detection effects of "interlayer merging" and "layer by layer merging" on the characteristic layer of the HDBottleneckCSP network.

Table 4. Effect comparison of "interlayer merging" and "layer by layer merging".

Model	mAP/%	Recall/%	Parameters
Layer by layer merging	90.9	92.8	894,842
Interlayer merging	91.7	93.6	717,645

It can be seen from Table 4 that, compared with "layer by layer merging", the "interlayer merging" used for feature map aggregation had fewer training parameters and a better detection effect. This was mainly because the "layer by layer merging" led to excessive duplication of the use of feature maps, which can easily cause feature redundancy and increase the difficulty of learning for the model. In addition, overly dense feature map aggregation increases the number of channels in the feature map, thus increasing the number of parameters and consuming more computing resources.

Figure 13. Waste detection with YOLOv5m-DSM in different scenarios.

4. Conclusions

This paper focused on waste detection for a robotic arm based on YOLOv5. In addition to object identification in general image analysis, a waste-sorting robotic arm not only needs to identify a target object but also needs to accurately judge its placement angle, so that the robotic arm can set the appropriate grasping angle. In order to address this need, we added an angular prediction network to the detection head to provide the grasping angle information for the waste object and proposed a dynamic smoothing algorithm for angle loss to enhance the model's angular prediction ability. In addition, we improved the method's feature extraction and aggregation abilities by optimizing the backbone and feature aggregation network of the model. The experimental results showed that the performance of the improved method in oriented waste detection was better than that of comparison methods; the average precision and recall rate were 93.9% and 94.8%, respectively, which were 11.6% and 7.6% higher than those of the original network, respectively.

Author Contributions: Conceptualization, W.Y., Y.X. and P.G.; methodology, W.Y. and Y.X.; software: W.Y. and P.G.; validation: W.Y.; writing—original draft preparation, W.Y.; writing—review and editing, W.Y., Y.X. and P.G. All authors have read and agreed to the published version of the manuscript.

Funding: This research was partially supported by Guangdong Key Discipline Scientific Research Capability Improvement Project: no. 2021ZDJS144; Projects of Young Innovative Talents in Universities of Guangdong Province: no. 2020KQNCX126; Key Subject Projects of Guangzhou Xinhua University: no. 2020XZD02; Scientific Projects of Guangzhou Xinhua University: no. 2020KYQN04; and the Plan of Doctoral Guidance: no. 2020 and no. 2021. Xie's work was supported by the Natural Science Foundation of China (no. 61972431).

Institutional Review Board Statement: Not applicable.

Informed Consent Statement: Not applicable.

Data Availability Statement: Not applicable.

Conflicts of Interest: The authors declare no conflict of interest.

References

1. Adedeji, O.; Wang, Z. Intelligent Waste Classification System Using Deep Learning Convolutional Neural Network. In Proceedings of the 2nd International Conference on Sustainable Materials Processing and Manufacturing (SMPM), Sun City, South Africa, 8–10 March 2019; pp. 607–612.
2. Alsubaei, F.S.; Al-Wesabi, F.N.; Hilal, A.M. Deep Learning-Based Small Object Detection and Classification Model for Garbage Waste Management in Smart Cities and IoT Environment. *Appl. Sci.* **2022**, *12*, 2281. [CrossRef]
3. Song, Q.; Li, S.; Bai, Q.; Yang, J.; Zhang, X.; Li, Z.; Duan, Z. Object Detection Method for Grasping Robot Based on Improved YOLOv5. *Micromachines* **2021**, *12*, 1273. [CrossRef]
4. Wu, Y.; Zhang, Y.; Li, M. Fine classification model for plastic waste based on FV-DCNN. *Transducer Microsyst. Technol.* **2021**, *40*, 118–120.
5. Chen, Z.-C.; Jiao, H.-N.; Yang, J.; Zeng, H.-F. Garbage image classification algorithm based on improved MobileNet v2. *J. Zhejiang Univ. (Eng. Sci.)* **2021**, *55*, 1490–1499.
6. Liu, Y.; Ge, Z.; Lv, G.; Wang, S. Research on Automatic Garbage Detection System Based on Deep Learning and Narrowband Internet of Things. In Proceedings of the 3rd Annual International Conference on Information System and Artificial Intelligence (ISAI), Suzhou, China, 22–24 June 2018.
7. Kang, Z.; Yang, J.; Li, G.; Zhang, Z. An Automatic Garbage Classification System Based on Deep Learning. *IEEE Access* **2020**, *8*, 140019–140029. [CrossRef]
8. Cui, W.; Zhang, W.; Green, J.; Zhang, X.; Yao, X. YOLOv3-DarkNet with adaptive clustering anchor box for garbage detection in intelligent sanitation. In Proceedings of the 2019 3rd International Conference on Electronic Information Technology and Computer Engineering (EITCE), Xiamen, China, 18–20 October 2019; pp. 220–225.
9. Xu, W.; Xiong, W.; Yao, J.; Sheng, Y. Application of garbage detection based on improved YOLOv3 algorithm. *J. Optoelectron. Laser* **2020**, *31*, 928–938.
10. Chen, W.; Zhao, Y.; You, T.; Wang, H.; Yang, Y.; Yang, K. Automatic Detection of Scattered Garbage Regions Using Small Unmanned Aerial Vehicle Low-Altitude Remote Sensing Images for High-Altitude Natural Reserve Environmental Protection. *Environ. Sci. Technol.* **2021**, *55*, 3604–3611. [CrossRef] [PubMed]
11. Majchrowska, S.; Mikolajczyk, A.; Ferlin, M.; Klawikowska, Z.; Plantykow, M.A.; Kwasigroch, A.; Majek, K. Deep learning-based waste detection in natural and urban environments. *Waste Manag.* **2022**, *138*, 274–284. [CrossRef] [PubMed]
12. Meng, J.; Jiang, P.; Wang, J.; Wang, K. A MobileNet-SSD Model with FPN for Waste Detection. *J. Electr. Eng. Technol.* **2022**, *17*, 1425–1431. [CrossRef]
13. Li, J.; Qiao, Y.; Liu, S.; Zhang, J.; Yang, Z.; Wang, M. An improved YOLOv5-based vegetable disease detection method. *Comput. Electron. Agric.* **2022**, *202*, 107345. [CrossRef]
14. Chen, Z.; Wu, R.; Lin, Y.; Li, C.; Chen, S.; Yuan, Z.; Chen, S.; Zou, X. Plant Disease Recognition Model Based on Improved YOLOv5. *Agronomy* **2022**, *12*, 365. [CrossRef]
15. Ling, L.; Tao, J.; Wu, G. Research on Gesture Recognition Based on YOLOv5. In Proceedings of the 33rd Chinese Control and Decision Conference (CCDC), Kunming, China, 22–24 May 2021; pp. 801–806.
16. Wang, Z.; Wu, L.; Li, T.; Shi, P. A Smoke Detection Model Based on Improved YOLOv5. *Mathematics* **2022**, *10*, 1190. [CrossRef]
17. Gao, P.; Lee, K.; Kuswidiyanto, L.W.; Yu, S.-H.; Hu, K.; Liang, G.; Chen, Y.; Wang, W.; Liao, F.; Jeong, Y.S.; et al. Dynamic Beehive Detection and Tracking System Based on YOLO V5 and Unmanned Aerial Vehicle. *J. Biosyst. Eng.* **2022**, *47*, 510–520. [CrossRef]
18. Yang, X.; Yan, J. Arbitrary-oriented object detection with circular smooth label. In Proceedings of the European Conference on Computer Vision, Glasgow, UK, 23–28 August 2020; pp. 677–694.

19. Rezatofighi, H.; Tsoi, N.; Gwak, J.; Sadeghian, A.; Reid, I.; Savarese, S.; Soc, I.C. Generalized Intersection over Union: A Metric and A Loss for Bounding Box Regression. In Proceedings of the 32nd IEEE/CVF Conference on Computer Vision and Pattern Recognition (CVPR), Long Beach, CA, USA, 16–20 June 2019; pp. 658–666.
20. Chen, K.; Zhu, Z.; Deng, X.; Ma, C.; Wang, H. Deep Learning for Multi-scale Object Detection: A Survey. *J. Softw.* **2021**, *32*, 1201–1227.
21. Hariharan, B.; Arbelaez, P.; Girshick, R.; Malik, J. Hyper columns for Object Segmentation and Fine-grained Localization. In Proceedings of the IEEE Conference on Computer Vision and Pattern Recognition (CVPR), Boston, MA, USA, 7–12 June 2015; pp. 447–456.
22. Shelhamer, E.; Long, J.; Darrell, T. Fully Convolutional Networks for Semantic Segmentation. *IEEE Trans. Pattern Anal. Mach. Intell.* **2017**, *39*, 640–651. [CrossRef] [PubMed]
23. Yang, S.; Ramanan, D. Multi-scale recognition with DAG-CNNs. In Proceedings of the IEEE International Conference on Computer Vision, Santiago, Chile, 11–18 December 2015; pp. 1215–1223.
24. Kaiming, H.; Xiangyu, Z.; Shaoqing, R.; Jian, S. Deep residual learning for image recognition. In Proceedings of the 2016 IEEE Conference on Computer Vision and Pattern Recognition (CVPR), Las Vegas, NV, USA, 27–30 June 2016; pp. 770–778.
25. Huang, G.; Liu, Z.; van der Maaten, L.; Weinberger, K.Q. Densely Connected Convolutional Networks. In Proceedings of the 30th IEEE/CVF Conference on Computer Vision and Pattern Recognition (CVPR), Honolulu, HI, USA, 21–26 July 2017; pp. 2261–2269.
26. Hu, J.; Shen, L.; Sun, G. Squeeze-and-Excitation Networks. In Proceedings of the 31st IEEE/CVF Conference on Computer Vision and Pattern Recognition (CVPR), Salt Lake City, UT, USA, 18–23 June 2018; pp. 7132–7141.
27. Woo, S.; Park, J.; Lee, J.-Y.; Kweon, I.S. CBAM: Convolutional Block Attention Module. In Proceedings of the 15th European Conference on Computer Vision (ECCV), Munich, Germany, 8–14 September 2018; pp. 3–19.
28. Hou, Q.; Zhou, D.; Feng, J.; Ieee Comp, S.O.C. Coordinate Attention for Efficient Mobile Network Design. In Proceedings of the IEEE/CVF Conference on Computer Vision and Pattern Recognition (CVPR), Nashville, TN, USA, 19–25 June 2021; pp. 13708–13717.
29. Lin, T.-Y.; Dollar, P.; Girshick, R.; He, K.; Hariharan, B.; Belongie, S. Feature Pyramid Networks for Object Detection. In Proceedings of the 30th IEEE/CVF Conference on Computer Vision and Pattern Recognition (CVPR), Honolulu, HI, USA, 21–26 July 2017; pp. 936–944.
30. Liu, S.; Qi, L.; Qin, H.; Shi, J.; Jia, J. Path Aggregation Network for Instance Segmentation. In Proceedings of the 31st IEEE/CVF Conference on Computer Vision and Pattern Recognition (CVPR), Salt Lake City, UT, USA, 18–23 June 2018; pp. 8759–8768.
31. Mingxing, T.; Ruoming, P.; Le, Q.V. EfficientDet: Scalable and Efficient Object Detection. In Proceedings of the 2020 IEEE/CVF Conference on Computer Vision and Pattern Recognition Workshops (CVPRW), Seattle, WA, USA, 14–19 June 2020; pp. 10778–10787.
32. Liu, W.; Anguelov, D.; Erhan, D.; Szegedy, C.; Reed, S.; Fu, C.Y.; Berg, A.C. Ssd: Single shot multibox detector. In Proceedings of the European Conference on Computer Vision, Amsterdam, The Netherlands, 11–14 October 2016; pp. 21–37.
33. Redmon, J.; Farhadi, A. Yolov3: An incremental improvement. *arXiv* **2018**, arXiv:1804.02767.
34. Ding, J.; Xue, N.; Long, Y.; Xia, G.-S.; Lu, Q.; Soc, I.C. Learning RoI Transformer for Oriented Object Detection in Aerial Images. In Proceedings of the 32nd IEEE/CVF Conference on Computer Vision and Pattern Recognition (CVPR), Long Beach, CA, USA, 16–20 June 2019; pp. 2844–2853.
35. Xu, Y.; Fu, M.; Wang, Q.; Wang, Y.; Chen, K.; Xia, G.S.; Bai, X. Gliding vertex on the horizontal bounding box for multi-oriented object detection. *IEEE Trans. Pattern Anal. Mach. Intell.* **2020**, *43*, 1452–1459. [CrossRef] [PubMed]
36. Yang, X.; Yan, J.C.; Feng, Z.M.; He, T.; Assoc Advancement Artificial, I. R(3)Det: Refined Single-Stage Detector with Feature Refinement for Rotating Object. In Proceedings of the AAAI Conference on Artificial Intelligence, Virtual, 2–9 February 2021; pp. 3163–3171.
37. Han, J.; Ding, J.; Li, J.; Xia, G.S. Align deep features for oriented object detection. *IEEE Trans. Geosci. Remote Sens.* **2021**, *60*, 1–11. [CrossRef]

Article

Two Novel Models for Traffic Sign Detection Based on YOLOv5s

Wei Bai [1], Jingyi Zhao [1], Chenxu Dai [1], Haiyang Zhang [2], Li Zhao [3], Zhanlin Ji [1,4,*] and Ivan Ganchev [4,5,6,*]

[1] College of Artificial Intelligence, North China University of Science and Technology, Tangshan 063210, China
[2] Department of Computing, Xi'an Jiaotong-Liverpool University, Suzhou 215000, China
[3] Research Institute of Information Technology, Tsinghua University, Beijing 100080, China
[4] Telecommunications Research Centre (TRC), University of Limerick, V94 T9PX Limerick, Ireland
[5] Department of Computer Systems, University of Plovdiv "Paisii Hilendarski", 4000 Plovdiv, Bulgaria
[6] Institute of Mathematics and Informatics—Bulgarian Academy of Sciences, 1040 Sofia, Bulgaria
* Correspondence: zhanlin.ji@ncst.edu.cn (Z.J.); ivan.ganchev@ul.ie (I.G.)

Abstract: Object detection and image recognition are some of the most significant and challenging branches in the field of computer vision. The prosperous development of unmanned driving technology has made the detection and recognition of traffic signs crucial. Affected by diverse factors such as light, the presence of small objects, and complicated backgrounds, the results of traditional traffic sign detection technology are not satisfactory. To solve this problem, this paper proposes two novel traffic sign detection models, called YOLOv5-DH and YOLOv5-TDHSA, based on the YOLOv5s model with the following improvements (YOLOv5-DH uses only the second improvement): (1) replacing the last layer of the 'Conv + Batch Normalization + SiLU' (CBS) structure in the YOLOv5s backbone with a <u>t</u>ransformer self-attention module (T in the YOLOv5-TDHSA's name), and also adding a similar module to the last layer of its neck, so that the image information can be used more comprehensively, (2) replacing the YOLOv5s coupled head with a <u>d</u>ecoupled <u>h</u>ead (DH in both models' names) so as to increase the detection accuracy and speed up the convergence, and (3) adding a <u>s</u>mall-object detection layer (S in the YOLOv5-TDHSA's name) and an <u>a</u>daptive anchor (A in the YOLOv5-TDHSA's name) to the YOLOv5s neck to improve the detection of small objects. Based on experiments conducted on two public datasets, it is demonstrated that both proposed models perform better than the original YOLOv5s model and three other state-of-the-art models (Faster R-CNN, YOLOv4-Tiny, and YOLOv5n) in terms of the mean accuracy (*mAP*) and *F1 score*, achieving *mAP* values of 77.9% and 83.4% and *F1 score* values of 0.767 and 0.811 on the TT100K dataset, and *mAP* values of 68.1% and 69.8% and *F1 score* values of 0.71 and 0.72 on the CCTSDB2021 dataset, respectively, for YOLOv5-DH and YOLOv5-TDHSA. This was achieved, however, at the expense of both proposed models having a bigger size, greater number of parameters, and slower processing speed than YOLOv5s, YOLOv4-Tiny and YOLOv5n, surpassing only Faster R-CNN in this regard. The results also confirmed that the incorporation of the T and SA improvements into YOLOv5s leads to further enhancement, represented by the YOLOv5-TDHSA model, which is superior to the other proposed model, YOLOv5-DH, which avails of only one YOLOv5s improvement (i.e., DH).

Keywords: computer vision; object detection; traffic sign detection; you only look once (YOLO); attention mechanism; feature fusion

MSC: 68W01; 68T01

Citation: Bai, W.; Zhao, J.; Dai, C.; Zhang, H.; Zhao, L.; Ji, Z.; Ganchev, I. Two Novel Models for Traffic Sign Detection Based on YOLOv5s. *Axioms* **2023**, *12*, 160. https://doi.org/10.3390/axioms12020160

Academic Editor: Oscar Humberto Montiel Ross

Received: 28 December 2022
Revised: 29 January 2023
Accepted: 31 January 2023
Published: 3 February 2023

1. Introduction

The detection and recognition of traffic signs play essential roles in the fields of assisted driving and automatic driving. Traffic signs are not only the main sources for drivers to obtain the necessary road information, but they also help adjust and maintain traffic flows [1]. However, in real-life scenarios, the influence of complex weather conditions and the existence of various categories of objects presented on the road—with a large proportion of these being small objects—have brought great challenges to the research on automatic detection and recognition of traffic signs.

There were two traffic sign detection and recognition techniques in the early days—one based on color features and the other based on shape features. Later, hybrid techniques emerged, e.g., [2], which considered both the color and geometric information of traffic signs during the feature extraction. The noise reduction and morphological processing made it easier to process images based on shapes, using the geometric information of a triangle, a circle, or a square commonly found in traffic signs, along with the RGB color information, in order to identify the images containing traffic signs. Although such a technique can detect the presence of traffic signs in images, it cannot distinguish between different classes of traffic signs.

With the emergence of deep learning, some models based on it have been applied for image classification and object detection, showing excellent performance, such as the two-stage detectors, represented by, e.g., the region-based convolutional neural networks (R-CNNs), and the single-stage detectors, represented by, e.g., You Only Look Once (YOLO) versions. R-CNN [3] was the first model applying convolutional neural networks (CNNs) for object detection. R-CNN generates candidate boxes first before detection to reduce the information redundancy, thus improving the detection speed. However, it zooms and crops images, resulting in a loss of original information. SPP-net [4] defined a spatial pyramid pooling (SPP) layer in front of the fully connected layer, which allowed one to input images of an arbitrary size and scale, thus not only breaking the constraint of fixed sizes of input images but also reducing the computational redundancy. Fast R-CNN [5] changed the original string structure of R-CNN into a parallel structure and absorbed the advantages of SPP-net, which allowed it not only to accelerate the object detection but also to improve the detection accuracy. However, if a large number of invalid candidate regions is generated, it would lead to a waste of computing power, whereas a small number of candidate regions would result in missed detection. Based on the above problems, Ren et al. proposed the concept of region proposal networks (RPNs) [6], which generates candidate regions through neural networks to solve the mismatch between the generated candidate regions and the real objects. However, these two-stage models were not superior in training and detection speed, so single-stage models, represented by the YOLO family, came into existence [7]. By creating the feature map of the input image, the learning category probability, and the boundary box coordinates of the entire image, YOLO sets the object detection as a simple regression problem. The algorithm only runs once, which of course reduces the accuracy, but allows achieving a higher processing speed than the two-stage object detectors, thus making it suitable for real-time detection of objects. The first version of YOLO, YOLOv1 [8], divides each given image into a grid system. Each grid detects objects by predicting the number of bounding boxes of the objects in the grid. However, if small objects in the image appear in clusters, the detection performance is not as sufficient. The second version, YOLOv2 [9], preprocesses the batch normalization based on the feature extraction network of DarkNet19 to improve the convergence of the network. Later, YOLOv3 [10] added logic regression to predict the score of each bounding box. It also introduced the method of Faster R-CNN giving priority to only one bounding box. As a result, YOLOv3 can detect some small objects. However, YOLOv3 cannot fit well with the ground truth. YOLOv4 [11] uses weighted real connections (WRCs), cross-mini-batch normalization (CmBN), self-adaptive training (SAT), and other methods, which allows it to not only keep suitable training and detection speed but also achieve better detection accuracy. YOLOv5 passes each batch of training data through a data loader, which performs three types of data enhancement—zooming, color space adjustment, and mosaic enhancement. From the five models produced to date based on YOLOv5, this paper proposes improvements to the YOLOv5s model, which uses two cross-stage partial connections (CSP) structures (one for the backbone network and the other for the neck) and a weighted non-maximum suppression (NMS) [12] to improve the detection accuracy of the occluded objects in images.

The two-stage object detectors, such as R-CNN, SPP-net, and Fast R-CNN mentioned above, are not suitable for real-time detection of objects due to their relatively low detection speed. As single-stage object detectors, the YOLO versions are obviously better than the two-stage detectors in terms of the detection speed achieved. However, their detection performance is not as efficient. To tackle this problem, this paper proposes two novel YOLOv5s-based traffic sign detection models, called YOLOv5-DH and YOLOv5-TDHSA, with the following improvements to YOLOv5s (YOLOv5-DH uses only the second improvement below), which constitute the main contributions of the paper:

1. Replacing the last layer of the 'Conv + Batch Normalization + SiLU' (CBS) structure in the YOLOv5s backbone with a transformer self-attention module (T in the YOLOv5-TDHSA's name), and also adding a similar module to the last layer of its neck, so that the image information can be used more comprehensively;
2. Replacing the YOLOv5s coupled head with a decoupled head (DH in the both models' names) so as to increase the detection accuracy and speed up the convergence;
3. Adding a small-object detection layer (S in the YOLOv5-TDHSA's name) and an adaptive anchor (A in the YOLOv5-TDHSA's name) to the YOLOv5s neck to improve the detection of small objects.

Based on results obtained from experiments conducted on two public datasets (TT100K and CCTSDB2021), the proposed YOLOv5-DH and YOLOv5-TDHSA models outperform the original YOLOv5s model along with three other state-of-the-art models (Faster R-CNN, YOLOv4-Tiny, YOLOv5n), as shown further in the paper.

The rest of the paper is organized as follows. Section 2 introduces the attention mechanisms, feature fusion networks, and detection heads commonly used in object detection models. Section 3 presents the main representatives of the two-stage and single-stage object detection models. Section 4 explains the YOLOv5s improvements used by the proposed models, including the transformer self-attention mechanism, the decoupled head, the small-object detection layer, and the group of adaptive anchor boxes. Section 5 describes the conducted experiments, and presents and discusses the obtained results. Finally, Section 6 concludes the paper.

2. Background

2.1. Attention Mechanisms

Attention is a data processing mechanism used in machine learning and extensively applied in different types of tasks such as natural language processing (NLP), image processing, and object detection [13]. The squeeze-and-exchange (SE) attention mechanism aims to assign different weights to each feature map and focuses on more useful features [14]. SE pools the input feature map globally, then uses a full connection layer and an activation function to adjust the feature map, thus obtaining the weight of the feature, which is multiplied with the input feature at the end. The disadvantage of SE is that it only considers the channel information and ignores the spatial location information. The convolutional block attention module (CBAM) solves this problem by first generating different channel weights, and then compressing all feature maps into one feature map to calculate the weight of the spatial features [15]. Currently, the self-attention [16] is one of the most widely used attention mechanisms due to its strong feature extraction ability and the support of parallel computing. The transformer self-attention mechanism, used by the YOLOv5-TDHSA model proposed in this paper, can establish a global dependency relationship and expand the receptive field of images, thus obtaining more features of traffic signs.

2.2. Multi-Scale Feature Fusion

The feature pyramid network (FPN) [17] utilized in Faster R-CNN and Mask R-CNN [18] is shown in Figure 1a. It uses the features of the five stages of the ResNet convolution groups C2–C6, among which C6 is obtained from a MaxPooling operation by directly applying $1 \times 1/2$ on C5. The feature maps P2–P6 are obtained after the FPN fusion, as follows: P6 is equal to C6, P5 is obtained through a 1×1 convolution followed by a 3×3 convolution, and P2–P4 are obtained through a 1×1 convolution followed by a fusion with the feature of the former $2 \times$ Upsample and a 3×3 convolution.

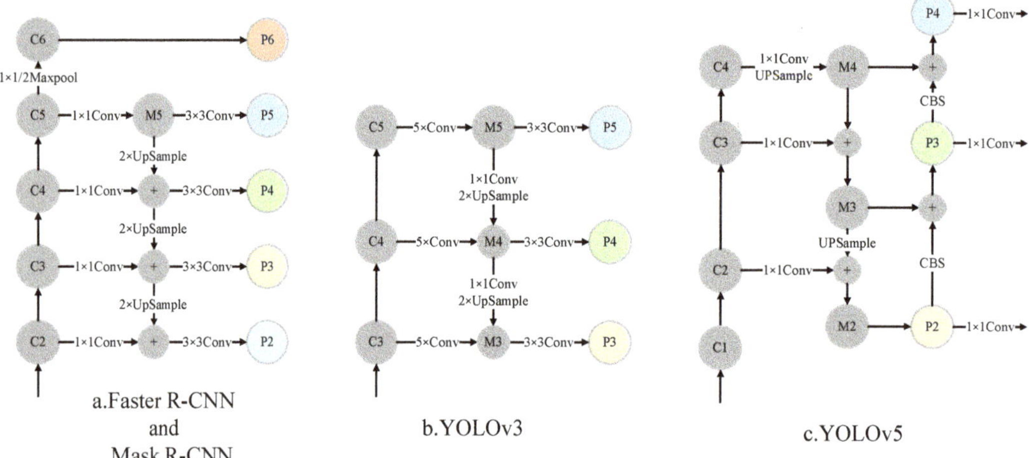

Figure 1. Different feature fusion structures.

The FPN in YOLOv3 is shown in Figure 1b. The features of C3, C4, and C5 are used. The features from C5 to P5 first pass through five layers of convolution, and then through one layer of 3×3 convolution. The features of P4 are obtained by connecting M5 (through 1×1 Conv + $2 \times$ Upsample) and C4 through five layers of convolution, and one layer of 3×3 convolution. The features of P3 are obtained by connecting M4 (through 1×1 Conv + $2 \times$ Upsample) and C3 through five layers of convolution, and one layer of 3×3 convolution.

The feature extraction network of YOLOv5 uses a 'FPN + Path Aggregation Network (PAN)' [19] structure, as shown in Figure 1c. PAN adds a bottom-up pyramid behind the FPN as a supplement. FPN conveys the strong semantic features from top to bottom, while PAN conveys strong positioning features from bottom to top. The specific operation of PAN includes first copying the last layer M2 of FPN as the lowest layer P2 of PAN, and then fusing M3 with the downsampled P2 to obtain P3. P4 is obtained through a feature fusion of M4 and downsampled P3. However, the feature extraction network does not work well for the detection of small objects. The feature fusion utilized by the YOLOv5-TDHSA model, proposed in this paper, is based on a small-object detection layer, making the detection of small objects more accurate. This is described in more detail in Section 4.3.

2.3. Detector Head

Since the head of YOLOv1 only generates two detection boxes for each grid, it is not suitable for both dense and small-object detection tasks. Its generalization ability is weak when the size ratio of the same-type objects is uncommon. The head of YOLOv2 improves the network structure and also adds an anchor box. YOLOv2 removes the last fully connected layer in YOLOv1, and uses convolution and anchor boxes to predict the detection box. However, since the use of convolution to downsample the feature map results in a loss of the fine-grained features, the model's detection of small objects is poor. Consequently, the passthrough layer structure has been introduced in the head of YOLOv2 to divide the feature map into four parts to preserve the fine-grained features. The head of

YOLOv3 introduces a multi-scale detection logic and utilizes a multi-label classification idea on the basis of YOLOv2. The loss function has been optimized as well. YOLOv4 adopts a multi-anchor strategy, different from YOLOv3. Any anchor box greater than the intersection over union (IoU) [20] threshold is regarded as a positive sample, thus ensuring that the positive samples ignored by YOLOv3 will be added to YOLOv4 to improve the detection accuracy of the model. The output of YOLOv5 has three prediction branches. The grid of each branch has three corresponding anchors. Instead of the IoU maximum matching method, YOLOv5 calculates the width–height ratio of the bounding box to the anchor of the current layer. If the ratio is greater than the parameter value set, this indicates that the matching degree is poor, which is considered as a background. The coupled detection head of YOLOv5s performs both the recognition and positioning tasks on a feature map simultaneously. However, these tasks have different focuses, making the final recognition accuracy low. The 'decoupled head' idea allows one to separate these two tasks and achieve better performance. Therefore, the models proposed in this paper use a decoupled head instead of the original YOLOv5s coupled head, which is described in more detail in Section 4.2.

3. Related Work

Over the past 20 years, the object detection models were divided into two categories: (1) traditional models (before 2012), such as V-J detection [21,22], HOG detection [23], DPM [24], etc., and (2) deep learning (DL) models, beginning with AlexNet [25]. The following subsections briefly present the DL object detection models, divided into two-stage and one-stage models, whose development route is illustrated in Figure 2.

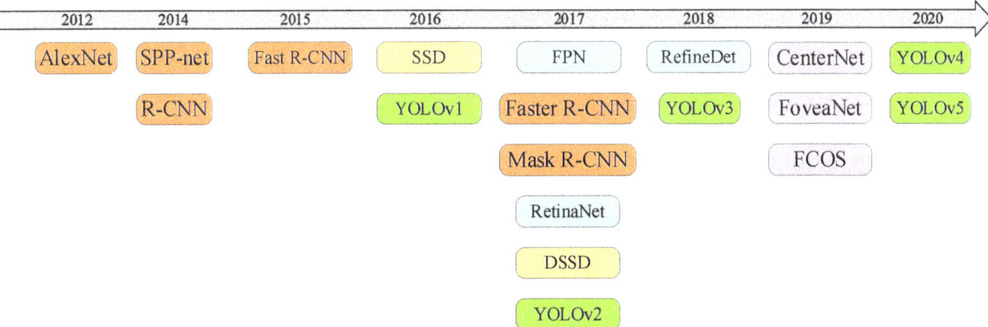

Figure 2. The development route of the DL object detection models.

3.1. Two-Stage Object Detection Models

Krizhevsky et al. proposed AlexNet as a CNN framework when participating (and winning the first place) in the ImageNet LSVRC 2012 competition. This model brought the climax to the development of deep learning.

Later, R-CNN emerged for object detection. However, R-CNN unifies the size of all candidate boxes, which causes a loss of the image content and affects the detection accuracy. Based on R-CNN, SPP-net, Fast R-CNN, Faster R-CNN, Mask R-CNN, and other models have been developed subsequently.

SPP-net was proposed in 2014. It inserts a spatial pyramid pooling layer between the CNN layer and fully connected layer, which allows it to solve the R-CNN loss of the image content caused by adjusting all candidate boxes to the same size. In order to find the location of each area in the feature map, the location information is added after the convolution layer. However, the time-consuming selective search (SS) [26] method is still used to generate the candidate areas.

On the basis of R-CNN, Fast R-CNN adds an RoI (region of interest) pooling layer and reduces the number of model parameters, thus greatly increasing the processing speed.

The method of SPP-net is used for reference, CNN is used to process the input images, and the serial structure of R-CNN is changed to a parallel structure, so that classification and regression can be carried out simultaneously, and the detection is accelerated.

In order to solve the problem that Fast R-CNN uses the SS method to generate candidate areas, Faster R-CNN uses an RPN to directly generate candidate areas, which enables the neural network to complete the detection task in an end-to-end fashion [27].

Based on Faster R-CNN, Mask R-CNN uses a fully constructive network (FCN). The model operates in two steps: (1) generating the candidate regions through an RPN, and (2) extracting the RoI features from candidate regions using RoIAlign (region of interest alignment) to obtain the probability of object categories and the location information of prediction boxes.

The two-stage object detection models are not suitable for real-time object detection because they require multiple detection and classification processes, which lowers the detection speed.

3.2. One-Stage Object Detection Models
3.2.1. YOLO

YOLO's training and detection are carried out in a separate network. The object detection is regarded as a process of solving a regression problem. As long as the input image passes through inference, the location information of the object and the probability of its category can be obtained [28]. Therefore, YOLO is particularly outstanding in terms of detection speed. There are different versions of YOLO proposed to date. Based on its fifth version, YOLOv5, five models have been produced, namely YOLOv5n, YOLOv5s, YOLOv5m, YOLOv5l, and YOLOv5x. The YOLOv5-DH and YOLOv5-TDHSA models, described in this paper, propose improvements to the YOLOv5s model, whose network structure is shown in Figure 3. A focus network structure is used at the beginning of the trunk to derive the value of every other pixel in an image. This is followed by four independent feature layers, which are stacked. At that point, the width and height information is concentrated on the channel, and the input channel is expanded four times.

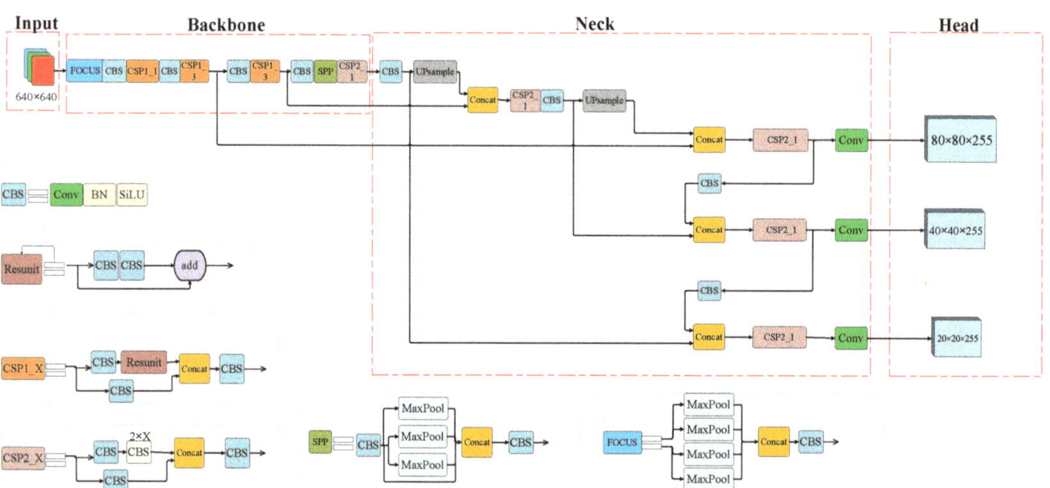

Figure 3. The structure of the YOLOv5s model.

YOLOv5s uses Mixup [29] and Mosaic for data enhancement, where Mosaic splices four images to enrich the background of the detected object. The data of the four images are processed at one time during a batch normalization computation.

In the backbone part, the model extracts features from the input image. The extracted features, through three feature layers, are used for the next network construction.

The main task of the neck part is to strengthen feature extraction and feature fusion, so as to combine feature information of different scales. In the Path Aggregation Network (PANet) structure, upsampling and downsampling operations are used to achieve feature extraction. When the input size is 640×640 pixels, the maximum scale of output feature is 80×80 pixels, so the minimum size of the detection frame is 8×8 pixels. However, when there are many smaller objects in the dataset, this will affect the detection accuracy. The proposed improvements of YOLOv5s in this regard are described in Section 4.3.

In the head part, the three feature layers, which have been strengthened, are regarded as a collection of feature points. This part is used to judge whether the feature points have objects corresponding to them. The YOLOv5s detection head is a coupled head which performs complete identification and location tasks on a feature map. However, recognition and location are two different tasks. Therefore, this paper proposes a branch structure to carry out recognition and location tasks separately. This improvement to the YOLOv5s structure is described in more detail in Section 4.2.

There have been some improvements of YOLOv5 recently proposed for traffic sign and traffic light recognition. For instance, Chen et al. [30] introduced a Global-CBAM attention mechanism for embedding into YOLOv5's backbone in order to enhance its feature extraction ability, and achieved sufficient balance between the channel attention and spatial attention for improving the target recognition. Due to this, the overall accuracy of the model was improved, especially for small-sized target recognition, and the mean accuracy (mAP) achieved was 6.68% higher than that before the improvement.

In order to solve the problem of using YOLOv5s for the recognition of small-sized traffic signs, Liu et al. [31] proposed to replace the original DarkNet-53 backbone of YOLOv5s with MobileNetV2 network for feature extraction, selecting Adam as the optimizer. The result of this was the reduction in the number of parameters by 65.6% and the computation amount by 59.1% on the basis of improving the mAP by 0.129.

Chen et al. [32] added additional multi-scale features to YOLOv5s to make it faster and more accurate in capturing traffic lights when these occupy a small area in images. In addition, a loop was established to update the parameters using a gradient of loss values. This led to mAP improvement (from 0.965 to 0.988) and detection time reduction (from 3.2 ms inference/2.5 ms to 2.4 ms inference/1.0 ms NMS per image).

3.2.2. SSD

The Single Shot MultiBox Detector (SSD) [33] is a one-stage object detection model proposed after YOLOv1. In order to improve YOLO's imperfection for small-object detection, SSD uses feature maps of different sizes and prior boxes of different sizes to further improve the regression rate and accuracy of the predicted box. The proportion of the prior frame size to the image is calculated as follows:

$$S_k = S_{min} + \frac{S_{max} - S_{min}}{m - 1}(k - 1), \tag{1}$$

where $k \in [1, m]$, m denotes the number of characteristic graphs, and S_{max} and S_{min} denote the maximum and minimum value of the ratio, respectively.

4. Proposed Improvements to YOLOv5s

This section describes the YOLOv5s improvements used by the models proposed in this paper. The decoupled head (DH) improvement is used by both proposed models, YOLOv5-DH and YOLOv5-TDHSA, whereas the other two improvements are used only by YOLOv5-TDHSA.

4.1. Transformer Self-Attention Mechanism

The transformer model was proposed by the Google team in June 2017 [34]. It has not only become the preferred model in the NLP field, but also showed strong potential in the field of image processing. The transformer abandons the sequential structure of Recurrent Neural Networks (RNNs) and adopts a self-attention mechanism to enable the model to parallelize training and make full use of the global information of training data.

The core mechanism of the transformer model is the self-attention depicted in Figure 4. The regular attention mechanism first calculates the attention distribution on all input information and then obtains the weighted average of the input information according to this attention distribution. Self-attention maps the input features to three new spaces for representation, namely Query (Q), Key (K), and Value (V). The correlation between Q and K is calculated as well, after which a *SoftMax* function is used to normalize the data and widen the gap between the data to enhance the attention. The weight coefficient and V are weighted and summed to obtain the attention value. The self-attention mechanism maps the features to three spatial representations, which allows one to avoid problems encountered when features are mapped to only one space. For example, if Q1 and Q2 are directly used to calculate the correlation, there will be no difference between the correlation between Q1 and Q2 and the correlation between Q2 and Q1. In this case, the expression ability of the attention mechanism will become weak. If K is introduced to calculate the correlation between the original data, it can reflect the difference between Q1 and K2 on one hand and Q2 and K1 on the other, which can also enhance the expression ability of the attention mechanism. Since the input of the next step is the attention weight obtained, it is not appropriate to use Q or K; thus, the third space, V, is introduced. Finally, the attention value is obtained through weighted summation.

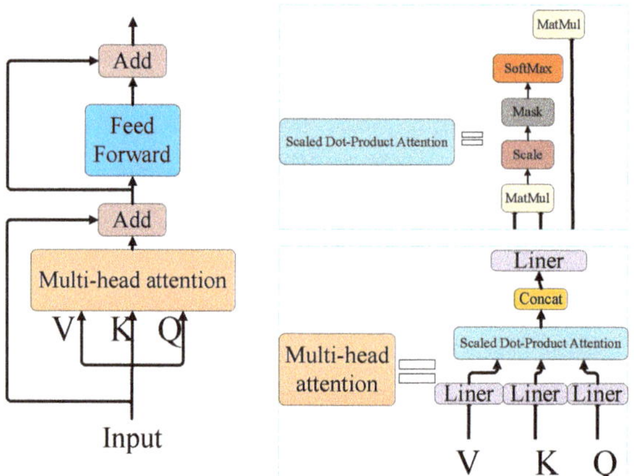

Figure 4. The module structure of the transformer self-attention mechanism.

However, the transformer model would significantly increase the amount of computation, resulting in higher training costs. The feature dimension is the smallest when the image features are transferred to the last layer of the network. At this moment, the influence on training the model would be the smallest if the transformer is added. Therefore, the proposed YOLOv5-TDHSA model uses the transformer only as a replacement of the CBS at the last layer of the backbone of the original YOLOv5s model, and also adds the transformer to the last layer of its neck.

4.2. Decoupled Head

After performing analytical experiments indicating that the coupled detection head may harm YOLO's performance, the authors of [35] recommend replacing the original YOLO's head with a decoupled one. This idea is taken on board by the models proposed in this paper to reduce the number of parameters and network depth, thus improving the model training speed and reducing the feature losses.

During the object detection, it is necessary to output the category/class and position information of the object. The decoupled head uses two different branches to output the category and position information separately as the recognition and positioning tasks have different focuses. The recognition focuses more on the existing class to which the extracted features are closer. The positioning focuses more on the location coordinates of the ground truth box so as to correct the parameters of the bounding box. YOLO's head uses a feature map to complete the two tasks of recognition and location in a convolution. Therefore, it does not perform as well as the decoupled head D1 shown in Figure 5, which is used by the models proposed in this paper. However, the decoupling process increases the number of parameters, thus affecting the training speed of the model. Therefore, in order to reduce the number of parameters, the feature first goes through a 1×1 convolution layer to reduce the dimension and then through two parallel branches with two 3×3 convolution layers. The first branch is used to predict the category. Since there are 45 categories in the TT100K dataset used in this paper, the channel dimension becomes 45 after a convolution operation and the processing of the *Sigmoid* activation function [36]. The second branch is mainly used to determine whether the object box is a foreground or background. As a result, the channel dimension becomes 1 after the convolution operation and *Sigmoid* activation function. There is also a third branch used to predict the coordinate information (x, y, w, h) of the object box. Therefore, after the convolution operation, the channel dimension becomes 4. Finally, the three outputs are integrated into $20 \times 20 \times 50$ feature information through *Concat* for the next operation. The decoupled heads D2, D3, and D4, shown in Figure 6, also follow the same steps to generate feature information of $40 \times 40 \times 50$, $80 \times 80 \times 50$, and $160 \times 160 \times 50$, respectively. The proposed YOLOv5-DH model only uses D1, D2, and D3 to replace the 'Head' part of the original YOLOv5s model (c.f., Figure 3).

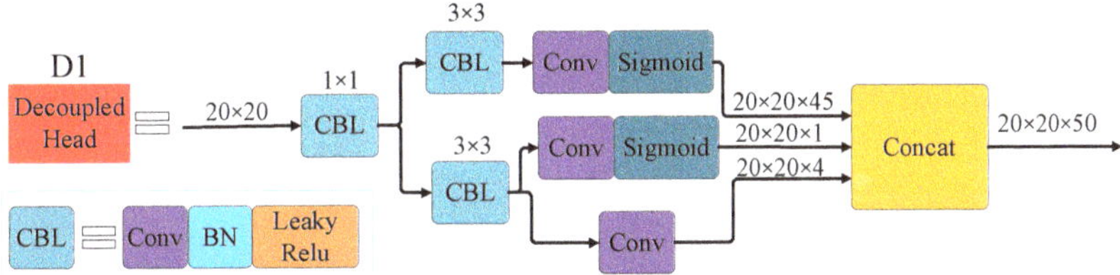

Figure 5. The structure of the decoupled head D1 used by the proposed models.

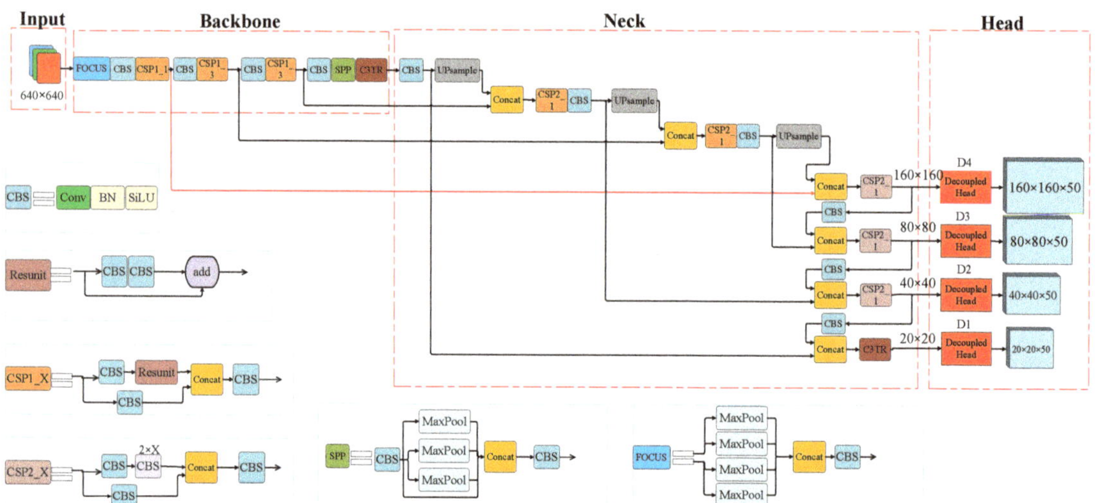

Figure 6. The structure of the proposed YOLOv5-TDHSA model.

4.3. Small-Object Detection Layer and Adaptive Anchor

During the detection of traffic signs, the changing distance between the shooting equipment and the object makes the size of traffic signs in the collected images different, which has a certain impact on the detection accuracy [37]. YOLOv5s solves this problem in the form of PANet. Taking an input image size of 640×640 pixels as an example, the feature information of the feature map output through the original model is $80 \times 80 \times 255$, $40 \times 40 \times 255$, and $20 \times 20 \times 255$, respectively. At this time, the grid sizes of the generated detection box are 8×8 pixels, 16×16 pixels, and 32×32 pixels, respectively. However, when there is a large number of objects with size smaller than 8×8 pixels in the dataset, the detection performance for these small objects is not acceptable. Furthermore, the feature pyramid pays more attention to the extraction and optimization of the underlying features. With increasing the depth of the network, some features at the top level will be lost, reducing the accuracy of the object detection.

To improve the detection of small objects, a branch structure is added to the PANet of YOLOv5 to maintain the same size of the input image. However, the neck part adds a $160 \times 160 \times 128$ feature information output. In other words, the feature map continues to expand by performing the convolution and upsampling on the feature map after layer 17. Meanwhile, the 160×160 pixels feature information obtained from layer 19 is fused with the layer 2 feature in the backbone at the layer 20 to make up for the feature loss during feature transmission. The addition of a small object detection layer in the network can ease the difficulty of small object detection. At the same time, it combines the features of the top level with those of the bottom level to supplement the features lost in the bottom level, thus improving the detection accuracy.

The network structure after the addition of the small-object detection layer is shown in Figure 6. A branch is added to connect layer 2 and layer 19 (the red solid line part). In this case, the added fourth output size is $160 \times 160 \times 128$. After the head decoupling, the feature information size is $160 \times 160 \times 50$. The minimum size of the generated detection box is 4×4 pixels, which improves the detection of small objects.

The original YOLOv5s network model has only three detection layers. As a result, there are three groups of anchor boxes corresponding to the feature maps at three different resolutions. In each group of anchor boxes, there are three different anchors. A total of nine anchors can be used to detect large, medium, and small objects. However, the YOLOv5-TDHSA model, proposed in this paper, deepens the network and adds an output layer of feature information. It uses a group of 12 anchor boxes, added to the original YOLOv5s model, to calculate the feature map at the new resolution. The ratio between an anchor and the width and height of each ground truth box is calculated, and the K-Means and genetic learning algorithms are used to obtain the best possible recall (BPR). When BPR is greater than 0.98, it indicated that the four groups of anchor boxes generated can be suitable for custom datasets.

The addition of the small-object detection layer and the group of adaptive anchor boxes allows us to significantly improve the detection accuracy of the proposed YOLOv5-TDHSA model, as demonstrated in the next section.

5. Experiments

5.1. Datasets

Two public datasets were used in the experiments conducted for the performance comparison of models. The first one was the Tsinghua-Tencent 100 K Chinese traffic sign detection benchmark [38], denoted as TT100K in [39]. It includes 100,000 high-definition images with large variations in illuminance and weather conditions, among which 10,000 images are annotated that contain 30,000 traffic sign instances (in total), each of which theoretically belongs to one of the 221 Chinese traffic sign categories. The images are taken from the Tencent Street View Map. Sample images are shown in Figure 7. However, there is a serious imbalance in the distribution of categories in this dataset, and even some categories do not have instances corresponding to them. Therefore, in the conducted experiments, similarly to [39], only categories with more than 100 traffic sign instances were used, resulting in 45 categories spread over 9170 images.

Figure 7. Sample images of the TT100K dataset.

The other dataset used in the experiments was the CCTSDB2021 Chinese traffic sign detection benchmark [40], which was built based on the CCTSDB2017 dataset [41,42] by adding 5268 annotated images of real traffic scenes and replacing images containing easily detected traffic signs with more difficult samples of a complex and changing detection environment. Three traffic sign classes are distinguished in CCTSDB2021, namely a warning, a mandatory, and a prohibitory traffic sign class, as shown in Figure 8. There are a total of 17,856 images, including 16,356 images in the training set and 1500 images in the test set. However, the weather environment attribute, which represents a great challenge for the object detection models, is only present in the images of the test set and not of the training set. Therefore, only these 1500 images, presenting greater difficulty to the detection of traffic signs contained in them, were used in the experiments.

Figure 8. Sample images of the CCTSDB2021 dataset, containing (**A**) warning traffic signs; (**B**) mandatory traffic signs; (**C**) prohibitory traffic signs.

In the experiments, as shown in Table 1, the 9170 TT100K images and 1500 CCTSDB021 images were separately divided (using the same ratio) into a training set (60% of the total number of images), a validation set (20%), and a test set (20%). The corresponding number of labels in each of these three sets is shown in Table 1.

Table 1. Splitting the datasets into training, validation, and test sets.

	TT100K Dataset	**CCTSDB2021 Dataset**
Training set	13,908 labels	1935 labels
Validation set	4636 labels	645 labels
Test set	4636 labels	645 labels

5.2. Experimental Environment

In the training process, the initial learning rate was set to 0.01, and a cosine annealing strategy was used to reduce it. 300 epochs were performed with the batch size set to 32. The experiments were conducted on a PC with a Windows 10 operating system, Intel (R) Core (TM) i7-10,700 CPU@2.90 GHz, NVIDIA GeForce RTX3090, and 24GB video memory, by using CUDA 11.1 for training acceleration, PyTorch 1.8.1 deep learning framework for training, and an input image size of 640×640 pixels, as shown in Table 2.

Table 2. Experimental environment's parameters.

Component	**Name/Value**
Operating system	Windows 10
CPU	Intel (R) Core (TM) i7-10,700
GPU	GeForce RTX3090
Video memory	24 GB
Training acceleration	CUDA 11.1
Deep learning framework for training	PyTorch 1.8.1
Input image size	640×640 pixels
Initial learning rate	0.01
Final learning rate	0.1
Training batch size	32

5.3. Evaluation Metrics

Evaluation metrics commonly used for the performance evaluation of object detection models include *precision, average precision (AP), mean average precision (mAP), recall, F1 score,* and *processing speed* measured in frames per second (fps).

Precision refers to the proportion of the true positive (*TP*) samples in the prediction results, as follows:

$$precision = \frac{TP}{TP + FP} \tag{2}$$

where *TP* denotes the number of images containing detected objects with IoU > 0.5, that is, the number of images containing positive samples that are correctly detected by the model; *FP* (false positive) represents the number of images containing detected objects with IoU ≤ 0.5.

Recall refers to the proportion of correct predictions in all positive samples, as follows:

$$recall = \frac{TP}{TP + FN}, \tag{3}$$

where *FN* (false negative) represents the number of images wrongly detected as not containing objects of interest.

The *average precision* (*AP*) is the area enclosed by the *precision–recall* curve and the X axis, calculated as follows:

$$AP = \int_0^1 p(r)dr, \tag{4}$$

where $p(r)$ denotes the precision function of recall r.

F1 score is the harmonic average of *precision* and *recall*, with a maximum value of 1 and a minimum value of 0, calculated as follows:

$$F1 = 2 \cdot \frac{precision \cdot recall}{precision + recall}. \tag{5}$$

The mean average precision (*mAP*) is the mean *AP* value over all classes of objects, calculated as follows:

$$mAP = \frac{\sum AP}{N_{classes}}, \tag{6}$$

where $N_{classes}$ denotes the number of classes.

5.4. Results

Based on the two datasets, experiments were conducted for performance comparison of the proposed YOLOv5-DH and YOLOv5-TDHSA models to four state-of-the-art models, namely R-CNN, YOLOv4-Tiny, YOLOv5n, and YOLOv5s. The size and number of parameters of models are shown in Table 3 and the duration of a single experiment conducted with each model is shown in Table 4. On the two datasets, TT100K and CCTSDB2021, five separate experiments were performed with each of the models compared. In each experiment, the same data were utilized for all models, generated by randomly splitting the used dataset into a training set, a validation set, and a test set, as per Table 1. The results obtained for each model were averaged over the five experiments in order to serve as the final evaluation of the model performance.

Table 3. The size and number of parameters of compared models.

Model	Size (MB)	Number of Parameters (Million)
Faster R-CNN	360.0	28.469
YOLOv4-Tiny	22.4	6.057
YOLOv5n	3.6	1.767
YOLOv5s	13.7	7.068
YOLOv5-DH	**22.8**	**11.070**
YOLOv5-TDHSA	**24.8**	**12.224**

Table 4. Single experiment duration of compared models.

Dataset	Faster R-CNN	YOLOv4-Tiny	YOLOv5n	YOLOv5s	YOLOv5-DH	YOLOv5-TDHSA
TT100K	47 h	37.5 h	30 h	32 h	33 h	35 h
CCTSDB 2021	8.5 h	4 h	0.8 h	1 h	2 h	2.5 h

Tables 5–10 show the *mAP* and *F1 score* results obtained in each experiment, conducted on the TT100K dataset, for each of the models compared. Table 11 shows the averaged *mAP* and *F1 score* results over the five experiments, along with the processing speed achieved, measured in frames per second (fps). The obtained results, shown in Table 11, demonstrate that on the TT100K dataset, both proposed models (YOLOv5-DH and YOLOv5-TDHSA) outperform all four state-of-the-art models in terms of *mAP* and *F1 score*, at the expense of having a bigger size, greater number of parameters, and slower processing speed (surpassing only Faster R-CNN). From the two proposed models, YOLOv5-TDHSA is superior to YOLOv5-DH in terms of both evaluation metrics (*mAP* and *F1 score*).

Table 5. Results of Faster R-CNN on TT100K dataset.

	Experiment 1	Experiment 2	Experiment 3	Experiment 4	Experiment 5
mAP (%)	52.9	53.6	54.1	53.4	52.6
F1 score	0.576	0.581	0.586	0.579	0.575

Table 6. Results of YOLOv4-TINY on TT100K dataset.

	Experiment 1	Experiment 2	Experiment 3	Experiment 4	Experiment 5
mAP (%)	57.7	62.8	63.1	64.6	63.2
F1 score	0.608	0.672	0.655	0.654	0.675

Table 7. Results of YOLOv5n on TT100K dataset.

	Experiment 1	Experiment 2	Experiment 3	Experiment 4	Experiment 5
mAP (%)	66.0	66.2	65.1	66.3	66.6
F1 score	0.651	0.645	0.639	0.646	0.641

Table 8. Results of YOLOv5s on TT100K dataset.

	Experiment 1	Experiment 2	Experiment 3	Experiment 4	Experiment 5
mAP (%)	74.5	75.6	75.2	75.3	75.1
F1 score	0.728	0.741	0.740	0.730	0.728

Table 9. Results of YOLOv5-DH on TT100K dataset.

	Experiment 1	Experiment 2	Experiment 3	Experiment 4	Experiment 5
mAP (%)	77.2	78.3	77.6	78.5	78.1
F1 score	0.762	0.771	0.762	0.772	0.769

Table 10. Results of YOLOv5-TDHSA on TT100K dataset.

	Experiment 1	Experiment 2	Experiment 3	Experiment 4	Experiment 5
mAP (%)	83.3	83.5	82.6	83.3	84.2
F1 score	0.819	0.811	0.797	0.810	0.816

Table 11. Results of compared models on TT100K dataset.

Model	F1Score	*mAP* (%)	Processing Speed (fps)
Faster R-CNN	0.579	53.3	40
YOLOv4-Tiny	0.653	62.3	160
YOLOv5n	0.644	66.0	111
YOLOv5s	0.733	75.1	100
YOLOv5-DH	**0.767**	**77.9**	**84**
YOLOv5-TDHSA	**0.811**	**83.4**	**77**

Tables 12–17 show the *mAP* and *F1 score* results obtained in each experiment, conducted on the CCTSDB2021 dataset, for each of the models compared. Table 18 shows the averaged *mAP* and *F1 score* results over the five experiments, along with the processing speed achieved. The obtained results, shown in Table 18, demonstrate that both proposed models (YOLOv5-DH and YOLOv5-TDHSA) outperform all four state-of-the-art models in terms of *mAP* and *F1 score* on this dataset as well, at the expense of having a bigger size, greater number of parameters, and slower processing speed (surpassing only Faster R-CNN). From the two proposed models, YOLOv5-TDHSA is again superior to YOLOv5-DH in terms of both evaluation metrics (*mAP* and *F1 score*).

Table 12. Results of Faster R-CNN on CCTSDB2021 dataset.

	Experiment 1	Experiment 2	Experiment 3	Experiment 4	Experiment 5
mAP (%)	61.9	48.7	45.7	46.0	61.1
F1 score	0.65	0.62	0.59	0.61	0.65

Table 13. Results of YOLOv4-TINY on CCTSDB2021 dataset.

	Experiment 1	Experiment 2	Experiment 3	Experiment 4	Experiment 5
mAP (%)	62.6	64.2	53.7	62.3	64.7
F1 score	0.66	0.68	0.62	0.65	0.67

Table 14. Results of YOLOv5n on CCTSDB2021 dataset.

	Experiment 1	Experiment 2	Experiment 3	Experiment 4	Experiment 5
mAP (%)	68.7	72.5	54.8	60.7	71.3
F1 score	0.72	0.72	0.60	0.63	0.74

Table 15. Results of YOLOv5s on CCTSDB2021 dataset.

	Experiment 1	Experiment 2	Experiment 3	Experiment 4	Experiment 5
mAP (%)	64.7	70.2	58.9	68.5	73.7
F1 score	0.70	0.72	0.63	0.69	0.76

Table 16. Results of YOLOv5-DH on CCTSDB2021 dataset.

	Experiment 1	Experiment 2	Experiment 3	Experiment 4	Experiment 5
mAP (%)	71.9	67.6	62.2	65.2	73.8
F1 score	0.71	0.70	0.68	0.68	0.76

Table 17. Results of YOLOv5-TDHSA on CCTSDB2021 dataset.

	Experiment 1	Experiment 2	Experiment 3	Experiment 4	Experiment 5
mAP (%)	69.1	73.4	62.1	69.7	74.7
F1 score	0.73	0.76	0.66	0.67	0.76

Table 18. Results of compared models on CCTSDB2021 dataset.

Model	F1Score	mAP (%)	Processing Speed (fps)
Faster R-CNN	0.62	52.7	28
YOLOv4-Tiny	0.66	61.5	162
YOLOv5n	0.68	65.6	83
YOLOv5s	0.70	67.2	77
YOLOv5-DH	**0.71**	**68.1**	70
YOLOv5-TDHSA	**0.72**	**69.8**	66

6. Discussion

The incorporation of the proposed improvements into YOLOv5 resulted in overall better traffic sign detection. This was confirmed by a series of experiments conducted for evaluating and comparing the performance of the proposed models (YOLOv5-DH and YOLOv5-TDHSA) to that of YOLOv5s and three other state-of-the-art models, namely Faster R-CNN, YOLOv4-Tiny, and YOLOv5n, based on two datasets—TT100K and CCTSDB2021. The obtained results clearly demonstrate that both proposed models outperform all four models, in terms of the *mean average precision (mAP)* and *F1 score*.

Although both proposed models are better than the two-stage detection Faster R-CNN model, in terms of the model's size, number of parameters, and processing speed, they still have some shortcomings in this regard compared with the one-stage detection models (YOLOv4-Tiny, YOLOv5n, YOLOv5s). Therefore, in the future, some lightweight modules will be introduced into the proposed YOLOv5-TDHSA model (which is superior to the other proposed model YOLOv5-DH) in order to reduce its size and number of parameters, and increase its processing speed.

To check if the proposed models are significantly different statistically from the compared state-of-the-art models, we applied the (non-parametric) Friedman test [43,44] with the corresponding post-hoc Bonferroni–Dunn test [45,46], which are regularly used for the comparison of classifiers (more than two) over multiple datasets.

First, using the Friedman test, we measured the performances of the models, used in the experiments described in the previous section, across both datasets. Basically, the Friedman test shows whether the measured average ranks of models are significantly different from the mean rank expected, by checking the null hypothesis (stating that all models perform the same and the observed differences are merely random), based on the following formula:

$$T_{x^2} = \frac{12N}{k(k+1)} \left(\sum_{i=1}^{k} r_i^2 - \frac{k(k+1)^2}{4} \right),$$ (7)

where k denotes the number of models, N denotes the number of datasets, and r_i. represents the average rank of the i-th model. In our case, $k = 6$ and $N = 2$.

Instead of Friedman's T_{x^2} statistic, we used the better Iman and Davenport statistic [47], which is distributed according to the F-distribution with $(k - 1)$ and $(k - 1)(N - 1)$ degrees of freedom, as follows:

$$T_F = \frac{(N-1)T_{x^2}}{N(k-1) - T_{x^2}}.$$ (8)

Using (8), we calculated the following values: $T_F = 34$ for *F1 score* and $T_F = \infty$ for *mAP*. As both these values are greater than the critical values of 3.45 and 5.05 for six models and two datasets, with confidence levels of $\alpha = 0.10$ and $\alpha = 0.05$, respectively, we rejected the null hypothesis and concluded that there are significant differences between the compared models.

Next, we proceeded with a post-hoc Bonferroni–Dunn test, in which the models were compared only to a control model and not between themselves [44,48]. In our case, we used the proposed YOLOv5-TDHSA model as a control model. The advantage of the Bonferroni–Dunn test is that it is easier to visualize because it uses the same Critical Difference (CD) for all comparisons, which can be calculated as follows [48]:

$$CD = q_\alpha \sqrt{\frac{k(k+1)}{6N}},$$ (9)

where q_α denotes the critical value for $\frac{\alpha}{k-1}$. When $k = 6$, $q_\alpha = 2.326$ for $\alpha = 0.10$, and $q_\alpha = 2.576$ for $\alpha = 0.05$ [48]. Then, the corresponding CD values, calculated according to (9), are equal to 4.352 and 4.819, respectively. Figure 9 shows the CD diagrams based on *F1 score* and *mAP*. As can be seen from Figure 9, the proposed YOLOv5-TDHSA model is significantly superior to Faster R-CNN on both evaluation metrics for both confidence levels, and achieves at least comparable performance to that of YOLOv4-Tiny on both evaluation metrics for both confidence levels, and to that of YOLOv5n on *F1 score* for both confidence levels. It is not surprising that the Bonferroni–Dunn test found YOLOv5-DH and YOLOv5-TDHSA similar to YOLOv5s, as both proposed models are based on it. Having incorporated only one YOLOv5s improvement into itself, naturally, YOLOv5-DH is reported by the Bonferroni–Dunn test as more similar to YOLOv5s than YOLOv5-TDHSA.

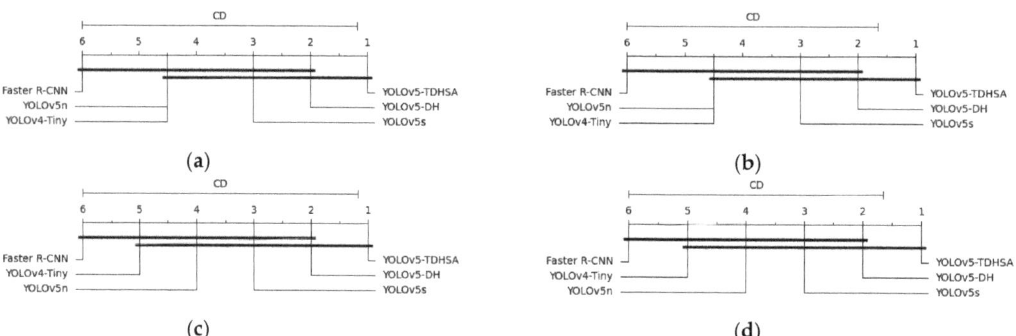

Figure 9. Critical difference (CD) comparison of YOLOv5-TDHSA (the control model) against other compared models with the Bonferroni–Dunn test, based on (**a**) *F1 score* with confidence level $\alpha = 0.05$, CD = 4.819; (**b**) *F1 score* with confidence level $\alpha = 0.10$, CD = 4.352; (**c**) *mAP* with confidence level $\alpha = 0.05$, CD = 4.819; (**d**) *mAP* with confidence level $\alpha = 0.10$, CD = 4.352 (any two models not connected by a thick black horizontal line are considered to have significant performance differences between each other).

7. Conclusions

We have proposed two novel models for accurate traffic sign detection, called YOLOv5-DH and YOLOv5-TDHSA, based on the YOLOv5s model with additional improvements. Firstly, a transformer self-attention module with stronger expression abilities was used in YOLOv5-TDHSA to replace the last layer of the 'Conv + Batch Normalization + SiLU' (CBS) structure in the YOLOv5s backbone. A similar module was added to the last layer of the YOLOv5-TDHSA's neck, so that the image information can be used more comprehensively. The features were mapped to the new three spaces for representation, thus improving the representation ability of the feature extraction. The multi-head mechanism used aims to realize the effect of multi-channel feature extraction. So, the transformer can increase the diversity of similarity computation between inputs and improve the ability of feature extraction. Secondly, a decoupled detection head was used in both proposed models to replace the YOLOv5s coupled head, which is responsible for the recognition and positioning on a feature map. As these two tasks have different focuses, resulting in a misalignment problem, the decoupled head uses two parallel branches—one responsible for the category recognition and the other responsible for positioning—which allows to improve the detection accuracy. However, as the decoupled head is not as fast as the coupled head, and due to the increase in the number of model parameters, the dimension was reduced through a 1×1 convolution before the decoupling to achieve balance between the speed and accuracy. Thirdly, for YOLOv5-TDHSA, a small-object detection layer was added to the YOLOv5s backbone and connected to the neck. At the same time, upsampling was used on the feature map of the neck to further expand the feature map. Supplemented by a group of adaptive anchor boxes, this new branch structure can not only ease the difficulty of small-object detection performed by YOLOv5-TDHSA, but can also compensate the feature losses caused by feature transmission with the increasing network depth.

Experiments conducted on two public datasets demonstrated that both proposed models outperform the original YOLOv5s model and three other state-of-the-art models (Faster R-CNN, YOLOv4-Tiny, YOLOv5n) in terms of the mean accuracy (*mAP*) and *F1 score*, achieving *mAP* values of 77.9% and 83.4% and *F1 score* values of 0.767 and 0.811 on the TT100K dataset, and *mAP* values of 68.1% and 69.8% and *F1 score* values of 0.71 and 0.72 on the CCTSDB2021 dataset, respectively, for YOLOv5-DH and YOLOv5-TDHSA. The results also confirm that the incorporation of the T and SA improvements into YOLOv5s leads to further enhancement, and a better performing model (YOLOv5-TDHSA), which is superior to the other proposed model (YOLOv5-DH) that avails of only one YOLOv5s improvement (i.e., DH).

Author Contributions: Conceptualization, Z.J., W.B.; methodology, W.B.; validation, J.Z., C.D.; formal analysis, H.Z., L.Z.; writing—original draft preparation, W.B.; writing—review and editing, I.G.; supervision, I.G.; project administration, J.Z. All authors have read and agreed to the published version of the manuscript.

Funding: This publication has emanated from research conducted with the financial support of the National Key Research and Development Program of China under grant no. 2017YFE0135700, the Tsinghua Precision Medicine Foundation under grant no. 2022TS003,and the MES by grant no. D01-168/28.07.2022 for NCDSC part of the Bulgarian National Roadmap on RIs.

Data Availability Statement: The data used in this study are openly available as per [38] and [40].

Conflicts of Interest: The authors declare no conflict of interest.

References

1. Saadna, Y.; Behloul, A. An overview of traffic sign detection and classification methods. *Int. J. Multimed. Inf. Retr.* **2017**, *6*, 193–210. [CrossRef]
2. Yıldız, G.; Dizdaroğlu, B. Traffic Sign Detection via Color and Shape-Based Approach. In Proceedings of 2019 1st International Informatics and Software Engineering Conference (UBMYK), Ankara, Turkey, 6–7 November 2019; pp. 1–5.
3. Shen, X.; Liu, J.; Zhao, H.; Liu, X.; Zhang, B. Research on Multi-Target Recognition Algorithm of Pipeline Magnetic Flux Leakage Signal Based on Improved Cascade RCNN. In Proceedings of 2021 3rd International Conference on Industrial Artificial Intelligence (IAI), Shenyang, China, 8–11 November 2021; pp. 1–6.
4. Wang, X.; Wang, S.; Cao, J.; Wang, Y. Data-driven based tiny-YOLOv3 method for front vehicle detection inducing SPP-net. *IEEE Access* **2020**, *8*, 110227–110236. [CrossRef]
5. Rani, S.; Ghai, D.; Kumar, S. Object detection and recognition using contour based edge detection and fast R-CNN. *Multimed. Tools Appl.* **2022**, *81*, 42183–42207. [CrossRef]
6. Ren, S.; He, K.; Girshick, R.; Sun, J. Faster R-CNN: Towards Real-Time Object Detection with Region Proposal Networks. *IEEE Trans. Pat. Analys. Mach. Intell.* **2017**, *39*, 1137–1149. [CrossRef] [PubMed]
7. Fan, J.; Huo, T.; Li, X. A Review of One-Stage Detection Algorithms in Autonomous Driving. In Proceedings of 2020 4th CAA International Conference on Vehicular Control and Intelligence (CVCI), Hangzhou, China, 18–20 December 2020; pp. 210–214.
8. Redmon, J.; Divvala, S.; Girshick, R.; Farhadi, A. You Only Look Once: Unified, Real-Time Object Detection. In Proceedings of Proceedings of the IEEE Conference on Computer Vision and Pattern Recognition, San Juan, PR, USA, 17–19 June 1997; pp. 779–788.
9. Redmon, J.; Farhadi, A. YOLO9000: Better, Faster, Stronger. In Proceedings of the IEEE Conference on Computer Vision and Pattern Recognition, San Juan, PR, USA, 17–19 June 1997; pp. 7263–7271.
10. Lv, N.; Xiao, J.; Qiao, Y. Object Detection Algorithm for Surface Defects Based on a Novel YOLOv3 Model. *Processes* **2022**, *10*, 701. [CrossRef]
11. Bochkovskiy, A.; Wang, C.-Y.; Liao, H.-Y.M. Yolov4: Optimal speed and accuracy of object detection. *arXiv* **2020**, arXiv:2004.10934.
12. Ma, W.; Zhou, T.; Qin, J.; Zhou, Q.; Cai, Z. Joint-attention feature fusion network and dual-adaptive NMS for object detection. *Knowl. Based Syst.* **2022**, *241*, 108213. [CrossRef]
13. Niu, Z.; Zhong, G.; Yu, H. A review on the attention mechanism of deep learning. *Neurocomputing* **2021**, *452*, 48–62. [CrossRef]
14. Hu, J.; Shen, L.; Sun, G. Squeeze-And-Excitation Networks. In Proceedings of the IEEE Conference on Computer Vision and Pattern Recognition, San Juan, PR, USA, 17–19 June 1997; pp. 7132–7141.
15. Woo, S.; Park, J.; Lee, J.-Y.; Kweon, I.S. Cbam: Convolutional block attention module. In Proceedings of the European Conference on Computer Vision (ECCV), Glasgow, UK, 23–28 August 2020; pp. 3–19.
16. Liang, H.; Zhou, H.; Zhang, Q.; Wu, T. Object Detection Algorithm Based on Context Information and Self-Attention Mechanism. *Symmetry* **2022**, *14*, 904. [CrossRef]
17. Lou, Y.; Ye, X.; Li, M.; Li, H.; Chen, X.; Yang, X.; Liu, X. Object Detection Model of Cucumber Leaf Disease Based on Improved FPN. In Proceedings of the 2022 IEEE 6th Advanced Information Technology, Electronic and Automation Control Conference (IAEAC), Beijing, China, 3–5 October 2022; pp. 743–750.
18. He, K.; Gkioxari, G.; Dollár, P.; Girshick, R. Mask R-Cnn. In Proceedings of the IEEE International Conference on Computer Vision, Venice, Italy, 22–29 October 2017; pp. 2961–2969.
19. Liu, S.; Qi, L.; Qin, H.; Shi, J.; Jia, J. Path Aggregation Network for Instance Segmentation. *arXiv* **2018**, arXiv:1803.01534.
20. Zheng, Z.; Wang, P.; Liu, W.; Li, J.; Ye, R.; Ren, D. Distance-IoU loss: Faster and better learning for bounding box regression. *Proc. AAAI Conf. Artif. Intell.* **2020**, *34*, 12993–13000. [CrossRef]
21. Zou, Z.; Shi, Z.; Guo, Y.; Ye, J. Object detection in 20 years: A survey. *arXiv* **2019**, arXiv:1905.05055. [CrossRef]

22. Xiao, Y.; Tian, Z.; Yu, J.; Zhang, Y.; Liu, S.; Du, S.; Lan, X. A review of object detection based on deep learning. *Multimed. Tools Appl.* **2020**, *79*, 23729–23791. [CrossRef]
23. Dixit, U.D.; Shirdhonkar, M.; Sinha, G. Automatic logo detection from document image using HOG features. *Multimed. Tools Appl.* **2022**, *82*, 863–878. [CrossRef]
24. Kim, J.; Lee, K.; Lee, D.; Jhin, S.Y.; Park, N. DPM: A novel training method for physics-informed neural networks in extrapolation. *Proc. AAAI Conf. Artif. Intell.* **2021**, *35*, 8146–8154. [CrossRef]
25. Krizhevsky, A.; Sutskever, I.; Hinton, G.E. Imagenet classification with deep convolutional neural networks. *Commun. ACM* **2017**, *60*, 84–90. [CrossRef]
26. Niepceron, B.; Grassia, F.; Moh, A.N.S. Brain Tumor Detection Using Selective Search and Pulse-Coupled Neural Network Feature Extraction. *Comput. Inform.* **2022**, *41*, 253–270. [CrossRef]
27. Du, L.; Zhang, R.; Wang, X. Overview of Two-Stage Object Detection Algorithms. *Proc. J. Phys. Conf. Ser.* **2020**, *1544*, 012033. [CrossRef]
28. Jiang, P.; Ergu, D.; Liu, F.; Cai, Y.; Ma, B. A Review of Yolo algorithm developments. *Procedia Comput. Sci.* **2022**, *199*, 1066–1073. [CrossRef]
29. Bouabid, S.; Delaitre, V. Mixup regularization for region proposal based object detectors. *arXiv* **2020**, arXiv:2003.02065.
30. Chen, Y.; Wang, J.; Dong, Z.; Yang, Y.; Luo, Q.; Gao, M. An Attention based YOLOv5 Network for Small Traffic Sign Recognition. In Proceedings of the 2022 IEEE 31st International Symposium on Industrial Electronics (ISIE), Anchorage, AK, USA, 1–3 June 2022; pp. 1158–1164.
31. Liu, X.; Jiang, X.; Hu, H.; Ding, R.; Li, H.; Da, C. Traffic Sign Recognition Algorithm Based on Improved YOLOv5s. In Proceedings of the 2021 International Conference on Control, Automation and Information Sciences (ICCAIS), Xi'an, China, 14–17 October 2021; pp. 980–985.
32. Chen, X. Traffic Lights Detection Method Based on the Improved YOLOv5 Network. In Proceedings of the 2022 IEEE 4th International Conference on Civil Aviation Safety and Information Technology (ICCASIT), Dali, China, 12–14 October 2022; pp. 1111–1114.
33. Chen, Z.; Guo, H.; Yang, J.; Jiao, H.; Feng, Z.; Chen, L.; Gao, T. Fast vehicle detection algorithm in traffic scene based on improved SSD. *Measurement* **2022**, *201*, 111655. [CrossRef]
34. Vaswani, A.; Shazeer, N.; Parmar, N.; Uszkoreit, J.; Jones, L.; Gomez, A.N.; Kaiser, Ł.; Polosukhin, I. Attention is all you need. In Proceedings of the Advances in Neural Information Processing Systems 30 (NIPS 2017), Long Beach, CA, USA, 4–9 December 2017.
35. Ge, Z.; Liu, S.; Wang, F.; Li, Z.; Sun, J. Yolox: Exceeding yolo series in 2021. *arXiv* **2021**, arXiv:2107.08430.
36. Ballantyne, G.H. Review of sigmoid volvulus. *Dis. Colon Rectum* **1982**, *25*, 823–830. [CrossRef] [PubMed]
37. Wang, J.; Chen, Y.; Gao, M.; Dong, Z. Improved YOLOv5 network for real-time multi-scale traffic sign detection. *arXiv* **2021**, arXiv:2112.08782. [CrossRef]
38. Zhu, Z.; Liang, D.; Zhang, S.; Huang, X.; Li, B.; Hu, S. Traffic-Sign Detection and Classification in the Wild. In Proceedings of the IEEE Conference on Computer Vision and Pattern Recognition, San Juan, PR, USA, 17–19 June 1997; pp. 2110–2118.
39. Chen, J.; Jia, K.; Chen, W.; Lv, Z.; Zhang, R. A real-time and high-precision method for small traffic-signs recognition. *Neural Comput. Appl.* **2022**, *34*, 2233–2245. [CrossRef]
40. Zhang, J.; Zou, X.; Kuang, L.-D.; Wang, J.; Sherratt, R.S.; Yu, X. CCTSDB 2021: A more comprehensive traffic sign detection benchmark. *Hum. Cent. Comput. Inf. Sci.* **2022**, *12*, 23.
41. Zhang, J.; Huang, M.; Jin, X.; Li, X. A real-time Chinese traffic sign detection algorithm based on modified YOLOv2. *Algorithms* **2017**, *10*, 127. [CrossRef]
42. Zhang, J.; Jin, X.; Sun, J.; Wang, J.; Sangaiah, A.K. Spatial and semantic convolutional features for robust visual object tracking. *Multimed. Tools Appl.* **2020**, *79*, 15095–15115. [CrossRef]
43. Friedman, M. The use of ranks to avoid the assumption of normality implicit in the analysis of variance. *J. Am. Stat. Assoc.* **1937**, *32*, 675–701. [CrossRef]
44. Friedman, M. A comparison of alternative tests of significance for the problem of m rankings. *Ann. Math. Stat.* **1940**, *11*, 86–92. [CrossRef]
45. Dunn, O.J. Multiple Comparisons among Means. *J. Am. Stat. Assoc.* **1961**, *56*, 52–64. [CrossRef]
46. Lin, Y.; Hu, Q.; Liu, J.; Li, J.; Wu, X. Streaming feature selection for multilabel learning based on fuzzy mutual information. *IEEE Trans. Fuzzy Syst.* **2017**, *25*, 1491–1507. [CrossRef]
47. Iman, R.L.; Davenport, J.M. Approximations of the critical region of the fbietkan statistic. *Commun. Stat. Theory Methods* **1980**, *9*, 571–595. [CrossRef]
48. Demšar, J. Statistical Comparisons of Classifiers over Multiple Data Sets. *J. Mach. Learn. Res.* **2006**, *7*, 1–30.

Article

Barrier Options and Greeks: Modeling with Neural Networks

Nneka Umeorah [1,*], Phillip Mashele [2], Onyecherelam Agbaeze [3] and Jules Clement Mba [4]

[1] School of Mathematics, Cardiff University, Cardiff CF24 4AG, UK
[2] School of Economic and Financial Sciences, University of South Africa, Pretoria 0003, South Africa; mashehp@unisa.ac.za
[3] College of Arts and Sciences, Troy University, Troy, AL 36082, USA; onagbaeze@alumni.troy.edu
[4] School of Economics, College of Business and Economics, University of Johannesburg, Johannesburg 2092, South Africa; jmba@uj.ac.za
* Correspondence: umeorahn@cardiff.ac.uk

Abstract: This paper proposes a non-parametric technique of option valuation and hedging. Here, we replicate the extended Black–Scholes pricing model for the exotic barrier options and their corresponding Greeks using the fully connected feed-forward neural network. Our methodology involves some benchmarking experiments, which result in an optimal neural network hyperparameter that effectively prices the barrier options and facilitates their option Greeks extraction. We compare the results from the optimal NN model to those produced by other machine learning models, such as the random forest and the polynomial regression; the output highlights the accuracy and the efficiency of our proposed methodology in this option pricing problem. The results equally show that the artificial neural network can effectively and accurately learn the extended Black–Scholes model from a given simulated dataset, and this concept can similarly be applied in the valuation of complex financial derivatives without analytical solutions.

Keywords: barrier options; Black–Scholes model; polynomial regression; random forest regression; machine learning; artificial neural network; option Greeks; data analysis

MSC: 91G20; 91G30; 62J05; 68T07

Citation: Umeorah, N.; Mashele, P.; Agbaeze, O.; Mba, J.C. Barrier Options and Greeks: Modeling with Neural Networks. *Axioms* **2023**, *12*, 384. https://doi.org/10.3390/axioms12040384

Academic Editor: Oscar Humberto Montiel Ross

Received: 19 December 2022
Revised: 29 March 2023
Accepted: 7 April 2023
Published: 17 April 2023

1. Introduction

The concept, techniques and applications of artificial intelligence (AI) and machine learning (ML) in solving real-life problems have become increasingly practical over the past years. The general aim of machine learning lies in attempting to 'learn' data and make some predictions from a variety of techniques. In the financial industry, they offer a more flexible and robust predictive capacity compared to the classical mathematical and econometric models. They equally provide significant advantages to the financial decision makers and market participants regarding the recent trends in financial modeling and data forecasting. The core applications of AI in finance are risk management, algorithmic trading, and process automation [1]. Hedge funds and broker dealers utilize AI and ML to optimize their execution. Financial institutions use the technologies to estimate their credit quality and evaluate their market insurance contracts. Both private and public sectors use these technologies to detect fraud, assess data quality, and perform surveillance. ML techniques are generally classified into supervised and non-supervised systems. A branch of ML (supervised) techniques that have been fully recognized is deep learning, as it provides and equips machines with practical algorithms needed to comprehend the fundamental principles, and pattern detection in a significant portion of data. The neural networks, the cornerstones of these deep learning techniques, evolved and developed in the 1960s. In the fields of quantitative finance, the neural networks are applied in the optimization of portfolios, financial model calibrations [2], high-dimensional futures [3], market prediction [4], and exotic options pricing with local stochastic volatility [5].

For the methodology employed in this paper, artificial neural networks (ANNs) are systems of learning techniques which focus on a cluster of artificial neurons forming a fully connected network. One aspect of the ANN is the ability to generally 'learn' to perform a specific task when fed with a given dataset. They attempt to replicate or mimic a mechanism which is observable in nature, and they gain their inspiration from the structure, techniques and functions of the brain. For instance, the brain is similar to a huge network with fully interconnected nodes (neurons, for example, the cells) through links, also referred to as synapses, biologically. The non-linearity feature is introduced to the network within these neurons as the non-linear activation functions are applied. The non-linearity aspect of the neural network tends to approximate any integrated function reasonably well. One significant benefit of the ANN method is that they are referred to as 'universal approximators'. This feature implies that they can fit any continuous function, together with functions having non-linearity features, even without the assumption of any mathematical relationship which connects the input and the output variables. Essentially, the ANNs are also fully capable of approximating the solutions to partial differential equations (PDE) [6], and they easily permit parallel processing, which facilitates evaluation processes on graphics processing units (GPUs) [7]. The presence of this universal approximator function is often a result of their typical architecture, training and prediction process.

Meanwhile, due to the practical significance of the use of financial derivatives, these instruments have sharply risen in more recent years. This has led to the development of sophisticated economic models, which tend to capture the dynamics of the markets, and there has also been an increase in proposing faster, more accurate and more robust models for the valuation process. The pricing of these financial contracts has significantly helped manage and hedge risk exposure in finance and businesses, improve market efficiencies and provide arbitrage opportunities for sophisticated market participants. The conventional pricing techniques of option valuation are theoretical, resulting in the formulation of analytical closed forms for some of these option types. In contrast, others rely heavily on numerical approximation techniques, such as Monte Carlo simulations, finite difference methods, finite volume methods, binomial tree methods, etc. These theoretical formulas are mainly based on assumptions about the behavior of the underlying prices of securities, constant risk-free interest rates, constant volatility, etc., and they have been duly criticized over the years. However, modifications have been made to the Black–Scholes model, thereby giving rise to such models as the mixed diffusion/pure jump models, displaced diffusion models, stochastic volatility models, constant elasticity of variance diffusion models, etc. On the other hand, neural networks (NNs) have proved to be emerging computing techniques that offer a modern avenue to explore the dynamics of financial applications, such as derivative pricing [8].

Recent years have seen a huge application of AI and ML, as they have been utilized greatly in diverse financial fields. They have contributed significantly to financial institutions, the financial market, and financial supervision. Li in [9] summarized the AI and ML development and analyzed their impact on financial stability and the micro-macro economy. In finance, AI has been utilized greatly in predicting future stock prices, and the concept lies in building AI models which utilize ML techniques, such as reinforcement learning or neural networks [10]. A similar stock price prediction was conducted by Yu and Yan [11]; they used the phase-space reconstruction method for time series analysis in combination with a deep NN long- and short-term memory networks model. Regarding applying neural networks to option pricing, one of the earliest research can be found in Malliaris and Salchenberger [12]. They compared the performance of the ANN in pricing the American-style OEX options (that is, options defined on Standard and Poor's (S&P) 100) and the results from the Black–Scholes model [13] with the actual option prices listed in the *Wall Street Journal* . Their results showed that in-the-money call options were valued significantly better when the Black–Scholes model was used, whereas the ANN techniques favored the out-of-the-money call option prices.

In pricing and hedging financial derivatives, researchers have incorporated the classical Black–Scholes model [13] into ML to ensure robust and more accurate pricing techniques. Klibanov et al. [14] used the method of quasi-reversibility and ML to predict option prices in corporations with the Black–Scholes model. Fang and George [15] proposed valuation techniques for improving the accuracy rate of Asian options by using the NN in connection with Levy approximation. Hutchinson et al. in [16] further priced the American call options defined on S&P 500 futures by comparing three ANN techniques with the Black–Scholes pricing model. Their results proved the supremacy of all three ANNs to the classical Black–Scholes model. Other comparative research studies on the ANN versus the Black–Scholes model are also applicable in pricing the following: European-style call options (with dividends) on the Financial Times Stock Exchange (FTSE) 100 index [17], American-style call options on Nikkei 225 futures [8], Apple's European call options [18], S&P 500 index call options with an addition of neuro-fuzzy networks [19], and in the pricing call options written on the Deutscher Aktienindex (DAX) German stock index [20]. Similar works on pricing and hedging options using the ML techniques can be found in [21–25].

Other numerical techniques, such as the PDE-based and the DeepBSDE-based (BSDE—-backward stochastic differential equations) methods, have also been employed in valuing the barrier options. For instance, Le et al. in [26] solved the corresponding option pricing PDE using the continuous Fourier sine transform and extended the concept of pricing the rebate barrier options. Umeorah and Mashele [27] employed the Crank–Nicolson finite difference method in solving the extended Black–Scholes PDE, describing the rebate barrier options and pricing the contracts. The DeepBSDE concept initially proposed by Han et al. in [28] converted high-dimensional PDE into BSDE, intending to reduce the dimensionality constraint, and they redesigned the solution of the PDE problem as a deep-learning problem. Further implementation of the BSDE-based using the numerical method with deep-learning techniques in the valuation of the barrier options is found in [29,30].

Generally, the concept of ANN can be classified into three phases: the neurons, the layers and the whole architecture. The neuron, which is the fundamental core processing unit, consists of three basic operations: summation of the weighted inputs, the addition of a bias to the input sum, and the computation of the output value via an activation function. This activation function is used after the weighted linear combination and implemented at the end of each neuron to ensure the non-linearity effect. The layers consist of an input layer, a (some) hidden layer(s) and an output layer. Several neurons define each layer, and stacking up various layers constitutes the entire ANN architecture. As the data transmission signals pass from the input layer to the output layer through the middle layers, the ANN serves as a mapping function among the input–output pairs [2]. After training the ANN in options pricing, computing the in-sample and out-of-sample options based on ANN becomes straightforward and fast [31]. Itkin [31] highlighted this example by pricing and calibrating the European call options using the Black–Scholes model.

This research is an intersection of machine learning, statistics and mathematical finance, as it employs recent financial technology in predicting option prices. To the best of our knowledge, this ML approach to pricing the rebate and zero-rebate barrier options has received less attention. Therefore, we aim to fill the niche by introducing this option pricing concept to exotic options. In the experimental section of this work, we simulate the barrier options dataset using the analytical form of the extended Black–Scholes pricing model. This is a major limitation of this research, and the choice was due to the non-availability of the real data. (A similar synthetic dataset was equally used by [32], in which they constructed the barrier option data based on the LIFFE standard European option price data by the implementation of the Rubenstein and Reiner analytic model. These datasets were used in the pricing of the up-and-out barrier call options via the use of a neural net model.) We further show and explain how the fully connected feed-forward neural networks can be applied in the fast and robust pricing of derivatives. We tuned different hyperparameters and used the optimal in the modeling and training of the NN. The performance of the optimal NN results is compared by benchmarking the results against other ML models, such as the random forest regression model and the polynomial regression model. Finally, we show how the barrier options and their Greeks can be trained and valued accurately under the extended Black–Scholes model. The major contributions of this research are classified as follows:

- We propose a non-parametric technique of barrier option valuation and hedging using the concept of a fully connected feed-forward NN.
- Using different evaluation metrics, we measure the performance of the NN algorithm and propose the optimal NN architecture, which prices the barrier options effectively in connection to some specified data-splitting techniques.
- We prove the accuracy and performance of the optimal NN model when compared to those produced by other ML models, such as the random forest and the polynomial regression, and extract the barrier option prices and their corresponding Greeks with high accuracy using the optimal hyperparameter.

The format of this paper is presented as follows: In Section 1, we provide a brief introduction to the topic and outline some of the related studies on the applications of ANN in finance. Section 2 introduces the concept of the Black–Scholes pricing model, together with the extended Black–Scholes pricing models for barrier options and their closed-form Greeks. Section 3 focuses on the machine learning models, such as the ANN, as well as its applications in finance, random forest regression and the polynomial regression models. In Section 4, we discuss the relevant results obtained in the course of the numerical experiments, and Section 5 concludes our research study with some recommendations.

2. Extended Black–Scholes Model for Barrier Options

The classical Black–Scholes model developed by Fischer Black and Myron Scholes is an arbitrage-free mathematical pricing model used to estimate the dynamics of financial derivative instruments. The model was initially designed to capture the price estimate of the European-style options defined under the risk-neutral measure. As a mathematical model, certain assumptions, such as the log-normality of underlying prices, constant volatility, frictionless market, continuous trading without dividends applied to stocks, etc., are made for the Black–Scholes model to hold [13]. Though the Black–Scholes model has been criticized over the years due to some underlying assumptions, which are not applicable in the real-world scenario, certain recent works are associated with the model [33–36]. Additionally, Eskiizmirliler et al. [37] numerically solved the Black–Scholes equation for the European call options using feed-forward neural networks. In their approach, they constructed a function dependent on a neural network solution, which satisfied the given boundary conditions of the Black–Scholes equation. Chen et al. [38] proposed a Laguerre neural network to solve the generalized Black–Scholes PDE numerically. They experimented with this technique on the European options and generalized option pricing models.

On the other hand, the valuation of exotic derivatives, such as the barrier options, has been extensively studied by many authors, mainly by imploring a series of numerical approximation techniques. Barrier options are typically priced using the Monte-Carlo simulations since their payoffs depend on whether the underlying price has/has not crossed the specified barrier level. The closed-form solutions can equally be obtained analytically using the extended Black–Scholes models [39], which shall be implemented as a benchmark of the exact price in this work. The structure of the model is described below.

2.1. Model Structure

Generally, the Black–Scholes option pricing formula models the dynamics of an underlying asset price S as a continuous time diffusion process given below:

$$\mathrm{d}S(t) = S(r\mathrm{d}t + \sigma \mathrm{d}B(t)), \tag{1}$$

where r is the risk-free interest rate, σ, the volatility and $B(t)$ is the standard Brownian motion at the current time t. Suppose $V(S, t)$ is the value of a given non-dividend paying

European call option. Then, under the pricing framework of Black and Scholes, $V(S,t)$ satisfies the following PDE:

$$\frac{\partial V(S,t)}{\partial t} + rS\frac{\partial V(S,t)}{\partial S} + \frac{\sigma^2 S^2}{2}\frac{\partial^2 V(S,t)}{\partial S^2} - rV(S,t) = 0, \tag{2}$$

subject to the following boundary and terminal conditions:

$$V(0,t) = 0, \forall\, t \in [0,T] \tag{3}$$
$$V(S,t) = S - Ke^{-r(T-t)} \text{ for } S \to \infty, \tag{4}$$
$$V(S,T) = \max\{S(T) - K, 0\}, \tag{5}$$

where K is the strike price and T is the time to expiration.

Since the barrier options are the focus of this study, the domain of the PDE in Equation (2) reduces to $\mathcal{D} = \{(S,t) : B \leq S \leq \infty; t \in [0,T]\}$ with the introduction of a predetermined level known as barrier B, and that feature distinguishes them from the vanilla European options. The boundary and terminal conditions above remain the same, with the exception of Equation (3), which reduces to $V(B,t) = 0$ for zero-rebate and $V(B,t) = R$ for the rebate barrier option. (In this paper, we shall consider the rebate paid at knock-out. The other type is the rebate paid at expiry, and in that case, Equation (3) becomes $V(B,t) = Re^{-r(T-t)}, \forall\, t \in [0,T]$.) The barrier options are either activated (knock-in options) or extinguished (knock-out options) once the underlying price attains the barrier level. The direction of the knock-in or the knock-out also determines the type of barrier options being considered, as this option is generally classified into up-and-in, up-and-out, down-and-in, and down-and-out barrier options. This paper will consider the down-and-out (DO) barrier options, both with and without rebates. For this option style, the barrier level is normally positioned below the underlying, and when the underlying moves in such a way that the barrier is triggered, the option becomes void and nullified (zero rebate). However, when the barrier is triggered, and the option knocks out with a specified payment compensation made to the option buyer by the seller, then we have the rebate barrier options. Under the risk-neutral pricing measure Q, the price of the down-and-out (DO) barrier options is given as

$$V(S,t) = \mathbb{E}^Q\left[e^{-r(T-t)}(S_T - K)^+\mathbb{I}\{\min_{0\leq t\leq T} S_t > B\}\right], \tag{6}$$

and the solution to the above is given in the following theorem.

Theorem 1. *Extended Black–Scholes for a DO call option (note that Equations (7) and (8) occurs when the strike price $K \geq B$. For $K < B$, we substitute $K = B$ into d_1 and d_3.) is given by [39]*

$$V(S,t) = SN(d_1) - Ke^{-r\tau}N(d_2) - \left[S\left(\frac{B}{S}\right)^{2\eta}N(d_3) - Ke^{-r\tau}\left(\frac{B}{S}\right)^{2\eta-2}N(d_4)\right] \tag{7}$$

$$\text{for } d_1 = \frac{\log\left(\frac{S}{K}\right) + \left(r + \frac{\sigma^2}{2}\right)\tau}{\sigma\sqrt{\tau}}, \quad d_3 = \frac{\log\left(\frac{B^2}{SK}\right) + \left(r + \frac{\sigma^2}{2}\right)\tau}{\sigma\sqrt{\tau}}, \quad d_5 = \frac{\log\left(\frac{B}{S}\right) + \left(r + \frac{\sigma^2}{2}\right)\tau}{\sigma\sqrt{\tau}},$$

where $\tau = T - t$, $d_{2,4} = d_{1,3} - \sigma\sqrt{\tau}$, $\eta = (2r + \sigma^2)(2\sigma^2)^{-1}$ and $N(x) = \int_{-\infty}^{x}\frac{1}{\sqrt{2\pi}}e^{-\frac{y^2}{2}}dy$ is the cumulative standard normal distribution function. In the presence of a rebate R, the option value becomes

$$V_R(S,t) = V(S,t) + R\left[\left(\frac{B}{S}\right)^{2\eta-1}N(d_5) + \left(\frac{S}{B}\right)N(d_5 - 2\eta\sigma\sqrt{\tau})\right] \tag{8}$$

2.2. Option Greeks

These refer to the sensitivities of option prices with respect to different pricing parameters. The knowledge and the application of option Greeks can equip investors with risk-minimization strategies, which will be applicable to their portfolios. Such knowledge is as vital as hedging the portfolio risk using any other risk management tools. For options that have an analytical form based on the Black–Scholes model or other closed-form models, the Greeks or the sensitivities are normally estimated from these formulas. In the absence of analytical option values, numerical techniques are employed to extract the Greeks. These Greeks are adopted from [40], and we only consider the delta (Δ_{DO}), gamma (Γ_{DO}) and the vega (ν_{DO}).

2.2.1. Delta

This measures the sensitivity of options values to changes in the underlying prices. The delta for the DO call options behaves like the delta of the European call options when the option is deep in-the-money, and it becomes very complicated as the underlying price approaches the barrier level:

$$\frac{\partial V(S,t)}{\partial S} = N(d_1) - \left(\frac{B}{S}\right)^{2\eta-2}\left\{-\frac{B^2}{S^2}N(d_3) + \frac{2\eta-2}{S}\left(\frac{B^2}{S}N(d_4) - Ke^{-r\tau}N(d_3)\right)\right\},$$

where d_1, d_3 and d_4 are given in Theorem 1.

2.2.2. Gamma

This measures the sensitivity of delta to a change in the underlying price, or the second partial derivative of the option value with respect to the underlying price:

$$\frac{\partial^2 V(S,t)}{\partial S^2} = = \frac{\phi(d_1)}{\sigma S\sqrt{\tau}} - \left(\frac{B}{S}\right)^{2\eta-2}\left\{\frac{(2\eta-2)(8\eta-7)}{S}\left(\frac{B^2}{S}N(d_4) - Ke^{-r\tau}N(d_3)\right)\right.$$
$$\left. + \frac{B^2}{S^2}\left(2N(d_3) + \frac{\phi(d_3)}{\sigma\sqrt{\tau}}\right) + 2(2\eta-2)\left(\frac{B^2}{S^2}N(d_3)\right)\right\},$$

where d_1, d_3 and d_4 are given in Theorem 1; also, $\phi(x) = \frac{1}{\sqrt{2\pi}}e^{\frac{-x^2}{2}}$ is the probability density function of the standard normal distribution.

2.2.3. Vega

This measures the sensitivity of options values to changes to volatility. It is calculated as

$$\frac{\partial V(S,t)}{\partial \sigma} = S\sqrt{\tau}\phi(d_1) - \left(\frac{B}{S}\right)^{2\eta-2}\left\{\sqrt{\tau}Ke^{-r\tau}\phi(d_4) - \frac{4r}{\sigma^3}\left(\frac{B^2}{S}N(d_4) - Ke^{-r\tau}N(d_3)\right)\ln\frac{B}{S}\right\}.$$

where d_1, d_3 and d_4 are given in Theorem 1; also, $\phi(x) = \frac{1}{\sqrt{2\pi}}e^{\frac{-x^2}{2}}$ is the probability density function of the standard normal distribution.

3. Machine Learning Models

Machine learning models, such as the ANN, polynomial regression model and random forest regression models, form the methodology in this research. Here, we briefly describe each of them and their financial application as they relate to the rebate barrier options problem. The numerical experiments are performed on an 11th Gen Intel(R) Core(TM) i7-1165G7 @ 2.80GHz processor, 16 GB RAM, 64-bit Windows 10 operating system and x64-based processor.

3.1. Artificial Neural Networks

This subsection utilizes the concept of ANN in the approximation of functions which describe the financial model in perspective. It will highlight the whole network environment and the multi-layer perceptron (MLP) idea. In connection with the application of ANN to option pricing, the concept lies first in the generation of the financial data (barrier option pricing data) and then employing the ANN to predict the option prices according to the trained model.

3.1.1. Network Environment

The research computations and the data processing are implemented using `Python` (version 3.9.7), which is an open-source programming tool. The ANN employed in the data analysis and construction of the model, as well as the training and validation, is implemented with `Keras` (https://keras.io/about/, accessed on 6 April 2023), which is a deep learning application programming interface, running concurrently with the machine learning platform `Tensorflow` (version 2.2.0).

3.1.2. Multi-Layer Perceptron

An MLP is a feed-forward ANN category comprising a minimum of three layers: the input layer, the hidden layer and the output layer. The MLP with as little as one hidden layer tends to approximate a large category of non-linear and linear functions with arbitrary accuracy and precision. Except for input nodes, every other node consists of neurons triggered by non-linear activation functions. During the training phase, an MLP employs the supervised learning techniques, also known as backpropagation, and in this section, we use the backpropagation network method, which is by far the most widespread neural network type.

Mathematically, consider an MLP network's configuration with first and second hidden layers $h_k^{(1)}$ and $h_k^{(2)}$, respectively, and input units x_k, where k denotes the number of the units. The non-linear activation function is written as $f(.)$, and we denote $f^{(1)}(.)$, $f^{(2)}(.)$ and $f^{(3)}(.)$ differently since the network layers can have various activation functions, such as the sigmoid (Sigmoid is defined by $f(z) = 1/(1 + \exp(-z))$, where z is the input to the neuron), hyperbolic tangent (Tanh is defined by $f(z) = 2\mathrm{sigmoid}(z) - 1$, where z is the input to the neuron), rectified linear unit (ReLU) (ReLU is defined by $f(z) = \max[0, z]$, where z is the input to the neuron), etc. The weights of the network are denoted by w_{jk}, the activation output value y_j, and the bias b_j, where j denotes the number of units in each layer. Thus, we have the following representation:

$$h_j^{(1)} = f^{(1)}\left(\sum_k w_{jk}^{(1)} x_k + b_j^{(1)} \right)$$

$$h_j^{(2)} = f^{(2)}\left(\sum_k w_{jk}^{(2)} h_k^{(1)} + b_j^{(2)} \right)$$

$$y_j = f^{(3)}\left(\sum_k w_{jk}^{(3)} h_k^{(2)} + b_j^{(3)} \right).$$

3.1.3. The Hyperparameter Search Space and Algorithm

This section further explains the hyperparameter optimization techniques, which aim to search for the optimal algorithm needed for our optimization problem. It is essential to note that the parameters of the NN are internal configuration variables that the models can learn. Examples are the weights and the bias. In contrast, the hyperparameters are external and cannot be learned from the data but are used to control the learning process and the structure of the NN. These parameters are set before the training process, and some examples include the activation function, batch size, epoch, learning rates, etc. The choice of hyperparameters hugely affects the accuracy of the network. As a result, different optimal methods, such as manual search, Bayesian optimization, random search and grid search, have been developed. We will employ the Keras tuner framework (https://keras.io/keras_tuner/, accessed on 6 April 2023), which encompasses some algorithms, such as the random search, hyperband and Bayesian optimization. For these three search algorithms, we choose the validation loss as the objective with the maximum search configuration of six trials. The following variables define the search space of our NN architecture:

(1) Width and depth: The depth refers to the number of hidden layers, and the width is the number of units in each hidden layer. Thus, the depth ranges from $[1, 4]$ with a step of 1, and the width ranges from $[32, 512]$ with a step of 32.
(2) Activation function: These are non-linear activation functions in the NN, and we only consider Sigmoid, ReLU and Tanh .
(3) Optimizer: This modifies the model parameters and is mostly used for weight adjustments to minimize the loss function. We only consider Adam, SGD, Adagrad and RMSprop.
(4) Learning and dropout rates: Learning rates control the speed at which the NN learns, and we set it at $[0.01, 0.001, 0.0001]$. The dropout is a regularization technique employed to improve the ability of NN to withstand overfitting. We set it at $[0.1, 0.5]$ with a step size of 0.1.

The activation functions are used in each layer, except the output layer. The network is trained with 45 epochs, 256 batch sizes and an early stopping callback on the validation loss with `patience` = 3. Since the option pricing model is a regression problem, our primary objective is to keep the mean squared error (MSE) of the predicted prices to a minimum. The essence of training a neural network entails minimizing the errors obtained during the regression analysis, and this is done by selecting a set of weights in both the hidden and the output nodes. Thus, to evaluate the performance of our ANN, we consider the MSE as the loss function used by the network and the mean absolute error (MAE) as the network metrics, which are given, respectively, as follows:

$$\text{MSE} = \frac{1}{N} \sum_{i=1}^{N} (V_i(S,t) - \hat{V}_i(S,t))^2$$

$$\text{MAE} = \frac{1}{N} \sum_{i=1}^{N} |V_i(S,t) - \hat{V}_i(S,t)|,$$

where N is the number of observations, $V_i(S,t)$ is the exact option values and $\hat{V}_i(S,t)$ is the predicted option values. Finally, we alternate the activation functions, optimizers, batch normalization and dropout rates to investigate the effect of the network training on the option valuation and avoid overfitting the models.

3.1.4. Data Splitting Techniques for the ANN

Data splitting is a fundamental aspect of data science, especially for developing data-based models. The dataset is divided into training and testing sets, and an additional set known as the validation can also be created. The training set is used mainly for training, and the model is expected to learn from this dataset while optimizing any of its parameters. The testing set contains the data which are used to fit and measure the model's performance. The validation set is mainly used for model evaluation. If the difference between the training set error and the validation set error is large, there is a case of over-fitting, as the model has high variance. This paper considers supervised learning in which the model is trained to predict the outputs of an unspecified target function. This function is denoted by a finite training set $\mathcal{F}$ consisting of inputs and corresponding desired outputs: $\mathcal{F} = \{[\vec{a_1}, \vec{x_1}], [\vec{a_2}, \vec{x_2}], \cdots [\vec{a_n}, \vec{x_n}]\}$, where n is the number of 2-tuples of input/output samples.

Train–Test Split

This paper considers the train–test split as 80:20 and a further 80:20 on the new train data to account for a validation dataset. Thus, 80% of the whole dataset will account for the training set and 20% for the test dataset. Additionally, 80% of the training set will be used as the actual training dataset and the remaining 20% for validation. After training, the final model should correctly predict the outputs and generalize the unseen data. Failure to accomplish this leads to over-training, and these two crucial conflicting demands between accuracy and complexity are known as the bias–variance trade-off [41,42]. A common approach to balance this trade-off is to use the cross-validation technique.

K-Fold Cross Validation

The *k*-fold cv is a strategy for partitioning data with the intent of constructing a more generalized model and estimating the model performance on unseen data. Denote the validation (testing) set as $\mathcal{F}_{te}$ and the training set as $\mathcal{F}_{tr}$. The algorithm (Algorithm 1) is shown below.

Algorithm 1 Pseudocode for the *k*-fold cross validation

Input the dataset $\mathcal{F}$, number of folds k and the error function (MSE)

1: **Data split :**
- Randomly split $\mathcal{F}$ into k independent subsets $\mathcal{F}_i, \mathcal{F}_2, \cdots, \mathcal{F}_k$ of same size.
- For $i = 1, 2, \cdots, k$: $\mathcal{F}_{te} \leftarrow \mathcal{F}_i$ and $\mathcal{F}_{tr} \leftarrow \mathcal{F} \setminus \{\mathcal{F}_i\}$.

2: **Fitting and Training:**
- Fit and train model on $\mathcal{F}_{tr}$ and evaluate model performance using $\mathcal{F}_{te}$ periodically: $\mathcal{R}_{te}(i)=$Error $(\mathcal{F}_{te})$.
- Terminate model training when the $\mathcal{R}_{te}(i)$ stop criterion is satisfied.

3: **Evaluation:**
- Evaluate the model performance using $\mathcal{R}_{te} = \frac{1}{k}\sum_{i=1}^{k} \mathcal{R}_{te}(i)$.

3.1.5. Architecture of ANN

This research considers a fully connected MLP NN in the option valuation for this research, which will consist of eight input nodes (in connection to the extended Black–Scholes for the rebate option parameters). There will be one output node (option value); the hidden layers and nodes will be tuned. There are two main models classified under the data-splitting techniques: Model A (train–test split) and Model B (5-fold cross-validation split). Each model is further subdivided into 3 according to the hyperparameter search algorithm. Thus, Models A1, A2, and A3 represent the models from the data train–test split for the hyperband algorithm, random search algorithm and Bayesian optimization algorithm, respectively. Additionally, Models B1, B2, and B3 represent the models from the k-fold cross-validation data split for the hyperband algorithm, random search algorithm and Bayesian optimization algorithm, respectively. Finally, Tables 1 and 2 present the post-tuning search details and the optimal model hyperparameters for the NN architecture, respectively.

Table 1. Trainable parameter search details.

Search Details	Model A1	Model A2	Model A3	Model B1	Model B2	Model B3
Best MAE score	0.07081	0.02794	0.01602	0.10141	0.03391	0.02911
Trainable parameters	254,305	218,145	113,345	138,945	80,961	4609
Total search time (secs)	620	714	642	642	491	754

Table 1 compares the search time taken by each of the algorithms in tuning the hyperparameters. We observed that the hyperband algorithm is highly efficient regarding the search time for the train–test split and the k-fold cross-validation models. The hyperband algorithm search time is generally less when compared to the random search and the Bayesian optimization algorithm. Furthermore, the Bayesian optimization provided the lowest MAE score and required fewer trainable parameters than the hyperband and the random search algorithm. This characteristic is equally observable for both models A and B. From the tuning, we can see that the Bayesian optimization effectively optimizes the hyperparameter when producing the lowest MAE, though it had the disadvantage of a higher search time. In contrast to the Bayesian optimization, the hyperband algorithm is optimal in terms of search time, despite having a higher MAE score. From the results section, the final comparison of optimality will be made in terms of the deviation from the actual values when all the models are used in the pricing process.

Table 2. Architecture of the ANN.

Hyperparameters	Model A1	Model A2	Model A3	Model B1	Model B2	Model B3
Activation Fn	Sigmoid	Tanh	Tanh	ReLU	Tanh	ReLU
Optimizer	Adam	Rmsprop	Adam	Adagrad	Rmsprop	Adam
Learning rate	0.001	0.0001	0.0001	0.1	0.0001	0.0001
Hidden layers	4	3	4	4	2	3
Layer 1 (dropout)	512(0.2)	288(0.2)	32(0.1)	384(0.3)	480(0.2)	512(0.1)
Layer 2 (dropout)	192(0.2)	480(0.3)	32(0.1)	352(0.2)	160(0.3)	32(0.1)
Layer 3 (dropout)	224(0.4)	160(0.5)	512(0.2)	448(0.4)	Nil	352(0.1)
Layer 4 (dropout)	480(0.5)	Nil	224(0.2)	192(0.4)	Nil	Nil

3.2. Random Forest Regression

Random forest combines tree predictors in such a way that each tree in the ensemble is contingent on the values of a randomly sampled vector selected from the training set. This sampled vector is independent and has similar distribution with all the trees in the forest [43]. The random forest regressor uses averaging to improve its predictive ability and accuracy.

Let $f(\mathbf{x}; \beta_n)$ be the collection of tree predictors where $n = 1, 2, \cdots, N$ denotes the number of trees. Here, $\mathbf{x}$ is the observed input vector from the random vector $\mathbf{X}$, and β_n are the independent and identically distributed random vectors. The random forest prediction is given by

$$\bar{f}(\mathbf{x}) = \frac{1}{N} f(\mathbf{x}; \beta_n),$$

where $\bar{f}(\mathbf{x})$ is the unweighted average over the tree collection $f(\mathbf{x})$. As the number of trees increases, the tree structure converges. This convergence explains why the random forest does not overfit, but instead, a limiting value of the generalization (or prediction) error is produced [43,44]. Thus, we have that as $n \to \infty$, the law of large numbers ensures that

$$\mathbb{E}_{\mathbf{X},Y}[Y - \bar{f}(\mathbf{X})]^2 \to \mathbb{E}_{\mathbf{X},Y}[Y - \mathbb{E}_\beta[\bar{f}(\mathbf{X}; \beta)]]^2$$

Here, Y is the outcome. The training data are assumed to be drawn independently from the joint distribution of $(\mathbf{X}, Y)$. In this research, we use the 80:20 train–test split techniques to divide the whole dataset into a training set and a testing set. Using the `RandomForestRegressor()` from the scikit-learn ML library, we initialize the regression model, fit the model, and predict the target values.

3.3. Polynomial Regression

Polynomial regression is a specific type of linear regression model which can predict the relationship between the independent variable to the dependent variable as an nth degree polynomial. In this research, we first create the polynomial features object using the `PolynomialFeatures()` from the scikit-learn ML library and indicate the preferred polynomial degree. We next use the 80:20 train–test split techniques to divide this new polynomial feature into training and testing datasets, respectively. Next, we construct the polynomial regression model, fit the model and predict the responses.

4. Results and Discussion

4.1. Data Structure and Description

For the ANN model input parameters, we generated 100000 sample data points and then used Equation (8) to obtain the exact price for the rebate barrier call options. These random observations will train, test and validate an ANN model to mimic the extended Black–Scholes equation. We consider the train–test split and the cross-validation split on the dataset and then measure these impacts on the loss function minimization and the option values. The generated samples consist of eight variables, that is $(S, K, B, R, T, \sigma, r, V_R)$, which are sampled uniformly, except the option price V_R, and following the specifications and logical ranges of each of the input variables (See Table 3). During the training process, we fed the ANN the training samples with the following inputs $(S, K, B, R, T, \sigma, r)$, where V_R is the expected output. In this phase, the ANN 'learns' the extended Black–Scholes model from the generated dataset, and the testing phase follows suit, from which the required results are predicted. Meanwhile, under the Black–Scholes framework, we assume that the stock prices follow a geometric Brownian motion, and we used GBM$(x = 150, r = 0.04, \sigma = 0.5, T = 1, N = 100{,}000)$ for the random simulation. Table 3 below shows the extended Black–Scholes parameters used to generate the data points, whereas Table 4 gives the sample data generated. The range for the rebate, strike and barrier is from the uniform random distribution, and they are multiplied by the simulated stock price to obtain the final range.

Table 3. Extended Black–Scholes range of parameters—rebate barrier.

Strike	Barrier	Rebate	Time	Volatility	Rate
$[0.4, 1]$	$[0.4, 1]$	$[0.01, 0.05]$	$[0.5, 1.5]$	$[0.1, 0.5]$	$[0.01, 0.05]$

Table 4. Sample training data for rebate barrier option pricing model.

Stock	Barrier	Strike	Rebate	Rate	Volatility	Time	Call Option
98.25745	51.92557	46.33445	0.01835	0.04355	0.22836	1.42274	54.61598
149.79728	96.91339	105.83799	0.04255	0.02863	0.27321	0.85166	47.22227
55.90715	38.39396	35.25565	0.01241	0.01989	0.34164	0.95034	20.32389
63.29343	41.03505	31.03422	0.04143	0.03072	0.21341	0.95572	32.80836
126.83153	116.65153	124.11145	0.02695	0.01888	0.45170	0.74936	9.60775
97.18410	75.27999	92.85564	0.01626	0.02186	0.39194	1.04861	15.67665
112.31872	112.30254	53.26221	0.01876	0.03539	0.43021	1.05975	0.04930

Statistics and Exploratory Data Analysis

In this section, we aim to summarize the core characteristics of our option dataset by analyzing and visualizing them. The descriptive statistics which summarize the distribution shape, dispersion and central tendency of the dataset are presented in Table 5. The following outputs were obtained: the number of observations or elements, mean, standard deviation, minimum, maximum and quartiles (25%, 50%, 75%) of the dataset. We observed that the distribution of the simulated stock is left skewed since the mean is lesser than the median, whereas the distributions of the option values, strike price and barrier levels are right skewed.

Table 5. Descriptive statistics for the rebate barrier.

	Count	Mean	Std	Min	25%	50%	75%	Max
Stock	100,000.0	100.914912	25.412662	49.601351	89.084095	100.055323	110.997594	171.710128
Strike	100,000.0	70.644521	25.269816	20.236066	51.255058	68.083559	86.709133	170.197316
Rebate	100,000.0	0.029946	0.011529	0.010000	0.019957	0.029952	0.039881	0.049999
Barrier	100,000.0	70.574580	25.264409	20.380974	51.162052	67.929254	86.572706	169.477720
Time	100,000.0	1.000314	0.288309	0.500003	0.751330	0.999858	1.250020	1.499976
Sigma	100,000.0	0.299688	0.115622	0.100001	0.199680	0.299840	0.400159	0.499991
Rate	100,000.0	0.030006	0.011557	0.010001	0.019994	0.029978	0.040008	0.050000
OptionV	100,000.0	27.183098	17.482423	0.025235	13.547223	24.047318	38.324748	103.946198

In Figure 1, we consider the visualization using the seaborn library in connection with the pairplot function to plot a symmetric combination of two main figures, that is, the scatter plot and the kernel density estimate (KDE). The KDE plot is a non-parametric technique mainly used to visualize the nature of the probability density function of a continuous variable. In our case, we limit these KDE plots to the diagonals. We focus on the relationship between the stock, strike, rebate and the barrier with the extended Black–Scholes price (OptionV) for the rebate barrier options. From the data distribution for the feature columns, we notice that the sigma, time and rate columns could be ignored. This is because the density distribution shows that these features are basically uniform, and the absence of any variation makes it very unlikely to improve the model performance. Suppose we consider this problem as a classification problem; then, no split on these columns will increase the entropy of the model.

On the contrary, however, if this was a generative model, then there would be no prior to updating given a uniform posterior distribution. Additionally, the model will learn a variate of these parameters since, by definition of the exact option price (referred to as OptionV) function, these are the parameters which can take on constant values. Another method to consider would be to take these parameters 'sine' functions as inputs to the model instead of the actual values. We observed from our analysis that this concept works, but there is not a significant improvement in model performance, which can be investigated in further research.

4.2. Neural Network Training

The first category (train dataset) is employed to fit the ANN model by estimating the weights and the corresponding biases. The model at this stage tends to observe and 'learn' from the dataset to optimize the parameters. In contrast, the other (test dataset) is not used for training but for evaluating the model. This dataset category explains how effective and efficient the overall ANN model is and the prediction probability of the model. Next and prior to the model training, we perform data standardization techniques to improve the performance of the proposed NN algorithm. The StandardScalar function of the Sklearn python library was used to standardize the distribution of values by ensuring that the distribution has a unit variance and a zero mean. During the compilation stage, we plot the loss (MSE) and the evaluation metrics (accuracy) values for both the train and validation datasets. We equally observe that the error difference between the training and the validation dataset is not large, and as such, there is no case of over- or under-fitting of the ANN models. Once the 'learning' phase of the model is finished, the prediction phase will set in. The performance of the ANN model is measured and analyzed in terms of the MSE and the MAE. Table 6 gives the evaluation metrics for both the out-sample prediction (testing dataset) and the in-sample prediction (training dataset).

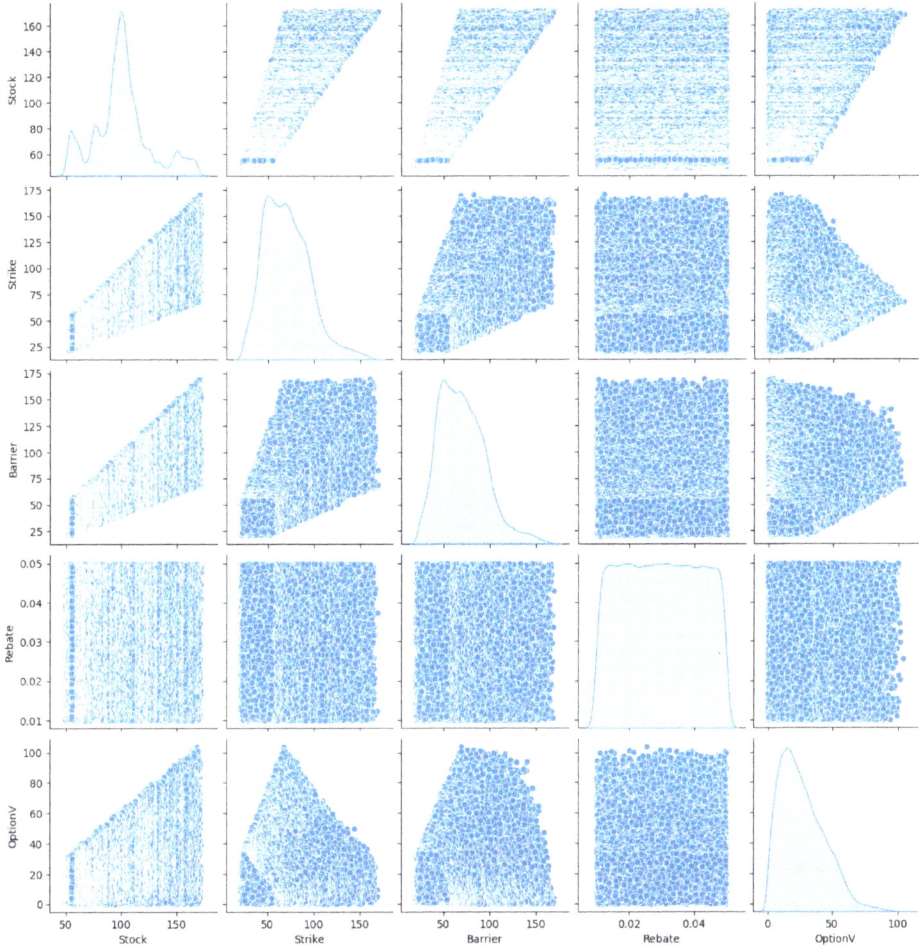

Figure 1. Visualization plot.

Table 6. Model evaluation for testing and training data (shows no over- or underfitting).

Models	Test		Train	
	Loss (MSE)	**Metrics (MAE)**	**Loss (MSE)**	**Metrics (MAE)**
Model A1	0.0214	0.0574	0.0249	0.0557
Model A2	0.0349	0.0987	0.0299	0.0929
Model A3	0.0446	0.1058	0.0423	0.1027
Model B1	0.0229	0.0601	0.0202	0.0584
Model B2	0.0425	0.0883	0.0394	0.0872
Model B3	0.0457	0.0782	0.0411	0.0762

Table 6 shows the model evaluation comparison for the train/test loss and accuracy. It is observed that the test loss is greater than the training loss, and the test accuracy is greater than the training accuracy for all the models. The differences in error sizes are not significant, and thus the chances of having an overfitting model are limited. Figures 2–5 show the training and validation (test) of the loss and MAE values for all the models when the models are fitted and trained on epoch = 45, batch size = 256, and verbose = 1. We visualize these graphs to ascertain whether there was any case of overfitting, underfitting or a perfect fitting of the model. In underfitting, the NN model fails to model the training data and learn the problem perfectly and sufficiently, leading to slightly poor performance on the training dataset and the holdout sample. Overfitting occurs mainly in complex models with diverse parameters, which happens when the model aims to capture all data points present in a specified dataset. In all the cases, we observe that the models show a good fit, as the training and validation loss are decreased to a stability point with an infinitesimal gap between the final loss values. However, the loss values for Model B3 followed by Model B2 are highly optimal in providing the best fit for the algorithm.

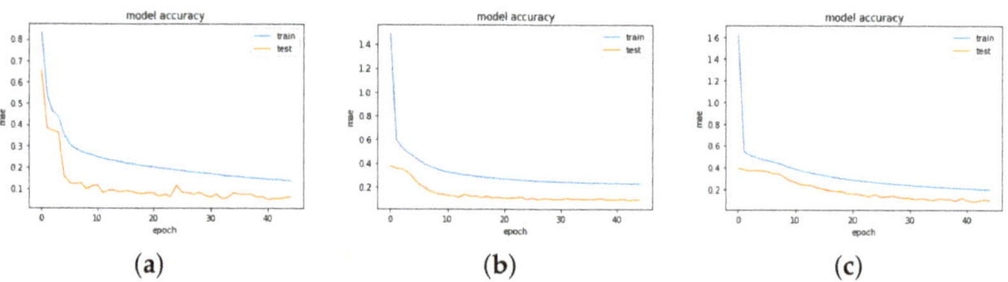

Figure 2. Train/test MAE values for Models A1, A2 and A3; (**a**) MAE—Model A1; (**b**) MAE—Model A2; (**c**) MAE—Model A3.

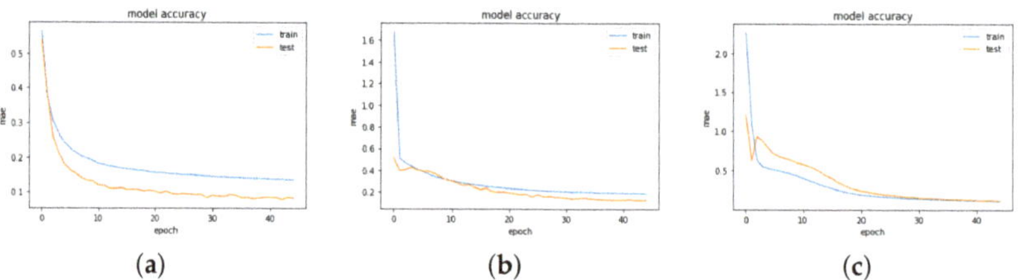

Figure 3. Train/Test MAE values for Models B1, B2 and B3; (**a**) MAE—Model B1; (**b**) MAE—Model B2; (**c**) MAE—Model B3.

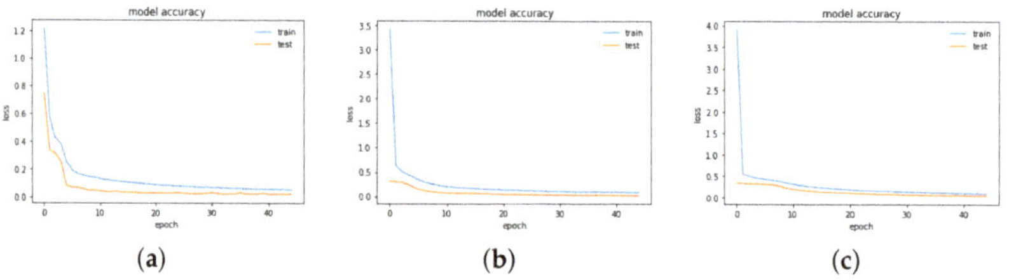

Figure 4. Train/Test LOSS values for Models A1, A2 and A3; (**a**) LOSS—Model A1; (**b**) LOSS—Model A2; (**c**) MAE—Model A3.

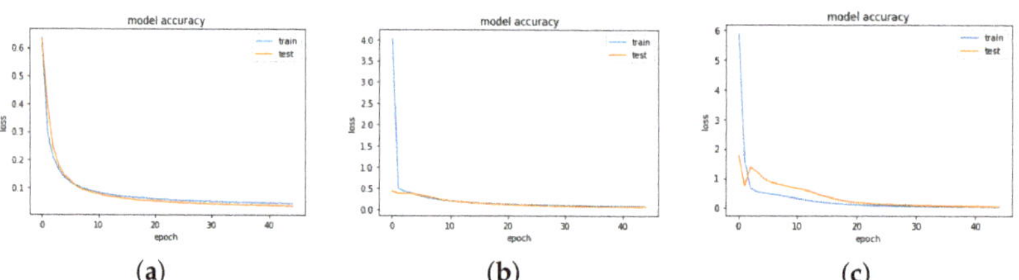

Figure 5. Train/Test LOSS values for Models B1, B2 and B3; (**a**) LOSS—Model B1; (**b**) LOSS—Model B2; (**c**) LOSS—Model B3.

Next, we display the plots obtained after compiling and fitting the models. The prediction is performed on the unseen data or the test data using the trained models. Figures 6 and 7 give the plot of the predicted values against the actual values, the density plot of the error values and the box plot of the error values for all six models.

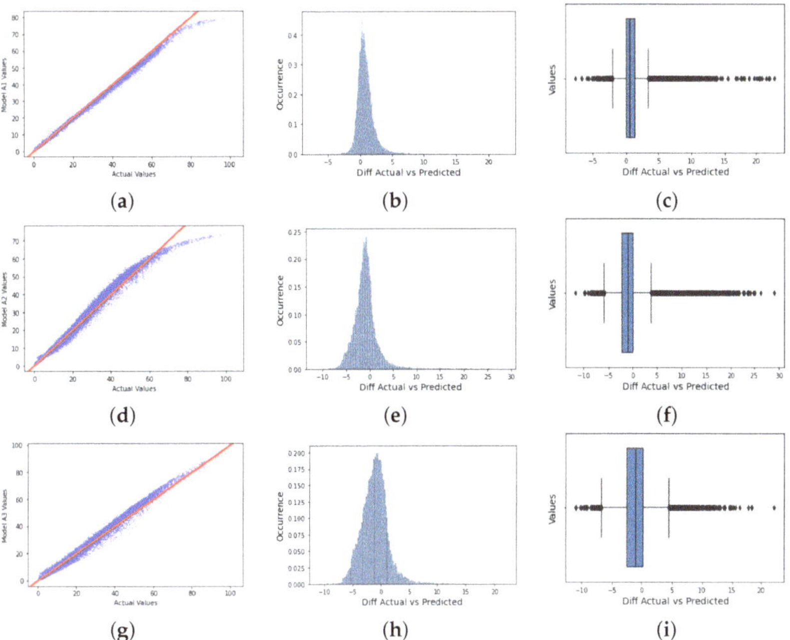

Figure 6. Option values visualization for Models A1, A2, A3; (**a**) Model A1: Regression plot; (**b**) Model A1: Histogram plot; (**c**) Model A1: Box plot; (**d**) Model A2: Regression plot; (**e**) Model A2: Histogram plot; (**f**) Model A2: Box plot; (**g**) Model A3: Regression plot; (**h**) Model A3: Histogram plot; (**i**) Model A3: Box plot.

The box plot enables visualization of the skewness and how dispersed the solution is. Model A2 behaved poorly, as this can be observed with the wide range of dispersion of the solution points, and the model did not fit properly. For a perfect fit, the data points are expected to concentrate along the 45 deg red line, where the predicted values are equal to the actual values. This explanation is applicable to Models A2 and A3, as there was no perfect alignment in the regression plots. We could retrain the neural network to improve this performance since each training can have different initial weights and biases. Further improvements can be made by increasing the number of hidden units or layers or using a larger training dataset. For the purpose of this research, we already performed the hyperparameter tuning, which solves most of the above suggestions. To this end, we focus on Model B, another training algorithm.

Models B3 and B1 provide a good fit when their performance is compared to the other models, though there are still some deviations around the regression line. The deviation of these solution data points is also fewer than in the other models. It is quite interesting to note that the solution data points of Models B1 and B3 are skewed to the left, as can be seen in the box plots. This could be a reason for their high performance compared to other models, such as A1, A2, and A3, which are positively skewed. However, this behavior would be worth investigating in our future work.

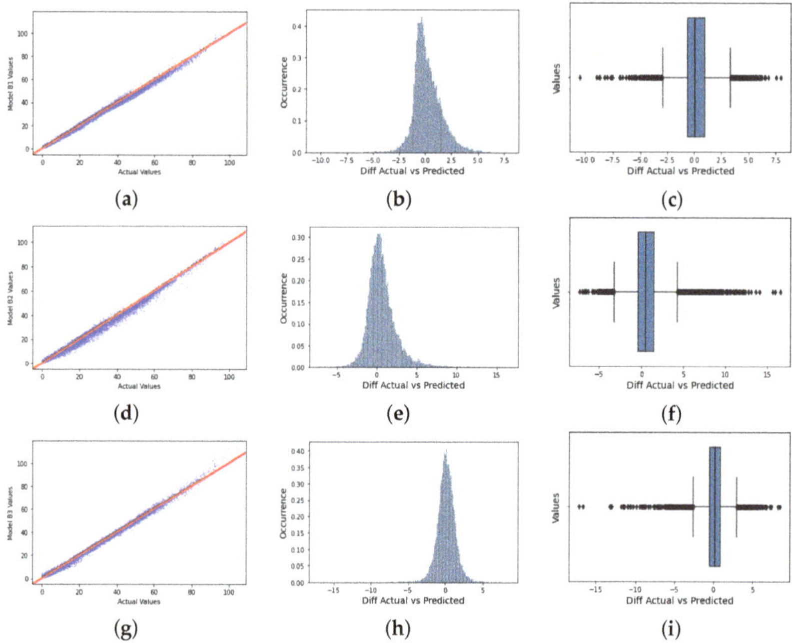

Figure 7. Option values visualization for Models B1, B2, B3; (**a**) Model B1: Regression plot; (**b**) Model B1: Histogram plot; (**c**) Model B1: Box plot; (**d**) Model B2: Regression plot; (**e**) Model B2: Histogram plot; (**f**) Model B2: Box plot; (**g**) Model B3: Regression plot; (**h**) Model B3: Histogram plot; (**i**) Model B3: Box plot.

Table 7 shows the error values in terms of the MSE, MAE, mean squared logarithmic error (MSLE), mean absolute percentage error (MAPE) and the R^2 (coefficient of determination) regression score. It also shows the models' comparison in terms of their computation speed, and it must be noted that the computation is measured in seconds. Mathematically, the MSLE and MAPE are given as

$$\text{MSLE} = \frac{1}{N} \sum_{i=1}^{N} [\log_e(1 + V_i(S,t)) - \log_e(1 + \hat{V}_i(S,t))]^2,$$

$$\text{MAPE} = \frac{100\%}{N} \sum_{i=1}^{N} \left| \frac{V_i(S,t) - \hat{V}_i(S,t)}{V_i(S,t)} \right|,$$

where N is the number of observations, $V_i(S,t)$ is the exact option values and $\hat{V}_i(S,t)$ is the predicted option values. For the MAPE, all the values lower than the threshold of 20% are considered 'good' in terms of their forecasting capacity [45]. Thus, all the models have good forecasting scores, with Model A1 possessing a highly accurate forecast ability. Similarly, the values for the MSLE measure the percentile difference between the log-transformed predicted and actual values. The lower, the better, and we can observe that all the models gave relatively low MSLE values, with Models A1 and B1 giving the lowest MSLE values.Please check that intended meaning is retained.

From Table 7, the R^2 measures the capacity of the model to predict an outcome in the linear regression format. Models B1 and B3 gave the highest positive values compared to the other models, and these high R^2-values indicate that these models are a good fit for our options data. It is also noted that for well-performing models, the greater the R^2, the smaller the MSE values. Model B3 gave the smallest MSE, with the highest R^2, compared to the least performed model A2, which had the largest MSE and the smallest R^2 score. The MAE measures the average distance between the predicted and the actual data, and the lower values of the MAE indicate higher accuracy.

Table 7. Error values and computation time for various NN models.

Models	Error Values					Comp Time (secs)
	R^2-Score	MAE	MSLE	MSE	MAPE	
Model A1	0.989839	1.128247	0.006056	3.143079	0.076672	93.22
Model A2	0.969974	2.080077	0.013861	9.175201	0.129603	76.93
Model A3	0.977491	1.991815	0.019283	6.840152	0.143385	60.21
Model B1	0.994019	0.990397	0.006361	1.821165	0.080418	51.48
Model B2	0.987908	1.328247	0.015289	3.681967	0.125486	51.85
Model B3	0.994371	0.932689	0.014919	1.714182	0.125429	31.17

Finally, we display the speed of the NN algorithm models in terms of their computation times, as shown in Table 7. The computation time taken by the computer to execute the algorithm encompasses the data splitting stage, standardization of set variables, ANN model compilation and training, fitting, evaluation and the prediction of the results. As noted in Models A1 and A2, the use of Sigmoid and Tanh activation functions accounted for higher computation time, and this is due to the presence of exponential functions, which need to be computed. Model A1 was the least performed in terms of the computation time, and Model B3 was the best, accounting for a 66.56% decrease in time. We observe that the computation time is reduced when the k-fold cross-validation split is implemented prior to the ANN model training, as compared to the traditional train–test split. This feature is evident as a further 41.62% decrease was observed when the average computation time for Model B was used against Model A.

The overall comparison of the tuned models is presented in Figure 8. Here, we rank the performance of each MLP model with regards to ST:TP, algorithm computation time, and finally, the errors spanning from the R^2 score, MAE and the MSE. The ST:TP ratio denotes the search time per one trainable parameter. The ranking is ascending, with 1 being the maximum preference and 6 being the least preference. From the results and regardless of the number of search times per one trainable parameter, we observe that Model B3 is optimal, followed by Model B1, and the lowest performing is Model A2. Hence, we can conclude that models which consist of the k-fold data split performed significantly well in the valuation of the rebate barrier options using the extended Black–Scholes framework.

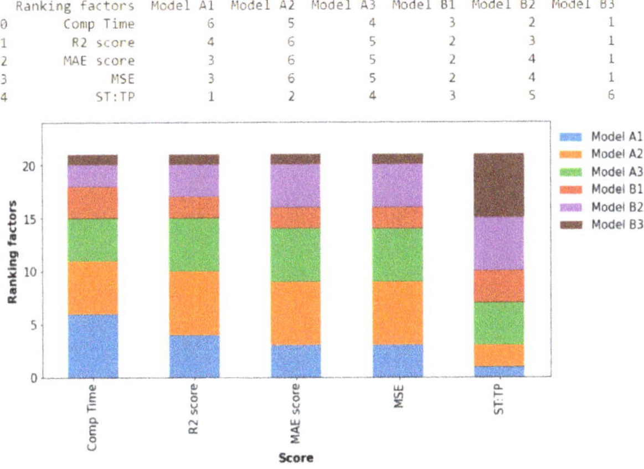

Figure 8. Ranking of models for optimality.

4.3. Analysis of Result Performance

One avenue to show the accuracy of our proposed model is to test the architecture on a non-simulated dataset for the rebate barrier options. At present, we are not able to obtain such real market data due to non-accessibility, and this is one of the limitations of the research. However, we compare the NN results with other machine learning models, such as the polynomial regression and the random forest regression on the same dataset. Both techniques are capable of capturing non-linear relationships that exist amongst variables.

Polynomial regression provides flexibility when modeling the non-linearity. Improved accuracy can be obtained when the higher-order polynomial terms are incorporated, and this feature makes it easier to capture the non-complex patterns in the dataset. It is also very fast when compared to both our proposed NN methodology and the random forest regression (Table 8). In this work, we only present the results obtained using the 2nd-, 3rd- and 4th-degree polynomial regressions. We observed that in terms of accuracy, polynomials of higher degrees gave rise to more accurate results and a significant reduction in their error components.

However, one of the issues facing the polynomial regression is model complexity; when the polynomial degree is high, the chances of model overfitting will be significantly high. Thus, we are faced with the trade-off between accuracy and over-fitting of the model. Regression random forest, on the other hand, combines multiple random decision trees, with each tree trained on a subset of data. We build random forest regression models using 10, 30, 50, and 70 decision trees, then we fit the model to the barrier options dataset, predict the target values, and then compute the error components. Finally, we compare these two models to the optimal model obtained with the NN results (Model B3), and we have the following table.

Table 8. Error values and computation time for Model B3, polynomial regression and random forest regression.

Models		Error values				Time (secs)
	R^2-Score	MAE	MSLE	MSE	MAPE	
Random Forest Regr.						
Decision trees (10)	0.990380	1.182121	0.016714	2.938727	0.145200	6.02
Decision trees (30	0.992352	1.056031	0.013461	2.293958	0.118368	15.56
Decision trees (50)	0.992449	1.037465	0.015827	2.299464	0.117032	26.34
Decision trees (70)	0.992825	1.022187	0.014146	2.211590	0.127853	36.02
Polynomial Regr.						
Polynomial order (2)	0.967764	2.156955	N/A	9.843225	0.327491	1.05
Polynomial order (3)	0.987900	1.269433	N/A	3.666175	0.206076	2.25
Polynomial order (4) [1]	0.996147	0.689323	N/A	1.177380	0.114092	3.75
Neural Network						
Model B3	0.994371	0.932689	0.014919	1.714182	0.125429	31.17

[1] We consider polynomials of order ≥ 4 to be higher-order, and this is because of the increase in their complexity. The accuracy of the 4th-order polynomial regression is actually higher than our proposed model, but the former has the issue of overfitting the data, which comes with a higher degree of polynomial regression. Additionally, the N/A in the MSLE cells is due to some negative values in the prediction set, which makes the logarithm of the values N/A.

Increasing the number of decision trees leads to more accurate results, and Oshiro et al. (2012) explained that the range of trees should be between 64 and 128. This feature will make it feasible to obtain a good balance between the computer's processing time, memory usage and the AUC (area under curve) [46]; we observed this feature in our research. The model was tested on 80, 100, 120, 140, 160, 180, and 200 decision trees, and we obtained the following coefficient of determination R^2 regression score (computation time): 0.9924 (34 secs), 0.9928 (52 secs), 0.9929 (62 secs), 0.9925 (75 secs), 0.9929 (83 secs), 0.9929 (89 secs) and 0.9926 (102 secs), respectively. We obtained the optimal decision tree to be between 110 and 120 with an R^2 score of 0.9929, and any other value below 110 will give rise to a less accurate result. Any value above 120 will not lead to any significant performance gain but will only lead to more computational cost.

Table 8 compares the performance of our optimal NN model to the random forest and the polynomial regressions. The performance is measured based on the error values and the computational time. The NN model performed better than the random forest regression regardless of the number of decision trees used, and this was obvious from the results presented in Table 8 above. On the other hand, polynomial regression of the 2nd and 3rd orders underperformed when compared to the NN model, but maximum accuracy was obtained when higher order (≥ 4) was used. This higher order posed a lot of complexity issues, which our optimal NN model does not face. Hence, more theoretical understanding is needed to further explain the phenomenon, and this current research does not account for it.

4.4. Option Prices and Corresponding Greeks

To compute the zero-rebate DO option prices and their corresponding Greeks, we simulate another set of data (1,000,000) in accordance with the extended Black–Scholes model, and the Table 9 below gives a subset of the full dataset after cleansing.

Table 9. Data subset of option values and Greeks.

S	B	K	R	r	σ	T	$V(t, S)$	Δ_{DO}	Γ_{DO}	v_{DO}
190.14286	80.0	100.0	0.0	0.05	0.25	1.0	95.04795	0.99811	0.00013	1.14425
83.61440	80.0	100.0	0.0	0.05	0.25	1.0	2.08913	0.48534	−0.00174	7.12088
108.12702	80.0	100.0	0.0	0.05	0.25	1.0	17.72886	0.74465	0.01029	30.44055
146.39879	80.0	100.0	0.0	0.05	0.25	1.0	51.77793	0.96780	0.00194	10.39359
121.23493	80.0	100.0	0.0	0.05	0.25	1.0	28.42419	0.86428	0.00678	24.95840

For the NN application, we used the hyperparameters of Model B3 to construct the NN architecture and train and predict the option values and their corresponding Greeks. The risks associated with the barrier options are complicated to manage and hedge due to their path-dependent exotic nature, which is more pronounced as the underlying approaches to the barrier level. For the Greeks considered here, we focus on predicting the delta, gamma and vega using the optimal NN model, and the following results were obtained.

Figure 9 shows the plot of the predicted and actual values of the DO option prices, together with the delta, gamma and vega values. For the option value, the DO call behaves like the European call when the option is far deep in-the-money, and this is because the impact of the barrier is not felt at that phase. The option value decreases and tends to zero as the underlying price approaches the barrier since the probability of the option being knocked out is very high. The in-the-money feature is equally reflected in the delta and gamma as they remain unchanged when the barrier is far away from the underlying. Here, the delta is one, and the gamma is zero.

Gammas for this option style are typically large when the underlying price is in the neighborhood of the strike price or even near the barrier, and it is the lowest for out-of-money options or knocked-out options. From Figure 9c, gamma tends to switch from positive to negative without switching from long to short options. The values of gammas are usually bigger than the gamma for the standard call option. These extra features pose a great challenge to risk managers during the rebalancing of portfolios. Lastly, vega measures the sensitivity of the option value with respect to the underlying volatility. It measures the change in option value based on a 1% change in implied volatility. Vega declines as the options approach the knock-out phase; it falls when the option is out-of-money and deep in-the-money, and it is maximum when the underlying is around the strike price. Overall, Figure 9a–d display how accurately Model B3 predicts the option values and their Greeks, as little or no discrepancies are observed in each dual plot.

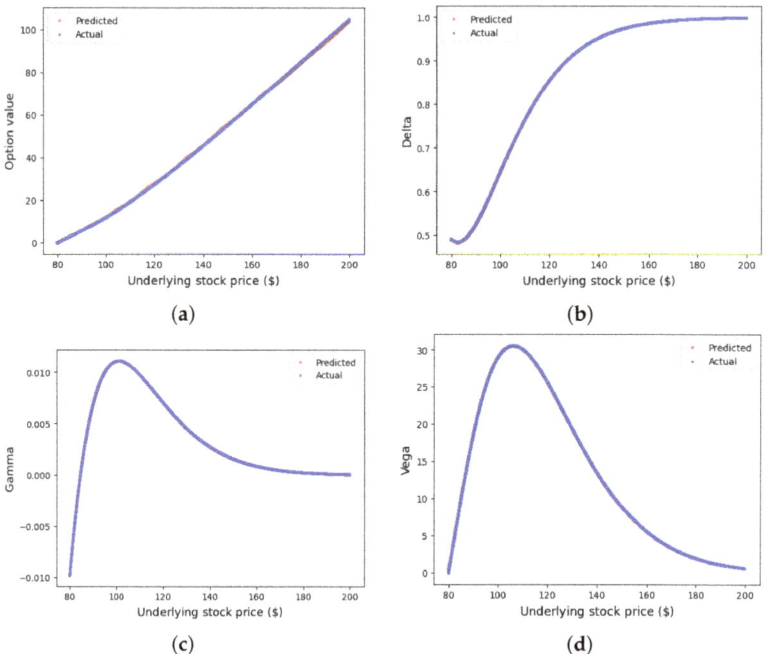

Figure 9. Option values and Greeks; (**a**) DO option value; (**b**) DO delta; (**c**) DO gamma; (**d**) DO vega.

5. Conclusions and Recommendations

This research suggested a more efficient and effective means of pricing the barrier call options, both with and without a rebate, by implementing the ANN techniques on the closed-form solution of these option styles. Barrier options belong to exotic financial options whose analytical solutions are based on the extended Black–Scholes pricing models. Analytical solutions are known to possess assumptions which are not often valid in the real world, and these limitations make them ideally imperfect in the effective valuation of financial derivatives. Hence, through the findings of this research, we were able to show that neural networks can be employed efficiently in the computation and the prediction of unbiased prices for both the rebate and non-rebate barrier options. This study showed that it is possible to utilize an efficient approximation method via the concept of ANN in estimating exotic option prices, which are more complex and often require expensive computational time. This research has provided an in-depth concept into the practicability of the deep learning technique in derivative pricing. This was made viable through some statistical and exploratory data analysis and analysis of the model training provided.

From the research, we conducted some benchmarking experiments on the NN hyperparameter tuning using the Keras interface and used different evaluation metrics to measure the performance of the NN algorithm. We finally estimated the optimal NN architecture, which prices the barrier options effectively in connection to some data-splitting techniques. We compared six models in terms of their data split and their hyperparameter search algorithm. The optimal NN model was constructed using the cross-validation data-split and the Bayesian optimization search algorithm, and this combination was more efficient than the other models proposed in this research. Next, we compared the results from the optimal NN model to those produced by other ML models, such as the random forest and the polynomial regression; the output highlights the accuracy and the efficiency of our proposed methodology in this option pricing problem.

Finally, hedging and risk management of barrier options are complicated due to their exotic nature, especially as the underlying is near the barrier. Our research extracted the barrier option prices and their corresponding Greeks with high accuracy using the optimal hyperparameter. The predicted and accurate results showed little or no difference, which explains our proposed model's effectiveness. For future research direction, more theoretical underpinning seems to be lacking in connection to the evaluation/error analysis for all the proposed models used in this research. Another limitation of this work is the use of a fully simulated dataset; it will suffice to implement these techniques on a real dataset to estimate the effectiveness. The third limitation of this research lies in the convergence analysis of the proposed NN scheme, and future research will address this issue. In addition, more research can be conducted to value these exotic barrier options from the partial differential perspective, that is, solving the corresponding PDE from this model using the ANN techniques and extending the pricing methodology to other exotic options, such as the Asian or the Bermudian options.

Author Contributions: These authors contributed equally to this work. All authors have read and agreed to the published version of the manuscript.

Funding: This research received no external funding.

Data Availability Statement: The data supporting this study's findings are available from the corresponding author upon reasonable request.

Acknowledgments: This work commenced while the first author was affiliated with the Center for Business Mathematics and Informatics, North-West University, Potchefstroom and the University of Johannesburg, both in South Africa. The authors wish to acknowledge their financial support in collaboration with the Deutscher Akademischer Austauschdienst (DAAD).

Conflicts of Interest: The author declares that they have no competing interests.

References

1. Aziz, S.; Dowling, M.M.; Hammami, H.; Piepenbrink, A. Machine learning in finance: A topic modeling approach. *European Financ. Manag.* **2022**, *28*, 744–770 [CrossRef]
2. Liu, S.; Borovykh, A.; Grzelak, L.A.; Oosterlee, C.W. A neural network-based framework for financial model calibration. *J. Math. Ind.* **2019**, *9*, 9. [CrossRef]
3. Beck, C.; Becker, S.; Grohs, P.; Jaafari, N.; Jentzen, A. Solving stochastic differential equations and Kolmogorov equations by means of deep learning. *arXiv* **2018**, arXiv:1806.00421.
4. Borovykh, A.; Bohte, S.; Oosterlee, C.W. Conditional time series forecasting with convolutional neural networks. *arXiv* **2017**, arXiv:1703.04691.
5. Babbar, K.; McGhee, W.A. A Deep Learning Approach to Exotic Option Pricing under LSVol; University of Oxford Working Paper. 2019. Available online: https://www.bayes.city.ac.uk/__data/assets/pdf_file/0007/494080/DeepLearningExoticOptionPricingLSVOL_KB_CassBusinessSchool_2019.pdf (accessed on 20 November 2022).
6. Sirignano, J.; Spiliopoulos, K. DGM: A deep learning algorithm for solving partial differential equations. *J. Comput. Phys.* **2018**, *375*, 1339–1364. [CrossRef]
7. Liu, S.; Oosterlee, C.W.; Bohte, S.M. Pricing options and computing implied volatilities using neural networks. *Risks* **2019**, *7*, 16. [CrossRef]
8. Yao, J.; Li, Y.; Tan, C.L. Option price forecasting using neural networks. *Omega* **2000**, *28*, 455–466. [CrossRef]
9. Li, C. The Application of Artificial Intelligence and Machine Learning in Financial Stability. In Proceedings of the International Conference on Machine Learning and Big Data Analytics for IoT Security and Privacy, Shanghai, China, 6–8 November 2020; pp. 214–219.
10. Lee, J.; Kim, R.; Koh, Y.; Kang, J. Global stock market prediction based on stock chart images using deep Q-network. *IEEE Access* **2019**, *7*, 167260–167277. [CrossRef]
11. Yu, P.; Yan, X. Stock price prediction based on deep neural networks. *Neural Comput. Appl.* **2020**, *32*, 1609–1628. [CrossRef]
12. Malliaris, M.; Salchenberger, L. A neural network model for estimating option prices. *Appl. Intell.* **1993**, *3*, 193–206. [CrossRef]
13. Black, F.; Scholes, M. The pricing of options and corporate liabilities. *J. Political Econ.* **1973**, *81*, 637–654. [CrossRef]
14. Klibanov, M.V.; Golubnichiy, K.V.; Nikitin, A.V. Application of Neural Network Machine Learning to Solution of Black-Scholes Equation. *arXiv* **2021**, arXiv:2111.06642.
15. Fang, Z.; George, K.M. Application of machine learning: An analysis of Asian options pricing using neural network. In Proceedings of the 2017 IEEE 14th International Conference on e-Business Engineering (ICEBE), Shanghai, China, 4–6 November 2017; pp. 142–149.
16. Hutchinson, J.M.; Lo A.W.; Poggio, T. A non-parametric approach to pricing and hedging derivative securities via learning networks. *J. Financ.* **1994**, *49*, 851–889. [CrossRef]

17. Bennell, J.; Sutcliffe, C. Black–Scholes versus artificial neural networks in pricing FTSE 100 options. *Intell. Syst. Acc. Financ. Manag. Int. J.* **2004**, *12*, 243–260. [CrossRef]
18. Yadav, K. Formulation of a rational option pricing model using artificial neural networks. In Proceedings of the SoutheastCon 2021, Virtual, 10–14 March 2021; pp. 1–8.
19. Ghaziri, H.; Elfakhani, S.; Assi, J. Neural, networks approach to pricing, options. *Neural Netw. World* **2000**, *1*, 271–277.
20. Anders, U.; Korn, O.; Schmitt, C. Improving the pricing of options: A neural network approach. *J. Forecast.* **1998**, *17*, 369–388. [CrossRef]
21. Buehler, H.; Gonon, L.; Teichmann, J.; Wood, B.; Mohan, B.; Kochems, J. *Deep Hedging: Hedging Derivatives under Generic Market Frictions using Reinforcement Learning*; SSRN Scholarly Paper ID 3355706; Social Science Research Network: Rochester, NY, USA, 2019.
22. Culkin, R.; Das, S.R. Machine learning in finance: The case of deep learning for option pricing. *J. Invest. Manag.* **2017**, *15*, 92–100.
23. De Spiegeleer, J.; Madan, D.B.; Reyners, S.; Schoutens, W. Machine learning for quantitative finance: Fast derivative pricing, hedging and fitting. *Quant. Financ.* **2018**, *18*, 1635–1643. [CrossRef]
24. Gan, L.; Wang, H.; Yang, Z. Machine learning solutions to challenges in finance: An application to the pricing of financial products. *Technol. Forecast. Soc. Change* **2020**, *153*, 119928. [CrossRef]
25. Hamid, S.A.; Habib, A. *Can Neural Networks Learn the Black-Scholes Model? A Simplified Approach*; Working Paper No. 2005–01; School of Business, Southern New Hampshire University: Manchester, NH, USA, 2005.
26. Le N.T.; Zhu, S.P.; Lu, X. An integral equation approach for the valuation of American-style down-and-out calls with rebates. *Comput. Math. Appl.* **2016**, *71*, 544–564. [CrossRef]
27. Umeorah, N.; Mashele, P. A Crank-Nicolson finite difference approach on the numerical estimation of rebate barrier option prices. *Cogent Econ. Financ.* **2019**, *7*, 1598835. [CrossRef]
28. Han, J.; Jentzen, A.; Weinan, E. Solving high-dimensional partial differential equations using deep learning. *Proc. Natl. Acad. Sci. USA* **2018**, *115*, 8505–8510. [CrossRef] [PubMed]
29. Ganesan, N.; Yu, Y.; Hientzsch, B. Pricing barrier options with DeepBSDEs. *arXiv* **2020**, arXiv:2005.10966.
30. Yu, B.; Xing, X.; Sudjianto, A. Deep-learning based numerical BSDE method for barrier options. *arXiv* **2019**, arXiv:1904.05921.
31. Itkin, A. Deep learning calibration of option pricing models: Some pitfalls and solutions. *arXiv* **2019**, arXiv:1906.03507.
32. Xu, L.; Dixon, M.; Eales, B.A.; Cai, F.F.; Read, B.J.; Healy, J.V. Barrier option pricing: Modelling with neural nets. *Phys. Stat. Mech. Its Appl.* **2004**, *344*, 289–293. [CrossRef]
33. Ghevariya, S. PDTM approach to solve Black Scholes equation for powered ML-Payoff function. *Comput. Methods Differ. Equ.* **2022**, *10*, 320–326.
34. Mehdizadeh, Khalsaraei, M.; Shokri, A.; Mohammadnia, Z.; Sedighi, H.M. Qualitatively Stable Nonstandard Finite Difference Scheme for Numerical Solution of the Nonlinear Black–Scholes Equation. *J. Math.* **2021**, *2021*, 6679484.
35. Rezaei, M.; Yazdanian, A. Pricing European Double Barrier Option with Moving Barriers Under a Fractional Black–Scholes Model. *Mediterr. J. Math.* **2022**, *19*, 1–16. [CrossRef]
36. Torres-Hernandez, A.; Brambila-Paz, F.; Torres-Martínez, C. Numerical solution using radial basis functions for multidimensional fractional partial differential equations of type Black–Scholes. *Comput. Appl. Math.* **2021**, *40*, 1–25. [CrossRef]
37. Eskiizmirliler, S.; Günel, K.; Polat, R. On the solution of the black–scholes equation using feed-forward neural networks. *Comput. Econ.* **2021**, *58*, 915–941. [CrossRef]
38. Chen, Y.; Yu, H.; Meng, X.; Xie, X.; Hou, M.; Chevallier, J. Numerical solving of the generalized Black-Scholes differential equation using Laguerre neural network. *Digit. Signal Process.* **2021**, *112*, 103003. [CrossRef]
39. Rich, D.R. The mathematical foundations of barrier option-pricing theory. *Adv. Futur. Options Res.* **1994**, *7*, 267–311.
40. Zhang, P.G. *Exotic Options: A Guide to Second Generation Options*; World Scientific: Singapore, 1997.
41. Kononenko, I.; Kukar, M. *Machine Learning and Data Mining*; Horwood Publishing: Chichester, UK, 2007
42. Reitermanová, Z. Data Splitting. In Proceedings of the WDS'10—19th Annual Conference of Doctoral Students, Prague, Czech Republic, 1–4 June 2010.
43. Breiman, L. Random forests. *Mach. Learn.* **2001**, *45*, 5–32. [CrossRef]
44. Segal, M.R. *Machine Learning Benchmarks and Random Forest Regression*; Division of Biostatistics, University of California: San Francisco, CA, USA, 2004.
45. Blasco, B.C.; Moreno, J.J.M.; Pol, A.P.; Abad, A.S. Using the R-MAPE index as a resistant measure of forecast accuracy. *Psicothema* **2013**, *25*, 500–506.
46. Oshiro, T.M.; Perez, P.S.; Baranauskas, J.A. How many trees in a random forest?. In Proceedings of the Machine Learning and Data Mining in Pattern Recognition: 8th International Conference, MLDM 2012, Berlin, Germany, 13–20 July 2012; pp. 154–168.

Article

Developing a Deep Learning-Based Defect Detection System for Ski Goggles Lenses

Dinh-Thuan Dang [1,2] and Jing-Wein Wang [3,*]

1   Department of Electronics Engineering, National Kaohsiung University of Science and Technology, Kaohsiung 80778, Taiwan
2   Department of Information Technology, Pham Van Dong University, Quang Ngai 57000, Vietnam
3   Institute of Photonics Engineering, National Kaohsiung University of Science and Technology, Kaohsiung 80778, Taiwan
*   Correspondence: jwwang@nkust.edu.tw

Abstract: Ski goggles help protect the eyes and enhance eyesight. The most important part of ski goggles is their lenses. The quality of the lenses has leaped with technological advances, but there are still defects on their surface during manufacturing. This study develops a deep learning-based defect detection system for ski goggles lenses. The first step is to design the image acquisition model that combines cameras and light sources. This step aims to capture clear and high-resolution images on the entire surface of the lenses. Next, defect categories are identified, including scratches, watermarks, spotlight, stains, dust-line, and dust-spot. They are labeled to create the ski goggles lenses defect dataset. Finally, the defects are automatically detected by fine-tuning the mobile-friendly object detection model. The mentioned defect detection model is the MobileNetV3 backbone used in a feature pyramid network (FPN) along with the Faster-RCNN detector. The fine-tuning includes: replacing the default ResNet50 backbone with a combination of MobileNetV3 and FPN; adjusting the hyper-parameter of the region proposal network (RPN) to suit the tiny defects; and reducing the number of the output channel in FPN to increase computational performance. Our experiments demonstrate the effectiveness of defect detection; additionally, the inference speed is fast. The defect detection accuracy achieves a mean average precision (mAP) of 55%. The work automatically integrates all steps, from capturing images to defect detection. Furthermore, the lens defect dataset is publicly available to the research community on GitHub. The repository address can be found in the Data Availability Statement section.

Keywords: ski goggles lenses; surface defect; automatic optical inspection; Faster-RCNN; fine-tune; MobileNetV3; FPN; RPN

MSC: 68T07, 68T20, 68T45

Citation: Dang, D.-T.; Wang, J.-W. Developing a Deep Learning-Based Defect Detection System for Ski Goggles Lenses. *Axioms* **2023**, *12*, 386. https://doi.org/10.3390/axioms12040386

Academic Editor: Oscar Humberto Montiel Ross

Received: 31 January 2023
Revised: 5 April 2023
Accepted: 13 April 2023
Published: 17 April 2023

1. Introduction

Winter sports such as skiing, snowboarding, and snowshoeing offer great enjoyment, and ski goggles are the necessary equipment to perform better in these activities. There are many advantages that can help protect the eyes from harmful ultraviolet rays, provide both facial and ocular safety protection, and offer color and contrast enhancement. Figure 1 shows some samples of ski goggles.

Figure 1. Some lens samples of ski goggles in different sizes, curvatures, and colors.

The most critical component of ski goggles is the lenses, which offer an unrivaled visual experience. In the manufacturing process of ski goggles, there are unavoidable defects [1,2] on the surface of the lenses. Therefore, lens manufacturers should implement a visual defect inspection system [3,4] to enhance product quality.

There are some challenging issues in the defect inspection for the lenses of ski goggles. Firstly, the lens surfaces are curved and vary in size, making it difficult to design an image acquisition system that captures their entire surface. As depicted in Figure 2, the usual lens samples exhibit distinct curvatures and sizes. Secondly, the lenses are coated with various tints, presenting a challenge in customizing the light source due to multiple reflections. Thirdly, some surface defects are extremely small, which are inconspicuous and subtle. Thus, the detection of such minor defects is particularly difficult and requires higher precision.

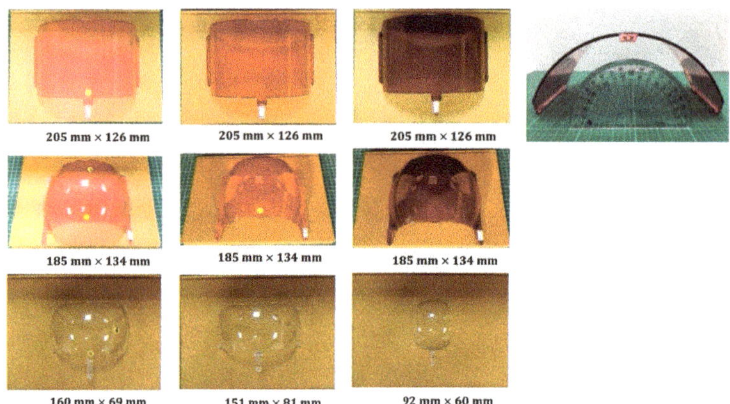

Figure 2. Some standard sizes of ski goggles lenses. The width ranges from 92 to 205 mm, and the height ranges from 60 to 126 mm.

Artificial intelligence (AI) technology can help organizations gain an edge over their competitors [5]. AI has proven especially beneficial for improving product quality and lowering costs. For manufacturers, AI promises benefits at every level of the value chain. The AI-based system can detect defects faster and more accurately than the human eye. A typical industrial visual inspection system based on deep learning incorporates several fundamental components to facilitate accurate and efficient product quality assessment. These components, critical to the system's operation, include:

- Image Acquisition Devices: High-resolution cameras [6,7], often fitted with specialized lenses, capture images of inspected items. These devices may employ various imaging technologies, such as monochrome, color, or infrared, contingent on the application's demands.
- Lighting: Customized illumination sources [8], including LED lights [9] or lasers, are employed to enhance the contrast and visibility of features under inspection. The choice and configuration of lighting are instrumental in achieving optimal image quality for precise defect detection.
- Deep Learning Algorithms: Deep learning-based defect detection typically requires training object detection models or alternative specialized architectures on the extensively labeled datasets of defect images. Object detection methodologies have been extensively applied in the detection of defects on the surfaces of industrial products, such as steel, plastic, wood, and silk [10–12]. The task of object detection in computer vision encompasses two primary functions: localization [13] and classification [14]. In traditional computer vision, classifiers [15] such as SVM, KNN, and K-means clus-

tering have played a vital role in categorizing classes. Meanwhile, object localization mainly employs fast template-matching-based algorithms [16].

In recent years, significant progress has been made in neural networks, machine learning, and deep learning. Classifiers such as ResNet [17], VGG [18], EfficientNet [19], and Vision Transformer [20] have achieved state-of-the-art results for classification tasks. Concurrently, object localization has been applied to anchor-based and anchor-free methods [21]. The anchor-based approach, known as the two-stage object detector, includes the Faster R-CNN [22] family. The anchor-free technique, referred to as the one-stage object detectors, comprises models such as YOLO [23] and FCOS [24].

The remarkable success in this field stems from the seamless integration of localization and classification tasks in deep learning. To gain a deeper insight, it is crucial to understand the Faster R-CNN architecture's automatic pipeline. The Faster R-CNN model systematically combines customizable neural sub-network blocks, including the backbone block, region proposal network [25], and ROI-head block.

Figure 3 illustrates the fundamental components of an optical initialization system, which include a light source, camera, and hardware to execute image processing algorithms.

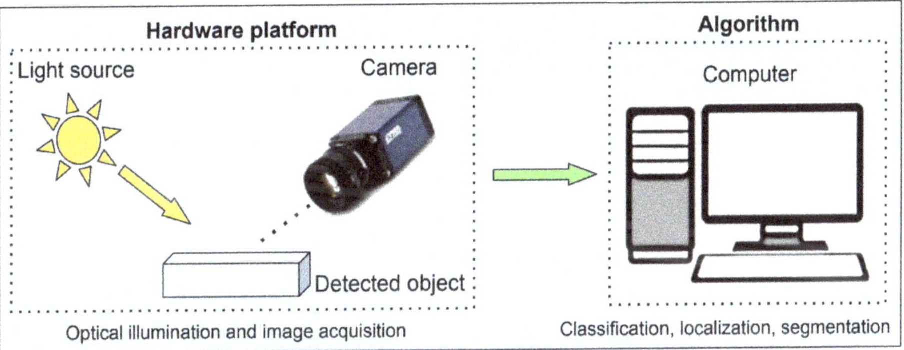

Figure 3. Some main components in the optical inspection system.

In light of the aforementioned challenges and emerging trends, this paper aims to develop an automatic defect detection system for ski goggles lenses, utilizing deep learning techniques. To accomplish this objective, the following steps are undertaken:

(1) Design of an image acquisition model that integrates cameras and light sources to effectively capture the entire surface of ski goggles lenses.
(2) Identification of lens defect categories and construction of a comprehensive ski goggles lens defect dataset.
(3) Fine-tuning of the integrated object detection model that combines Faster R-CNN, FPN, and MobileNetV3 by implementing the following modifications: replacement of the default ResNet50 backbone with a combination of MobileNetV3 and feature pyramid network (FPN) to optimize computational efficiency and performance; adjustment of the region proposal network (RPN) hyperparameters to accommodate the detection of minuscule defects; and a reduction of the output channel count in the FPN to enhance computational performance without sacrificing accuracy.

By executing these steps, the paper presents a novel deep learning-based approach for detecting defects on ski goggles lenses, demonstrating potential applicability to various manufacturing quality control scenarios.

The structure of the paper is organized as follows. Section 1 provides an introduction to the study. Section 2 introduces the image acquisition technique, data labeling for defects, and the defect detection method. Sections 3 and 4 present the research results and subsequent discussions, respectively. Finally, Section 5 offers concluding remarks and summarizes the paper's findings.

2. Materials and Methods

The deep learning-based defect detection system for ski goggles lenses presented in this study is depicted in Figure 4. The system comprises two modules: the image acquisition module and the defect detection module. The former is responsible for capturing images of ski goggles lenses, extracting regions of interest, and labeling data. The latter trains the customized model using input data and detects defects during inference.

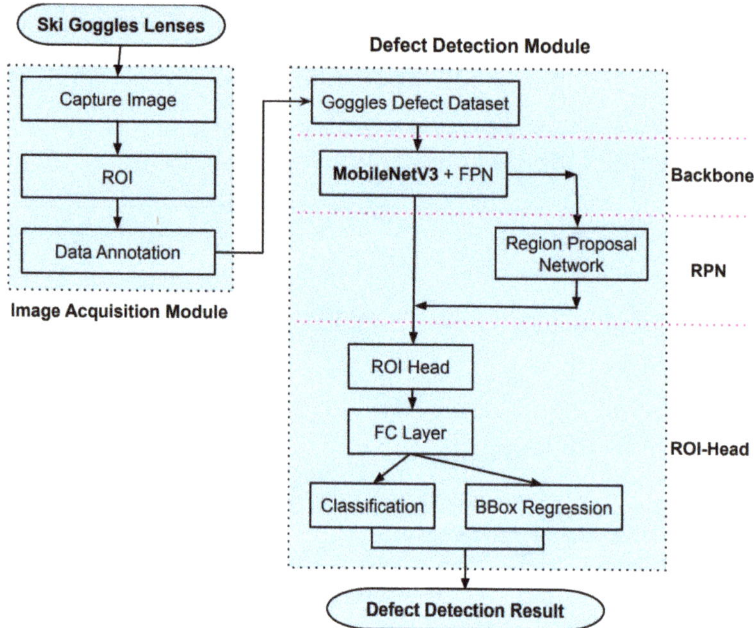

Figure 4. The flowchart of the proposed method. There are two modules for processing images. The first one is the image acquisition to capture the raw image, extract regions of interest, and data labeling. The second module is defect detection, which involves training data and inferencing defects. This module combines Faster R-CNN, MobileNetV3, and FPN to create the customized end-to-end model. It is compatible with data of the ski goggles lenses.

Image Acquisition Module: This module is responsible for obtaining high-quality images of ski goggles lenses. An optimized image acquisition setup, which combines cameras and light sources, is employed to ensure the entire surface of the lenses is captured with minimal glare and distortion. The regions of interest are extracted from the captured images, and the data are meticulously labeled to identify and categorize defects present in the lenses.

Defect Detection Module: A customized object detection model, based on the Faster R-CNN architecture, is designed and integrated into this module. The model involves replacing the backbone of Faster R-CNN with MobileNetV3 and integrating FPN for efficient feature extraction and multi-scale representation. The customized model is trained using the labeled input data and subsequently employed for detecting defects during the inference phase.

2.1. Image Acquisition Module

This module aims to collect accurate and high-quality images from the surface of the ski goggles lenses. It also prepares the well-formatted data for the next module.

2.1.1. Capture Image

The first part of the module is image capture, which consists of cameras, light sources, and ski goggles lens samples. To design the image capture system for the ski goggles, lenses need to overcome some challenges mentioned in Section 1. The surface of the lens is broad, and one camera cannot cover the whole of the lens surface. Therefore, we designed the image acquisition system using five cameras. Each camera will focus on each region marked on the lens, as in Figure 5.

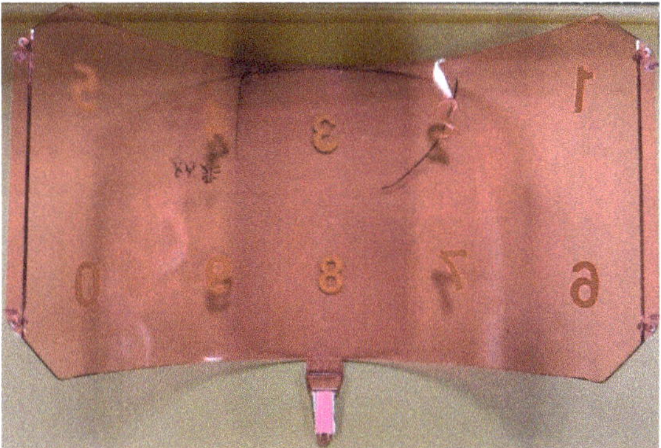

Figure 5. The ski goggles' lens sample. It is wide and curved; thus, we mark its surface to be easily controlled by cameras.

Furthermore, the surface is also curved, so we designed the custom light source as in Figure 6. The curvature of the light source is similar to the curvature of the lens, which helps to reflect uniform rays over the lens surface. The custom light source has five pieces of flat LED lights connected by an angle of 125 degrees. Figure 7 describes the detailed design diagram of the image acquisition system. The ray of each flat LED piece transits through the lens to the cameras opposite, respectively. Figure 8 depicts the actual pieces of equipment when deployed.

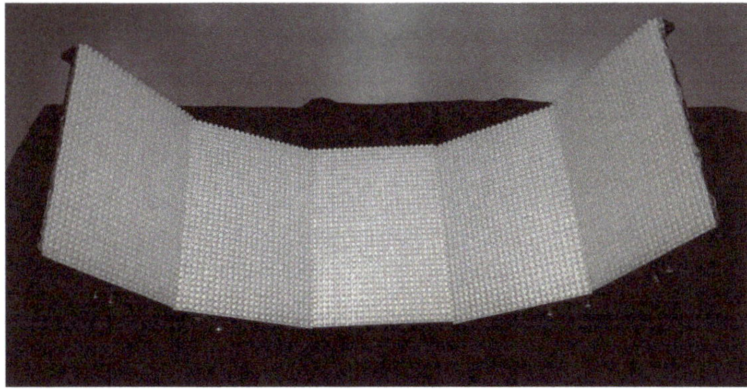

Figure 6. The most that a custom light source meets the curvature of the ski goggles lens. Five dot matrix LED modules are connected by an angle of 125°.

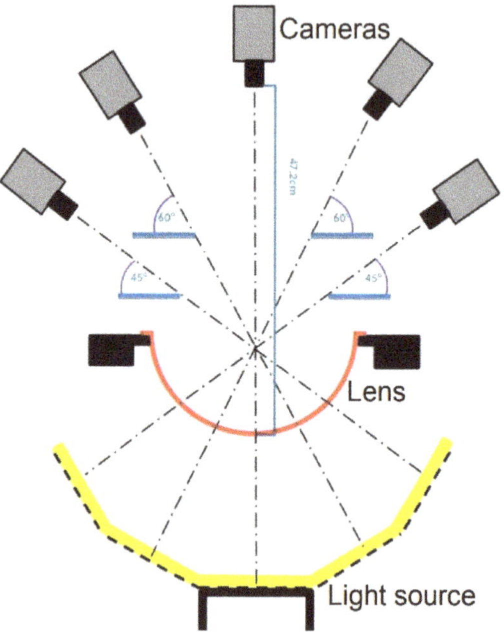

Figure 7. Design diagram of the image acquisition system. Five cameras are placed at the top. The custom light source (yellow) is placed at the bottom. The ski goggles' lens sample (red) is positioned in the middle and held on two sides.

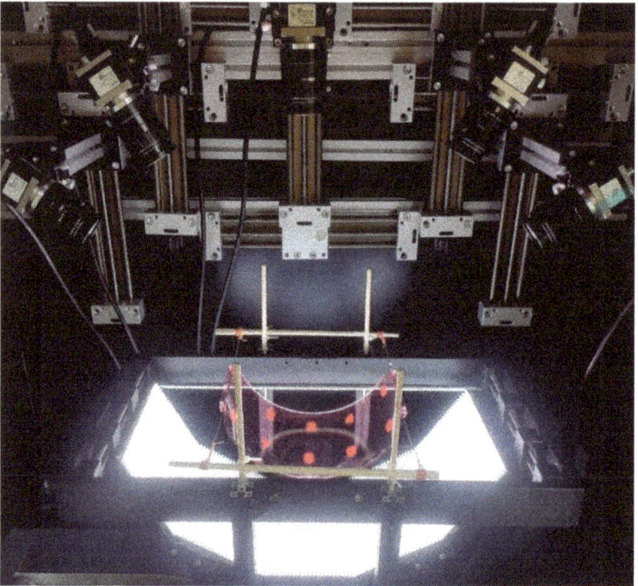

Figure 8. The actual model of image acquisition. The bottom is a custom light source that shines through the ski goggles lens surface to the cameras. The computer controls the five cameras through the acquisition card. The developed program will access the card's interface to capture images simultaneously.

The parameters of a camera, such as resolution, sensor, pixel size, and frame rate, are significantly influential in designing the distance from the camera lens to the inspected object. Table 1 lists the parameters of the image acquisition system.

Table 1. The equipment description of the image acquisition system.

Equipment	Producer	Specification
Camera	Basler	Model acA4112-30uc, sensor Sony IMX352, resolution 12 mp, pixel size 3.45 × 3.45 μm, frame rate 30 fps.
Acquisition Card	Basler	USB 3.0 Interface Card PCIe, Fresco FL1100, 4HC, x4, 4Ports. Data transfer with rates of up to 380 MB/s per port.
Vision Lens	Tokina	Model TC3520-12MP, image format 4/3 inch, mount C, focal length 35 mm, aperture range F2.0-22.
Light Source	Custom	The custom-designed light source comprises five-dot matrix LED modules that are connected by an angle of 125°.
Computer	Asus	Windows 10 Pro; hardware based on: mainboard Asus Z590-A, CPU Intel I7-11700K, RAM 16G, VGA gigabyte RTX 3080Ti 12 GB.

For ease of visualization, Figure 9 shows the input and output of the system. Inputs are lens samples. The system captures its surface and outputs images of the lens surface. We also developed a Python program to simultaneously control five cameras and automatically crop the areas of interest (ROI).

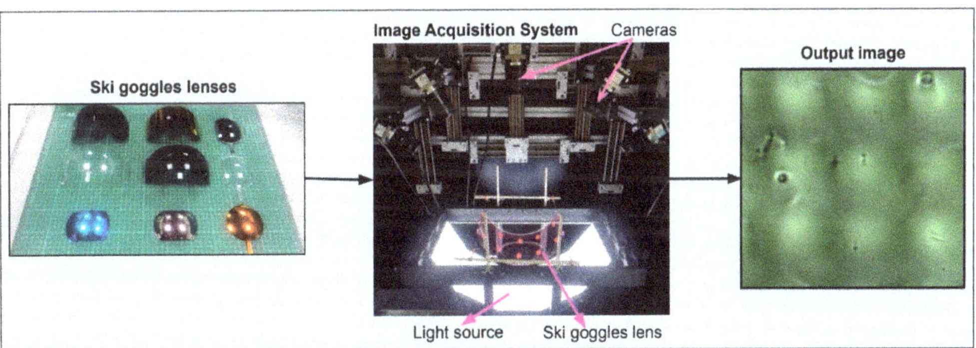

Figure 9. Input and Output of image acquisition. Input is some lens samples of ski goggles, and output is raw images from five cameras.

2.1.2. Regions of Interest

The system uses five cameras to capture the whole of the lens surface, and each camera only focuses on a portion of the lens. There are, however, limitations in the experiment, such as that the cameras can capture the overlapping or out-boundary parts. Therefore, we need to generate the ROIs from each raw image so that when stitched together, they become the image of the whole lens. The first line of Figure 10 shows five natural photos taken from the cameras, each containing redundant portions such as overlaps or areas outside of the lens. The five below images are the results of creating ROIs, respectively. We developed a program to capture images and generate ROIs seamlessly. The program was inherited from the Pypylon package of Basler and PyTorch framework.

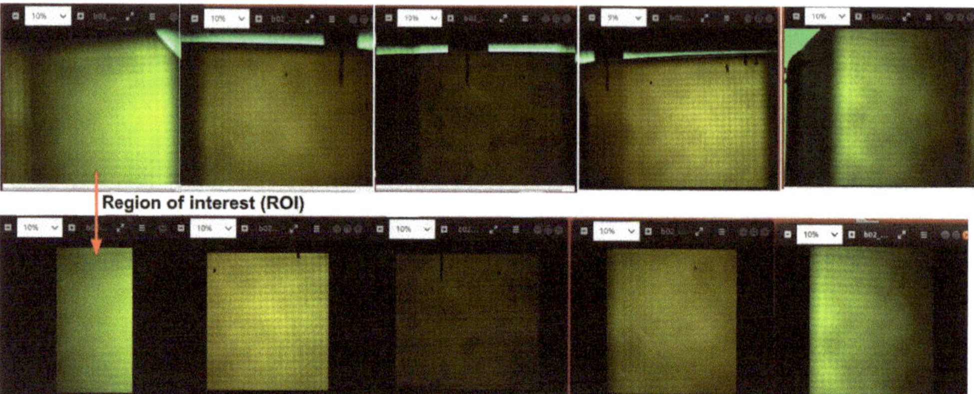

Figure 10. Crop regions of interest from raw images. Five cameras capture five images in the first row. To facilitate data labeling, extracting parts of good images is necessary. The **bottom** row is five regions of interest.

2.1.3. Data Labeling

Based on our practical experience with the imaging system and discussing with the ski goggles manufacturer, we conclude that there are the following common types of defects on the surface of ski goggles lenses: scratch, watermark, spotlight, stain, dust-line, and dust-spot. Figure 11 illustrates the detailed defect types.

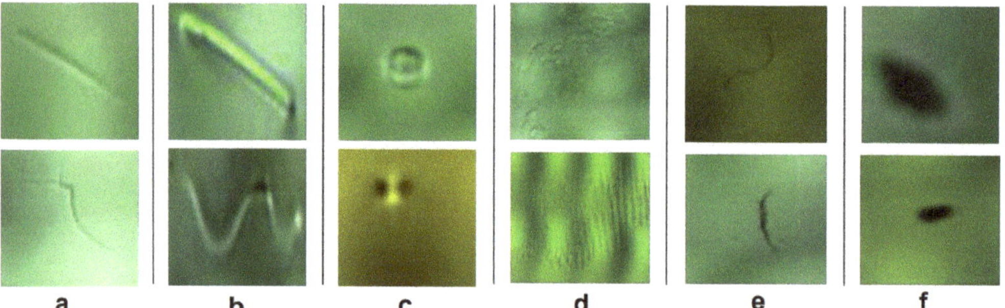

Figure 11. Some defect samples on the surface of ski goggles lenses. Column (**a**): scratch defects, (**b**): watermark, (**c**): spotlight, (**d**): stain, (**e**): dust-line, (**f**): dust-spot.

Because detect detection is based on supervised deep learning methods, the image data need to be labeled for the training phase. We used the LabelMe tool [26] to mark the defect regions with bounding boxes. Figure 12 illustrates the defect-labeling interface using the label tool.

From the 37 ski goggles lens pieces provided by the manufacturer, the image acquisition system captured and created a total of 654 images of 1330×800 pixels in size. We carry out defect labeling for the defect detection task. As outlined in Table 2, the count of labeled defects constitutes the initial dataset.

It is crucial to acknowledge that the distribution of defects in the dataset is imbalanced, with dust-line being the most prevalent (7292 instances) and watermark being the least prevalent (120 instances). This imbalance may result in a biased model. Consequently, the flip technique is employed to generate supplementary synthetic data. The quantity of underrepresented defect categories, including spotlight, stain, and watermark, is expected to increase. As depicted in Figure 13, the synthetic image is generated from a small batch of defects and backgrounds. A total of 200 synthetic images are generated and subsequently labeled. Table 3 presents the statistics regarding the categories of defects in the synthetic dataset.

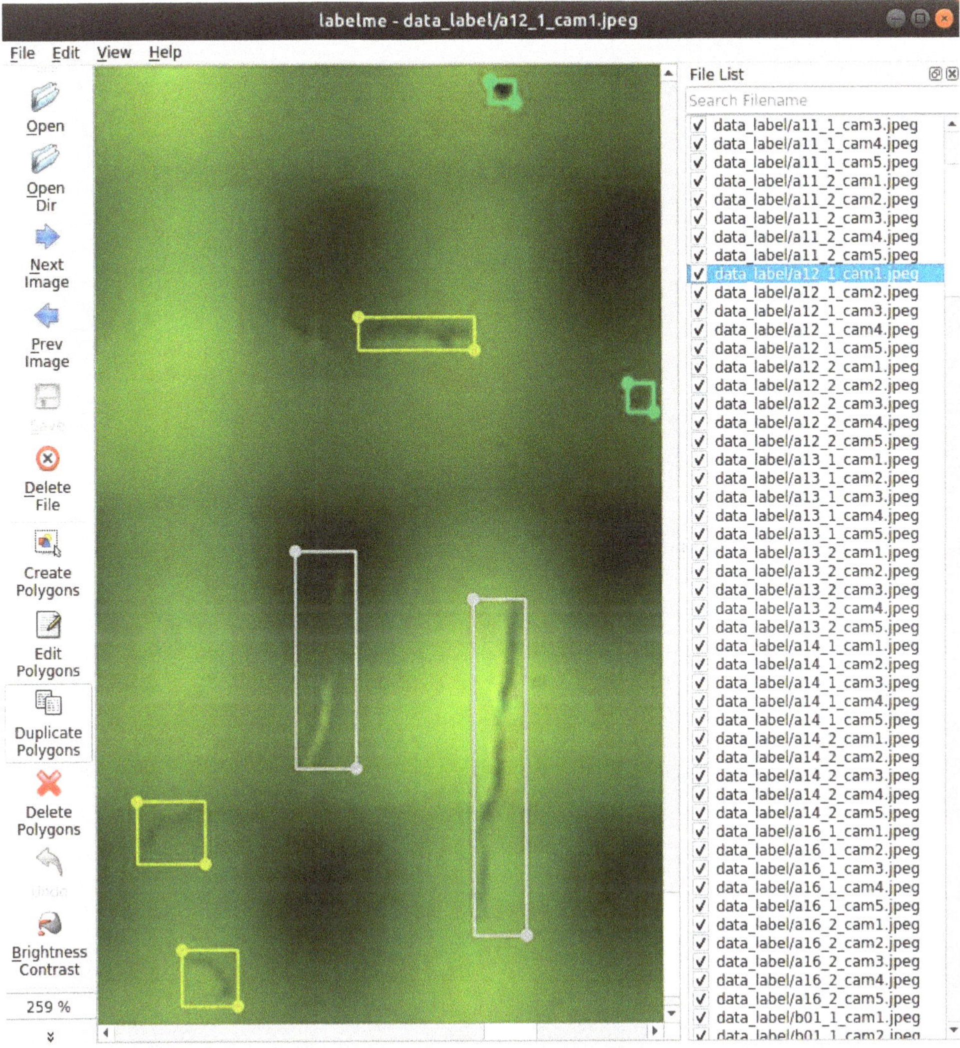

Figure 12. The GUI of LabelMe: the image annotation tool used to label defects on the surface of ski goggles lenses.

Table 2. The defect detection dataset of ski goggles lenses: defect type and its respective quantity.

Type	Defects	Type	Defects	Type	Defects
scratch	1972	spotlight	229	dust-line	7292
watermark	120	stain	281	dust-spot	1898
Total	11,792				

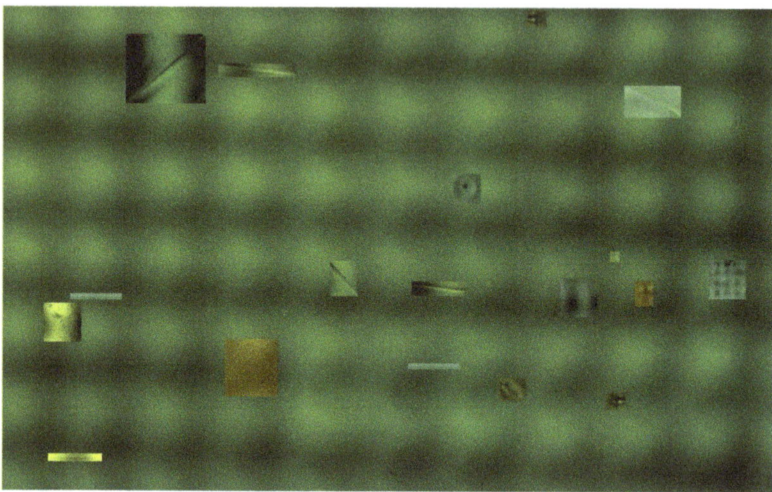

Figure 13. Synthetic image generated from the flip technique.

Table 3. The statistics of the number of defects in the synthetic dataset.

Type	Defects	Type	Defects	Type	Defects
scratch	0	spotlight	1093	dust-line	0
watermark	973	stain	1328	dust-spot	0
Total	3394				

Imbalanced and underrepresented data are a common challenge in collecting real-world data, which may impact detection results. To address this issue, a synthetic dataset is generated and subsequently merged with the initial dataset, forming a combined dataset as detailed in Table 4.

Table 4. The statistics of the defect categories from the combined dataset, which merges the initial and synthetic datasets.

Type	Defects	Type	Defects	Type	Defects
scratch	1972	spotlight	1322	dust-line	7292
watermark	1093	stain	1609	dust-spot	1898
Total	3394				

For a comprehensive and in-depth analysis of the dataset, Table 5 enumerates the number of images corresponding to each defect category.

Table 5. The number of images containing each defect category is extracted from the JSON file containing the labels.

Defect Type	Scratch	Watermark	Spotlight	Stain	Dust-Line	Dust-Spot
Images	447	199	352	316	546	612
Instances	1972	1093	1322	1609	7292	1898

The next section will describe the defect detection model, which is trained and utilized for inference using the aforementioned dataset.

2.2. Defect Detection Module

Finding a suitable object detection model for each data type is difficult. Faster R-CNN architecture is the two-state object detector which has proven to have high accuracy and be end-to-end trainable. Our work is to fine-tune this architecture by integrating the MobileNetV3 [27] backbone and feature pyramid network for extracting multi-scale features [28]. MobileNetv3 model is a lightweight neural network suitable for devices with a limited computational resource budget. Furthermore, we also reduce the number of channels to reduce latency in inference.

The following sections will cover the overview method of supervised machine learning theory, and the overview of the integrated Faster-RCNN architecture is shown in Figure 4. The backbone, RPN, and ROI-Head are the three main sub-networks in the defect detection module. First, the backbone block combines MobilenetV3 and FPN to extract multi-scale feature maps. Second, the RPN will create and propose the candidate defect regions. Finally, the ROI-Head block will locate the position of defects and classify them. Related theories, such as the bounding box regression, binary classification, multiclass classification, and assigning the boxes to the level of feature maps, are also discussed in detail.

2.2.1. Object Detection Problem Setting Based on Supervised Learning Approach

There are various machine-learning paradigms, such as supervised, unsupervised, and reinforcement learning. Because of the tasks related to detecting and classifying defects, we apply the supervised learning approach. This direction is related to the input data, labels, generative networks, loss functions, and measure metrics. This section describes the basic theory of supervised learning.

Description: When given an image, determine whether or not there are instances of objects from predefined classes and, if present, return the bounding box of each instance.

Input: A collection of N annotated images X_{train} and a label set Y_{train}.

$$X_{train} = \{x_1, x_2, \ldots, x_N\} \tag{1}$$

$$Y_{train} = \{y_1, y_2, \ldots, y_N\} \tag{2}$$

where y_i is annotation in image x_i, and each y_i has M_i objects belong to C classes.

$$y_i = \left\{ (b_1^i, c_1^i), (b_2^i, c_2^i), \ldots, (b_{M_k}^i, c_{M_k}^i) \right\} \tag{3}$$

where b_j^i and c_j^i denote the bounding box of jth object in x_i and the class, respectively.

Algorithm: Optimize the loss function L of classification L_{cls} and bounding-box regression $L_{box-reg}$:

$$L = L_{cls} + L_{box-reg} \tag{4}$$

Formally, L_{box} is based on the sum of squared errors (SSE) loss function, and L_{cls} is based on the cross-entropy loss function. The loss function is optimized by training the neural network after a specific amount of epochs.

Prediction: For x_{test}^i, the prediction result is y_{pred}^i,

$$y_{pred}^i = \left\{ (b_{pred_1}^i, c_{pred_1}^i, p_{pred_1}^i), (b_{pred_2}^i, c_{pred_2}^i, p_{pred_2}^i), \ldots \right\} \tag{5}$$

where $b_{pred_j}^i$, $c_{pred_j}^i$, $p_{pred_j}^i$ are results of the bounding box, object class, and reliability. For filtering the object detection results, we use a predefined threshold that compares the reliability.

Evaluation metric: The primary metric used to evaluate the object detection algorithms' performance is the mean average precision (mAP). This metric considers the prediction of correct category labels and accuracy location. There are two main performance evaluation criteria: precision (P) and recall ($Recall$). The statistic of true positives (TP), false positives (FP), true negatives (TN), and false negatives (FN) are needed to measure the P and $Recall$ values of the network model in the testing phase. The intersection-over-union (IoU) is a critical concept to determine whether the test results are correct or not. TP, FP, TN, and FN values depend on the IoU threshold. The formula of IoU is defined in Equation (6).

$$IoU(b_{pred}, b_{lb}) = \frac{Area(b_{pred} \cap b_{lb})}{Area(b_{pred} \cup b_{lb})} \tag{6}$$

The P and $Recall$ of each category of one image can be calculated as follows:

$$P_{C_{ij}} = \frac{TP_{C_{ij}}}{TP_{C_{ij}} + FP_{C_{ij}}} \tag{7}$$

$$Recall_{C_{ij}} = \frac{TP_{C_{ij}}}{TP_{C_{ij}} + FN_{C_{ij}}} \tag{8}$$

where $P_{C_{ij}}$ and $Recall_{C_{ij}}$ represent *Precision* and *Recall* of category C_{ij} in the jth image, respectively. The average precision (AP) of the category C_i can be calculated:

$$AP_{C_i} = \frac{1}{m} \sum_{j=1}^{m} P_{C_{ij}} \tag{9}$$

The dataset has multiple categories, the mAP of the entire category can be calculated as follows:

$$mAP = \frac{1}{n} \sum_{i=1}^{n} AP_{C_{ij}} \tag{10}$$

There are also many other criteria, but in this work's scope, the performance evaluation is measured mainly by the mAP metric.

2.2.2. Backbone: Feature Extractor Based on MobileNetV3 and Feature Pyramid Networks

MobileNetV3: It is important to emphasize the integration of the MobileNetV3 model into the faster R-CNN architecture by its suitability for optic inspection systems [29]. Most hardware of the inspection systems are low resource use cases, therefore, mobile-friendly models should be applied to reduce latency. MobileNetV3 backbone plays the role of a feature extractor in object detectors. MobileNetV1 [30] proposed depth-wise separable convolution to reduce the number of parameters to improve computation efficiency, and MobileNetV2 [31] introduced the inverted residual block to expand to a higher-dimensional feature space internally to make more efficient layer structures. MobileNetV3 inherited advances of V1 and V2; it deployed the Squeeze-and-Excitation [32] in the inverted residual bottleneck and flexibly used the h-swish nonlinearity to significantly improve the accuracy of neural networks.

The inverted residual is the main building block of the MobileNetV3 network. The block follows a narrow–wide–narrow approach by input–output channels. As an example in Figure 14, the input channel is 24, the space expansion channel is 72, and the output channel is 40. The inverted residual block uses a combination of the expand convolution, the depth-wise convolution, the squeeze-excitation block, and the projection convolution, as in Figure 14.

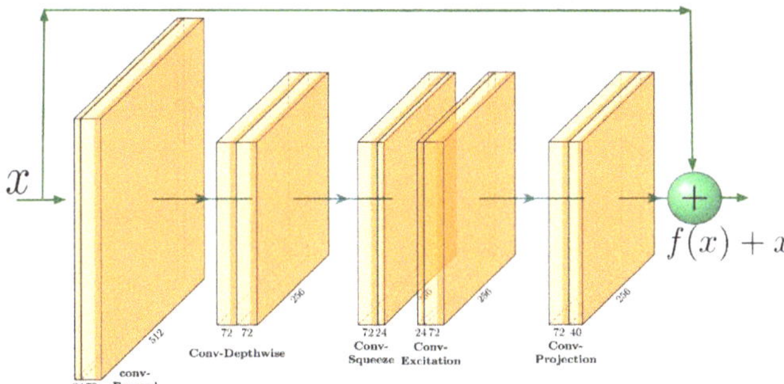

Figure 14. This is an instant of the inverted-residual block in MobileNetV3 architecture. First is a convolutional expand layer that widens channels from 24 to 72. Second is the convolutional depth-wise layer for better efficiency than traditional convolution. Its input and output channels are equal to 72, and the striking attribute of convolution halves the resolution. Next is the squeeze and excitation module to improve the power of features in the network. The final convolutional projection layer presents features in the lower dimension space, from 72 to 40.

FPN [33]: The object detection field has many more innovative algorithms, but current image data have become much more challenging, for instance, small object detection issues with only a few pixels. It is hard to extract the information about small objects in feature maps. FPN proposes a method to improve small object detection performance. It is an essential component that exploits the features of small objects on different levels of feature maps. FPN is an extended idea of pyramidal feature hierarchy that its architecture is a combination between top–down pathway, bottom–up pathway, and lateral connections.

As in Figure 15, the backbone architecture combines the MobilenetV3 and FPN. It is to extract multi-scale feature maps from the input image. The input is fed to MobilenetV3, which has 15 inverted residual blocks. The output is the multi-scale feature maps $\{C_1, C_2, C_3, C_4, C_5\}$, which are the input for FPN. FPN convolute and upsample C_i to output the better quality multi-scale feature maps $\{P_1, P_2, P_3, P_4, P_5\}$.

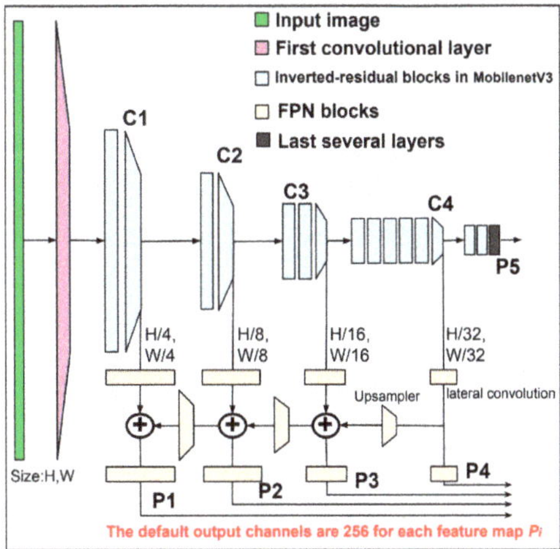

Figure 15. Backbone: the feature pyramid network and MobileNetV3 backbone together. The input is the image of size H, W. Firstly, The MobileNetV3 extracts the image to many multi-scale feature maps $\{C_1, C_2, C_3, C_4, C_5\}$. Secondly, the multi-scale output of MobileNetV3 is the input for FPN. The final result is the feature maps at multiple levels $\{P_1, P_2, P_3, P_4, P_5\}$.

The output channels of FPN present the multi-scale feature maps $\{P_1, P_2, P_3, P_4, P_5\}$. This hyper-parameter is vital to guarantee the quality of feature maps. Its default value is 256 for the large benchmark dataset. With the customized dataset, we will fine-tune the number of the output channels to obtain better performance. The results are shown in Section 3.2.

The output feature maps from FPN are $\{P_1, P_2, P_3, P_4, P_5\}$, which are also the input to the feature pyramid network and the ROI-Head. In the next section, we will describe these two blocks in detail.

2.2.3. RPN: Region Proposal Network

Detecting the position of objects is one of the two main tasks in object detection. The theoretical basis for initializing the temporary object position remains more challenging. In classical computer vision, selective search [34], multiscale combinatorial grouping [35], and CPMC [36] apply a strategy based on grouping super-pixels. EdgeBoxes [37] and objectness in Windows [38] use the window scoring technique. In deep learning-based computer vision, Shaoqing et al. [22] propose region proposal networks to create the box anchors to filter the potential positions. Anchor boxes are defined by two parameters: the wide range of scales and the aspect ratios.

RPN initialized a set of anchor boxes on each image or feature map by two hyper-parameters: scales and aspect ratio. They have a large impact on the final accuracy. Hence, we try to exploit them for optimal results. Figure 16 illustrates the creation of anchors on an image. The left image is an original, consisting of the red defect labels. The right imageinitializes a set of anchor boxes with scales $\{32^2, 64^2, 128^2, 256^2, 512^2\}$ and a range of aspect ratios $\{1:2, 1:1, 2:1\}$. Anchors are white, black, yellow, green, and blue rectangular boxes in the right image.

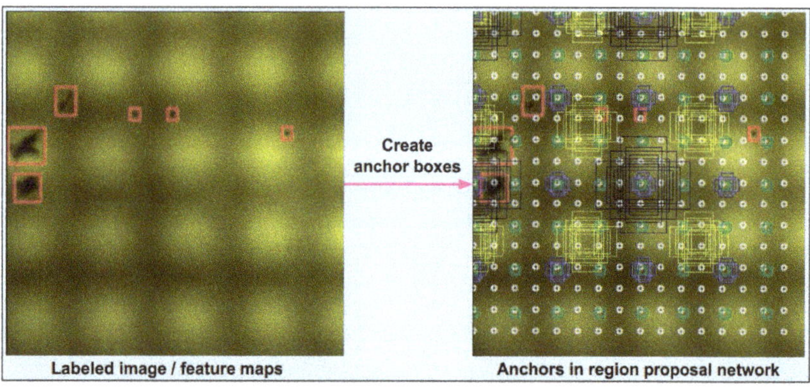

Figure 16. Illustrate how to create the box anchors in the region proposal network. Left is the image containing some red ground-truth boxes. RPN generates the reference boxes called "anchors" to map to ground-truth boxes. The multi-scale anchors are generated on the right image at various positions. They are the rectangular boxes marked with white, black, yellow, green, and blue colors.

The number of anchors generated is copious. RPN now tries to find anchors similar to the ground boxes (labels). The metric that determines whether an anchor is similar to the ground boxes is the IoU calculation. The pre-defined IoU thresholds are set to label the anchors as foreground, background, or ignored. If IoU is larger than the first threshold (typically 0.7), the anchor is assigned to one of the ground-truth boxes and labeled as foreground ('1'). If IoU is smaller than the second threshold (typically 0.3), the anchor is either labeled as a background ('0') or otherwise ignored ('−1').

In practice, the majority of anchors are background (the label is "0"), and so it is difficult to learn the foreground anchors due to the label imbalance. To solve the imbalance issue, the target number of foreground boxes N and the target number of background M are pre-defined.

At this time, we have the labeled anchor set and a target set, as shown in Figure 17. The RPN should learn to find rules to recognize the exact locations and shapes of ground-truth boxes. This issue is known as bounding-box regression, which will be presented in the next section.

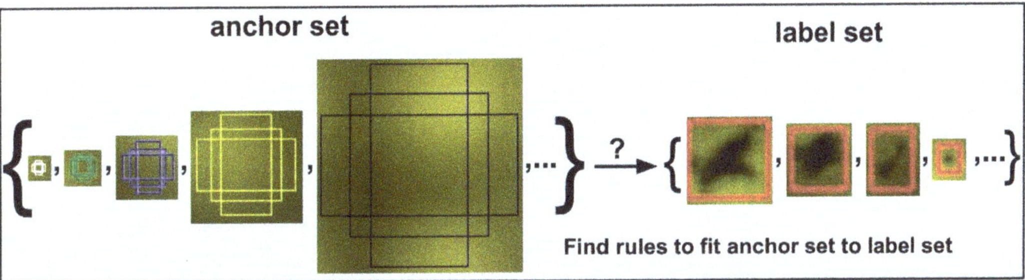

Figure 17. Box regression transforms the proposal anchor set to ground-truth label set.

The role of RPN is to propose the potential regions that contain the defects. To achieve this task, training the RPN is to regress the anchor boxes to the defect regions and classify anchors as the labels "1" or "0". Because of the large number of anchors, we only choose some of the quality anchors called *proposals* for the next sub-network. The loss function of RPN includes the L_1 [20] loss function for bounding-box regression and binary cross-entropy loss function for classifying the anchor as ground-truth or background.

$$L_{RPN} = L_{box-reg-RPN} + L_{binary-cls} \tag{11}$$

The following section details the bounding box regression and its loss function.

2.2.4. Bounding Box Regression

Bounding box regression will find some rules to scale-invariant transform a bounding box (anchor) to another bounding box (ground-truth/defect). The best idea is to consider the relationship between the center coordinates, where their width–height dimensions are significant. This section describes the formula between the ground-truth box and anchor. Figure 18 illustrates the parameters involved in transforming an anchor (the blue dotted-line rectangle) into a ground-truth box (the green rectangle) during the training phase, with the parameters are calculated using Equation (12).

$$\begin{aligned} \delta_x &= (b_x - a_x)/a_w, & \delta_y &= (b_y - a_y)/a_h \\ \delta_w &= \log(b_w/a_w), & \delta_h &= \log(b_h/a_h) \end{aligned} \tag{12}$$

where "a" and "b" denote anchor box and ground-truth box, respectively. Each one is represented by a 4-tuple in the form of (x, y, w, h), where (x, y) is the center coordinate and (w, h) is the width and height dimension. The regressor f aims to predict the transformation δ from the anchor a to the target ground-truth box b, represented as follows:

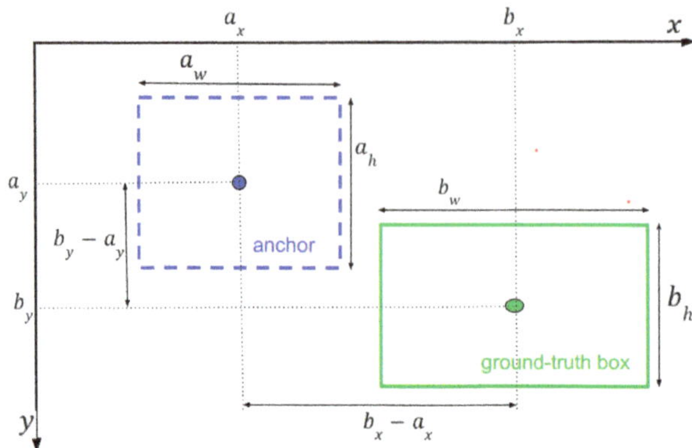

Figure 18. Illustration of transformation δ from the anchor a to the ground-truth box b. The formula is in Equation (12).

The image feature, denoted as x, is used as the input for the regressor f. Consiquently, the output is a prediction represented by $\hat{\delta} = f(x)$. The training process will minimize the bounding-box loss function:

$$L(\hat{\delta}, \delta) = \sum_{p \in \{x,y,w,h\}} L_1^{smooth}(\hat{\delta}_p - \delta_p), \tag{13}$$

where the function $L_1^{smooth}(.)$ is the robust L_1 loss defined in Equation (14).

$$L_1^{smooth}(t) = \begin{cases} 0.5t^2 \, if \, |t| < 1 \\ |t| - 0.5 \, otherwise \end{cases} \tag{14}$$

To calculate the final prediction box coordinates, the regressed anchor is inferred based on the inverse transformation of Equation (15) as follows:

$$\begin{aligned} a_x^{pred} &= \hat{\delta}_x a_w + a_x, \quad a_y^{pred} = \hat{\delta}_y a_h + a_y \\ a_w^{pred} &= a_w \exp(\hat{\delta}_w), \quad a_h^{pred} = a_h \exp(\hat{\delta}_h) \end{aligned} \tag{15}$$

The final summary is as follows: the bounding-box regressor f is a neural network with the input T, which are the image or feature maps, and the label is δ. A prediction is $\hat{\delta} = f(T)$. The training process will optimize the loss function $L_1^{smooth}(\delta - \hat{\delta})$. With the formulas δ, L_1^{smooth} and $\hat{\delta}$ in Equations (12) and (14), and formula $\hat{\delta} = f(T)$, respectively.

2.2.5. ROI-Head

ROI-Head converts the selected proposals on each feature map into a small fixed window (usually 7×7 pixels), and next is fed to the linear neural network to regress the bounding boxes and classify defects.

The inputs of ROI-Head are: The feature maps $\{P2, P3, P4, P5\}$ from the backbone block; the proposal boxes from the RPN block; and the label of defects. The ratio of foreground and background boxes will be customized to accelerate the training. The proposals with higher IoU than the threshold are counted as foreground and the others as background. This step will choose the best k proposals based on the IoU metric.

Before entering the ROI process, the top-k proposals are assigned to each level P_i of the appropriate feature maps based on the formula in Equation (16).

$$L_{P_i} = floor(k_0 + \log_2(\frac{\sqrt{w * h}}{canonical\_box\_size})) \tag{16}$$

where k_0 is the reference value, which is generally set to 4; w and h are the the width and height of the ROI area, respectively; *and* canonical_box_size is the canonical box size in pixels, set to 224, corresponding to the size of the pre-training image of the ImageNet dataset.

The ROIAlignV2 [37] process crops the rectangular regions on the feature maps specified by the proposal boxes. The linear neural network feeds the results of ROI to regress the proposals to ground boxes and classify the defect type.

When training the ROI-Head network, the loss function sums up the cost of classification $L_{defect-cls}$ and bounding-box regression $L_{box-reg-ROI}$, as in Equation (17).

$$L_{\text{detector}} = L_{defect-cls} + L_{box-reg-ROI} \tag{17}$$

where $L_{defect-cls}$ is the defect classification loss function that computes the cross-entropy loss; and $L_{box-reg-ROI}$ is the smooth L_1 loss as in Equation (13).

2.2.6. End-to-End Learning

The RPN block needs the cost to classify anchors as background or foreground (binary cross-entropy) and find proposals for the candidate locations of defects (bounding-box regression). Meanwhile, the ROI-Head network also incurs a cost to classify the defect type (cross-entropy) and locate the defects' position (bounding-box regression). Therefore, the network can be trained in an end-to-end manner using the multi-task loss function as follows:

$$L = L_{RPN} + L_{detector} \tag{18}$$

where L_{RPN} and $L_{detector}$ are based on the formulae in Equations (7) and (13), respectively.

3. Results

3.1. Experimental Setting

Defects are labeled and converted to COCO format to be compatible with object detection models. All images are resized to 1333 px for long edge and 800 px for short edge. We split the dataset into the training and test sets by a ratio of 80:20. In the first step, we train and test on some of the standard defect detection architectures such as two-stage object detectors (Faster-RCNN) and one-stage object detectors (Retina, FCOS). All models are implemented using PyTorch Vision's default configuration [39]. Table 6 displays the experimental outcomes obtained from training Faster R-CNN-based models with ResNet50, MobileNetV3-large, and MobileNetV3-small backbones, as well as the RetinaNet model, and the FCOS model, using the initial dataset. In the second step, for increment accuracy, we fine-tune the hyper-parameter in RPN, while for computational efficiency, we reduce the output channel of FPN in the backbone. The final result is presented in the following section.

Table 6. Comparison of defect detection between each architecture trained on the ski goggles defect dataset without any hyper-parameters adaptation.

Architecture	BACKBONE	IoU Metric			SPEED (S/IT)	
		AP	AP$_{50}$	AP$_{75}$	TRAIN	TEST
	RESNET50	56.3	78.5	63.3	0.528	0.126
Faster-RCNN	MOBILE-LARGE	41.3	72.8	38.1	0.127	0.059
	MOBILE-SMALL	10.0	25.1	08.4	0.086	0.045
FCOS	RESNET50	59.6	78.6	64.0	0.352	0.126
RetinaNet	RESNET50	10.2	25.2	05.9	0.331	0.140

The models are trained with 2 GPUs with a batch size of 8 for 26 epochs using SGD optimizer. The learning rate is initialized to 0.02 and learning ratio step at 16 and 22. Computer configuration is a CPU AMD Ryzen5 3600X, 64G RAM, 2 GPUs Gigabyte 2060 6G.

The method evaluates detection results based on the standard COCO-style average precision measured at *IoU* thresholds ranging from 0.5 to 0.75.

3.2. Defect Detection Results

To have a defect detection result baseline for the defect dataset on the surface of ski goggles lenses, we train and test different architectures and backbones. Parameters of Faster-RCNN-ResNet50, Faster-RCNN-Mobile-large, Faster-RCNN-Mobile-small, FCOS-Resnet50, and Retina-Resnet50 models are 41.1 M, 18.9 M, 15.8 M, 31.85 M, and 32.05 M, respectively. Table 6 shows the linear result of the larger architecture (more parameter) having better accuracy, but slower computational efficiency (speed, s/it). The balance between accuracy and computational efficiency is an issue in automatic optical inspection as hardware characteristics are compact. We recognize that the Faster-RCNN-Mobile-large model has gained a balanced result in terms of accuracy and computational efficiency. From this, we decide to fine-tune the Faster R-CNN with backbone Mobile-large to achieve a better result.

Faster R-CNN has proven to be a state-of-the-art object detector with high accuracy and flexible modular ability. Therefore, it can be integrated into some sub-network to improve performance. We implement the Faster R-CNN-based detector that uses an FPN-style backbone that extracts features from different convolutions of the MobileNetV3 model. The advance of MobileNetV3 block helps to improve speed; alternatively, FPN presents the invariant of feature maps, leading to an improvement in the small defect detection. However, Faster R-CNN has a drawback due to the complicated computation in creating anchor boxes. Its hyper-parameters in RPN are often sensitive to the final detection performance. The above disputation leads to fine-tuning Faster R-CNN to archive high performance.

First, we fine-tune the output channel of FPN to improve the network's speed. All feature maps extracted from the MobileNetV3 network have their output projected down to the number of channels by the FPN block. The default number of the output channel is 256. This parameter is finetuned within the value set $\{256, 128, 96, 64\}$ to obtain the best possible performance.

Second, we fine-tune the *anchor scale factor* in RPN to improve the accuracy. The *anchor scales* affect the handling of the bounding boxes of different sizes. Its invalid value setting causes the imbalance between negative and positive samples in training. The default value of *anchor scales* in the Faster R-CNN model is $\{32^2, 64^2, 128^2, 256^2, 512^2\}$. We augment two values: $\{16^2, 32^2, 64^2, 128^2, 256^2\}$ and $\{8^2, 16^2, 32^2, 64^2, 128^2\}$ in the RPN block.

Table 7 shows the aggregate results of fine-tuning the *output channel number* in FPN and the *anchor scales* in RPN. When the output channel of FPN decreases, the training and testing speed improves, while accuracy slightly decreases. Observing the efficiency of anchor scales, configuration $\{16^2, 32^2, 64^2, 128^2, 256^2\}$ achieved better results than the other two configurations, $\{32^2, 64^2, 128^2, 256^2, 512^2\}$ and $\{8^2, 16^2, 32^2, 64^2, 128^2\}$, with the same channel as FPN. With the channel reduction in FPN from 256 to 128, and the replacement of the anchor scale $\{32^2, 64^2, 128^2, 256^2, 512^2\}$ by $\{16^2, 32^2, 64^2, 128^2, 256^2\}$, the accuracy is 55.0 mAP, which is close to the best accuracy, while the AP_s metric achieved the best accuracy with 47.0. From this result, we choose the optimistic parameter set with the output channel equal to 128 and the *anchor scales* equal to $\{16^2, 32^2, 64^2, 128^2, 256^2\}$ for balance in computational efficiency and accuracy.

As mentioned in Section 2.1.3, the combined dataset (CoDS) arises from the fusion of initial (InDS) and synthetic datasets. Comparing these datasets is crucial to illustrate the efficacy in addressing a few data and imbalances. Table 8 depicts the evaluation outcomes for both datasets when training the Faster R-CNN model with optimal parameters.

Table 7. Defect detection results from fine-tuning the output channel of FPN and the anchor scale in RPN.

FPN	RPN	IOU METRIC		SPEED (S/IT)	
Out Channel	Anchor Scales	MAP	APS	TRAIN	TEST
256	$\{8^2, 16^2, 32^2, 64^2, 128^2\}$	55.3	46.4	0.4864	0.1161
256	$\{16^2, 32^2, 64^2, 128^2, 256^2\}$	49.2	42.8	0.4867	0.1179
128	$\{8^2, 16^2, 32^2, 64^2, 128^2\}$	53.6	45.4	0.3040	0.1133
128	$\{16^2, 32^2, 64^2, 128^2, 256^2\}$	55.0	47.0	0.3080	0.1074
96	$\{8^2, 16^2, 32^2, 64^2, 128^2\}$	46.7	39.6	0.2857	0.1094
96	$\{16^2, 32^2, 64^2, 128^2, 256^2\}$	51.8	42.7	0.2860	0.1046
64	$\{8^2, 16^2, 32^2, 64^2, 128^2\}$	47.6	38.3	0.2517	0.0993
64	$\{16^2, 32^2, 64^2, 128^2, 256^2\}$	51.4	46.0	0.2520	0.0968

Table 8. The comparative assessment of the initial dataset and the combined dataset using COCO metrics.

Model	COCO Metric	DATASET	
		INITIAL DS (INDS)	COMBINED DS (CODS)
Faster R-CNN with the MobileV3 Backbone	AP	51.4	55.1
	AP_{50}	71.5	75.8
	AP_{75}	40.7	47.4
The output channel number of FPN is 64	AP_S	46.0	48.2
	AP_m	47.3	50.3
	AP_l	59.1	63.4
The anchor scales of RPN $\{16^2, 32^2, 64^2, 128^2, 256^2\}$	AR_1	31.3	32.6
	AR_{10}	50.9	53.8
	AR_{100}	55.6	59.4
	AR_S	45.2	49.7
	AR_m	61.6	54.6
	AR_l	60.4	64.9

Figure 19 shows some results on the test set of the lens defect dataset. The red rectangular boxes mark defects; the above label is the defect category.

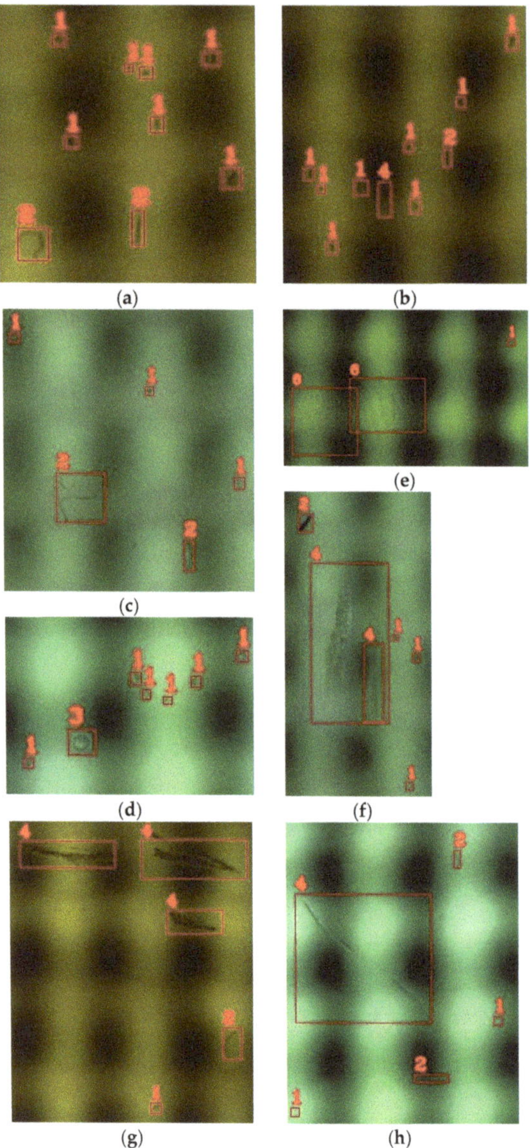

Figure 19. Selected examples of defect detection results. Defects are marked by the red rectangular boxes and the label above is the defect category. (**a–c**) display a variety of defect types, including dust-spot and dust-line. (**d**) showcases the spotlight defect type, while (**e**) highlights the stain defect type. (**f–h**) and h feature the scratch defect type.

4. Discussion

The data presented in Tables 6 and 7 have been computed using the detection evaluation metrics employed in the COCO detection challenge. A comparative analysis of object detection models, such as Faster R-CNN, FCOS, and RetinaNet, was conducted based on the data provided in Table 6. The Faster R-CNN model serves as a baseline for comparison due to its widespread use in object detection tasks. In terms of precision, FCOS outperforms Faster R-CNN with a 44.31% improvement in AP, a 7.97% enhancement in AP50, and a 67.98% increase in AP75. However, FCOS exhibits slower training and testing speeds, with a 177.17% reduction in training and a 113.56% decrement in testing relative to Faster R-CNN. Conversely, RetinaNet shows a significantly lower precision performance, with a 75.30% reduction in AP, a 65.35% decrease in AP50, and an 84.51% decline in AP75, while also demonstrating slower training and testing speeds (160.63% and 137.29% slower, respectively), in comparison to Faster R-CNN.

In the context of an optical inspection system, it is crucial to prioritize solutions that demand minimal hardware resources while maintaining satisfactory performance levels. A comprehensive analysis of various object detection models, including Faster R-CNN, FCOS, and RetinaNet, reveals that Faster R-CNN emerges as the most suitable candidate for deployment. Despite FCOS exhibiting superior overall precision, it significantly necessitates more hardware resources owing to its slower training and testing speeds, with a 177.17% reduction in training and a 113.56% decrement in testing compared to Faster R-CNN. In contrast, Faster R-CNN strikes an optimal balance between performance and resource efficiency, featuring notable precision performance and faster training and testing speeds in comparison to FCOS and RetinaNet. Consequently, Faster R-CNN is the preferable choice for deployment when considering hardware resource constraints.

Additionally, in Table 6, we dive into the performance analysis of some backbones of the Faster R-CNN model. Comparing speeds between ResNet50, MobileNet-Large, and MobileNet-Small as Faster R-CNN backbones showcases the trade-offs between performance and resource efficiency. Although the ResNet50 backbone outperforms the MobileNet-Large backbone in terms of AP, AP50, and AP75, it is notably slower in both training and testing phases, with training taking 315.75% longer and testing experiencing a 113.56% increase in duration. In contrast, MobileNet-Small provides the quickest training and testing speeds among the backbones, at 32.28% and 23.73% faster speeds than MobileNet-Large. However, this speed advantage comes at the cost of significantly reduced performance metrics. MobileNet-Large balances performance and speed, maintaining competitive performance metrics while achieving relatively faster training and testing speeds than ResNet50.

The optimal model for this context combines the strengths of Faster R-CNN, FPN, and MobileNetV3. The next step in optimizing this model is to fine-tune two parameters: the *output channel number* in FPN and the *anchor scales* in RPN. As shown in Table 7, the best-performing use-case has an FPN *output channel* of 256 and RPN *anchor scales* of $\{8^2, 16^2, 32^2, 64^2, 128^2\}$, with a *mAP* score that is 12.28% higher than the second-best use-case. Furthermore, the best-performing use-case has an AP_S score of 46.4, which is 8.62% higher than the second-best use-case. In cases where speed is prioritized, the fastest use-case is the one with an FPN *output channel* of 64 and RPN *anchor scales* of $\{16^2, 32^2, 64^2, 128^2, 256^2\}$. Compared to the best-performing use-case, this use-case is 48.78% faster in training and 16.28% faster in testing.

Table 8 compares two InDS and CoDS datasets utilizing various COCO metrics. Dataset CoDS exhibits superior performance in the majority of metrics when compared to dataset InDS. Notably, CoDS surpasses InDS in average precision (AP) with a 7.2% increase, as well as in size-based subcategories (AP_s, AP_m, and AP_l), with enhancements ranging from 4.8% to 7.3%. Regarding average recall (AR), dataset CoDS exceeds InDS in most categories, except for the medium object size category (AR_m), where InDS outperforms CoDS by 11.4%. In summary, the combined dataset demonstrates improved performance relative to the initial dataset.

Imbalanced datasets and scarce data are common challenges when training deep learning models with real-world data collection. CoDS performs better than InDS, thereby illustrating that generating additional synthetic data is an effective approach to addressing these challenges.

Despite the valuable contributions of this study, certain limitations should be acknowledged: the number of labels in the defect dataset is relatively small, which may impact the generalizability of the findings; the approach to addressing unbalanced and rare datasets is not extensively explored; the study does not employ newer defect detection methods within deep learning to analyze the dataset; and a limited range of metrics is considered for a detailed evaluation of defects, suggesting that additional performance measures could provide further insight. Future research may address these limitations by expanding the dataset, exploring more comprehensive solutions for unbalanced and rare data, incorporating cutting-edge defect detection techniques, and utilizing a wider array of evaluation metrics.

5. Conclusions

This paper solved the defect detection problem on the surface of ski goggles lenses based on the deep learning approach. The work has achieved the design of the image capture system that has five cameras cover the entire curved surface of the lenses, which enables it to capture images automatically from all angles at the same time. This work also presents the development of a surface defect detection dataset for ski goggles lenses, contributing to the diversification of surface data sources in the deep learning-based defect detection field. The defect detection result achieved excellent performance by fine-tuning the reasonable hyper-parameters of the Faster-RCNN modular architecture by replacing the ResNet backbone with MobileNetV3 and FPN to better extract feature maps, by reducing the number of the output channel of FPN to increase the computational performance, and by adjusting the anchor scale factor hyper-parameter in RPN, leading to better accuracy. This work is helpful for automatic optical inspection systems because of its limited hardware resource. The experimental results have reinforced the hypothesis for correctly choosing the Faster-RCNN defect detection architecture and fine-tuning the hyper-parameters. In the future, we will improve the dataset and make it publicly available to the research community.

Author Contributions: D.-T.D. developed the hardware system and software coding, and wrote the original draft. J.-W.W. guided the research direction and edited the paper. All authors discussed the results and contributed to the final manuscript. All authors have read and agreed to the published version of the manuscript.

Funding: This research was funded in part by MOST 109-2218-E-992-002 from the Ministry of Science and Technology, Taiwan.

Data Availability Statement: The datasets used in this study can be accessed through the following link: https://github.com/ddthuan/goggles (accessed on 30 January 2023).

Acknowledgments: The authors would like to thank Foresight Optical Co., Ltd., Taiwan, for their joint development of the inspection system.

Conflicts of Interest: The authors declare no conflict of interest. This article is authorized for the first author to continue with all his related study works.

References

1. Dang, D.-T.; Wang, J.-W.; Lee, J.-S.; Wang, C.-C. Defect Classification System for Ski Goggle Lens. In Proceedings of the 2021 International Symposium on Intelligent Signal Processing and Communication Systems (ISPACS), Hualien, Taiwan, 16–19 November 2021; pp. 1–3.
2. Le, N.T.; Wang, J.-W.; Wang, C.-C.; Nguyen, T.N. Novel Framework Based on HOSVD for Ski Goggles Defect Detection and Classification. *Sensors* **2019**, *19*, 5538. [CrossRef] [PubMed]
3. Luo, Q.; Fang, X.; Su, J.; Zhou, J.; Zhou, B.; Yang, C.; Liu, L.; Gui, W.; Tian, L. Automated Visual Defect Classification for Flat Steel Surface: A Survey. *IEEE Trans. Instrum. Meas.* **2020**, *69*, 9329–9349. [CrossRef]
4. Luo, Q.; Fang, X.; Liu, L.; Yang, C.; Sun, Y. Automated Visual Defect Detection for Flat Steel Surface: A Survey. *IEEE Trans. Instrum. Meas.* **2020**, *69*, 626–644. [CrossRef]
5. Peres, R.S.; Jia, X.; Lee, J.; Sun, K.; Colombo, A.W.; Barata, J. Industrial Artificial Intelligence in Industry 4.0—Systematic Review, Challenges and Outlook. *IEEE Access* **2020**, *8*, 220121–220139. [CrossRef]
6. Hirschberg, I.; Bashe, R. Charge Coupled Device (CCD) Imaging Applications—A Modular Approach. In *Applications of Electronic Imaging Systems*; Franseen, R.E., Schroder, D.K., Eds.; SPIE: Bellingham, WA, USA, 1978; Volume 0143, pp. 11–18.

7. Martins, R.; Nathan, A.; Barros, R.; Pereira, L.; Barquinha, P.; Correia, N.; Costa, R.; Ahnood, A.; Ferreira, I.; Fortunato, E. Complementary Metal Oxide Semiconductor Technology with and on Paper. *Adv. Mater.* **2011**, *23*, 4491–4496. [CrossRef] [PubMed]

8. Mersch, S. Overview of Machine Vision Lighting Techniques. In *Optics, Illumination, and Image Sensing for Machine Vision*; Svetkoff, D.J., Ed.; SPIE: Bellingham, WA, USA, 1987; Volume 0728, pp. 36–38.

9. Sieczka, E.J.; Harding, K.G. Light source design for machine vision. In *Optics, Illumination, and Image Sensing for Machine Vision VI*; Svetkoff, D.J., Ed.; SPIE: Bellingham, WA, USA, 1992; Volume 1614, pp. 2–10.

10. He, Z.; Liu, Q. Deep Regression Neural Network for Industrial Surface Defect Detection. *IEEE Access* **2020**, *8*, 35583–35591. [CrossRef]

11. Li, G.; Shao, R.; Wan, H.; Zhou, M.; Li, M. A Model for Surface Defect Detection of Industrial Products Based on Attention Augmentation. *Comput. Intell. Neurosci.* **2022**, *2022*, 9577096. [CrossRef] [PubMed]

12. Zhang, D.; Hao, X.; Liang, L.; Liu, W.; Qin, C. A novel deep convolutional neural network algorithm for surface defect detection. *J. Comput. Des. Eng.* **2022**, *9*, 1616–1632. [CrossRef]

13. He, Y.; Zhu, C.; Wang, J.; Savvides, M.; Zhang, X. Bounding Box Regression with Uncertainty for Accurate Object Detection. In Proceedings of the 2019 IEEE/CVF Conference on Computer Vision and Pattern Recognition (CVPR), Long Beach, CA, USA, 15–19 June 2019; pp. 2883–2892.

14. Touvron, H.; Vedaldi, A.; Douze, M.; Jegou, H. Fixing the train-test resolution discrepancy. In *Advances in Neural Information Processing Systems*; Wallach, H., Larochelle, H., Beygelzimer, A., d'Alché-Buc, F., Fox, E., Garnett, R., Eds.; Curran Associates, Inc.: Red Hook, NY, USA, 2019; Volume 32. Available online: https://proceedings.neurips.cc/paper/2019/file/d03a857a23b5285736 c4d55e0bb067c8-Paper.pdf (accessed on 5 November 2022).

15. Okwuashi, O.; Ndehedehe, C.E. Deep support vector machine for hyperspectral image classification. *Pattern Recognit.* **2020**, *103*, 107298. [CrossRef]

16. Jiao, J.; Wang, X.; Deng, Z.; Cao, J.; Tang, W. A fast template matching algorithm based on principal orientation difference. *Int. J. Adv. Robot. Syst.* **2018**, *15*, 1729881418778223. [CrossRef]

17. He, K.; Zhang, X.; Ren, S.; Sun, J. Deep Residual Learning for Image Recognition. In Proceedings of the 2016 IEEE Conference on Computer Vision and Pattern Recognition (CVPR), Las Vegas, NV, USA, 27–30 June 2016; pp. 770–778.

18. Simonyan, K.; Zisserman, A. Very Deep Convolutional Networks for Large-Scale Image Recognition. In Proceedings of the 3rd International Conference on Learning Representations (ICLR 2015), San Diego, CA, USA, 7–9 May 2015; Conference Track Proceedings. Bengio, Y., LeCun, Y., Eds.; 2015.

19. Tan, M.; Le, Q. EfficientNet: Rethinking Model Scaling for Convolutional Neural Networks. In Proceedings of Machine Learning Research, Proceedings of the 36th International Conference on Machine Learning, Long Beach, CA, USA, 10–15 June 2019; Chaudhuri, K., Salakhutdinov, R., Eds.; PMLR: London, UK, 2019; Volume 97, pp. 6105–6114.

20. Dosovitskiy, A.; Beyer, L.; Kolesnikov, A.; Weissenborn, D.; Zhai, X.; Unterthiner, T.; Dehghani, M.; Minderer, M.; Heigold, G.; Gelly, S.; et al. An Image is Worth 16×16 Words: Transformers for Image Recognition at Scale. *arXiv* **2021**, arXiv:2010.11929.

21. Zhang, S.; Chi, C.; Yao, Y.; Lei, Z.; Li, S.Z. Bridging the Gap Between Anchor-Based and Anchor-Free Detection via Adaptive Training Sample Selection. In Proceedings of the 2020 IEEE/CVF Conference on Computer Vision and Pattern Recognition (CVPR), Seattle, WA, USA, 13–19 June 2020; pp. 9756–9765.

22. Ren, S.; He, K.; Girshick, R.; Sun, J. Faster R-CNN: Towards Real-Time Object Detection with Region Proposal Networks. *IEEE Trans. Pattern Anal. Mach. Intell.* **2017**, *39*, 1137–1149. [CrossRef]

23. Redmon, J.; Divvala, S.; Girshick, R.; Farhadi, A. You Only Look Once: Unified, Real-Time Object Detection. In Proceedings of the 2016 IEEE Conference on Computer Vision and Pattern Recognition (CVPR), Las Vegas, NV, USA, 27–30 June 2016; pp. 779–788.

24. Tian, Z.; Shen, C.; Chen, H.; He, T. FCOS: A Simple and Strong Anchor-Free Object Detector. *IEEE Trans. Pattern Anal. Mach. Intell.* **2022**, *44*, 1922–1933. [CrossRef] [PubMed]

25. Ma, D.W.; Wu, X.J.; Yang, H. Efficient Small Object Detection with an Improved Region Proposal Networks. *IOP Conf. Ser. Mater. Sci. Eng.* **2019**, *533*, 012062. [CrossRef]

26. Wada, K. Labelme: Image Polygonal Annotation with Python; GitHub Repository. 2018. Available online: https://github.com/wkentaro/labelme (accessed on 1 October 2022).

27. Howard, A.; Sandler, M.; Chen, B.; Wang, W.; Chen, L.-C.; Tan, M.; Chu, G.; Vasudevan, V.; Zhu, Y.; Pang, R.; et al. Searching for MobileNetV3. In Proceedings of the 2019 IEEE/CVF International Conference on Computer Vision (ICCV), Seoul, Republic of Korea, 27 October–2 November 2019; pp. 1314–1324.

28. Huang, W.; Liao, X.; Zhu, L.; Wei, M.; Wang, Q. Single-Image Super-Resolution Neural Network via Hybrid Multi-Scale Features. *Mathematics* **2022**, *10*, 653. [CrossRef]

29. Ding, K.; Niu, Z.; Hui, J.; Zhou, X.; Chan, F.T.S. A Weld Surface Defect Recognition Method Based on Improved MobileNetV2 Algorithm. *Mathematics* **2022**, *10*, 3678. [CrossRef]

30. Howard, A.G.; Zhu, M.; Chen, B.; Kalenichenko, D.; Wang, W.; Weyand, T.; Andreetto, M.; Adam, H. MobileNets: Efficient Convolutional Neural Networks for Mobile Vision Applications. *arXiv* **2017**, arXiv:1704.04861.

31. Sandler, M.; Howard, A.; Zhu, M.; Zhmoginov, A.; Chen, L.-C. MobileNetV2: Inverted Residuals and Linear Bottlenecks. In Proceedings of the 2018 IEEE/CVF Conference on Computer Vision and Pattern Recognition, Salt Lake City, UT, USA, 18–23 June 2018; pp. 4510–4520.

32. Hu, J.; Shen, L.; Sun, G. Squeeze-and-Excitation Networks. In Proceedings of the 2018 IEEE/CVF Conference on Computer Vision and Pattern Recognition, Salt Lake City, UT, USA, 18–23 June 2018; pp. 7132–7141.
33. Lin, T.-Y.; Dollár, P.; Girshick, R.; He, K.; Hariharan, B.; Belongie, S. Feature Pyramid Networks for Object Detection. In Proceedings of the 2017 IEEE Conference on Computer Vision and Pattern Recognition (CVPR), Honolulu, HI, USA, 21–26 July 2017; pp. 936–944.
34. Uijlings, J.R.R.; van de Sande, K.E.A.; Gevers, T.; Smeulders, A.W.M. Selective Search for Object Recognition. *Int. J. Comput. Vis.* **2013**, *104*, 154–171. [CrossRef]
35. Arbeláez, P.; Pont-Tuset, J.; Barron, J.; Marques, F.; Malik, J. Multiscale Combinatorial Grouping. In Proceedings of the Computer Vision and Pattern Recognition, Columbus, OH, USA, 23–28 June 2014.
36. Carreira, J.; Sminchisescu, C. CPMC: Automatic Object Segmentation Using Constrained Parametric Min-Cuts. *IEEE Trans. Pattern Anal. Mach. Intell.* **2012**, *34*, 1312–1328. [CrossRef] [PubMed]
37. Zitnick, C.L.; Dollár, P. Edge Boxes: Locating Object Proposals from Edges. In Proceedings of the Computer Vision—ECCV 2014: 13th European Conference, Zurich, Switzerland, 6–12 September 2014; Fleet, D., Pajdla, T., Schiele, B., Tuytelaars, T., Eds.; Springer International Publishing: Cham, Switzerland, 2014; pp. 391–405.
38. Lampert, C.H.; Blaschko, M.B.; Hofmann, T. Beyond sliding windows: Object localization by efficient subwindow search. In Proceedings of the 2008 IEEE Conference on Computer Vision and Pattern Recognition, Anchorage, AK, USA, 23–28 June 2008; pp. 1–8.
39. Paszke, A.; Gross, S.; Massa, F.; Lerer, A.; Bradbury, J.; Chanan, G.; Killeen, T.; Lin, Z.; Gimelshein, N.; Antiga, A.; et al. PyTorch: An Imperative Style, High-Performance Deep Learning Library. In *Advances in Neural Information Processing Systems*; 2019; Volume 32. Available online: https://proceedings.neurips.cc/paper/2019/file/bdbca288fee7f92f2bfa9f7012727740-Paper.pdf (accessed on 6 July 2022).

Article

A Unified Learning Approach for Malicious Domain Name Detection

Atif Ali Wagan [1], Qianmu Li [1,*], Zubair Zaland [2], Shah Marjan [2], Dadan Khan Bozdar [3], Aamir Hussain [4], Aamir Mehmood Mirza [3] and Mehmood Baryalai [3]

[1] School of Computer Science and Engineering, Nanjing University of Science and Technology, Nanjing 210094, China
[2] Department of Software Engineering, Balochistan University of Information Technology Engineering and Management Sciences, Quetta 87300, Pakistan
[3] Department of Computer Science, Balochistan University of Information Technology Engineering and Management Sciences, Quetta 87300, Pakistan
[4] Department of Computer Science, Muhammad Nawaz Shareef University of Agriculture Multan, Multan 60000, Pakistan
* Correspondence: qianmu@njust.edu.cn

Abstract: The DNS firewall plays an important role in network security. It is based on a list of known malicious domain names, and, based on these lists, the firewall blocks communication with these domain names. However, DNS firewalls can only block known malicious domain names, excluding communication with unknown malicious domain names. Prior research has found that machine learning techniques are effective for detecting unknown malicious domain names. However, those methods have limited capabilities to learn from both textual and numerical data. To solve this issue, we present a novel unified learning approach that uses both numerical and textual features of the domain name to classify whether a domain name pair is malicious or not. The experiments were conducted on a benchmark domain names dataset consisting of 90,000 domain names. The experimental results show that the proposed approach performs significantly better than the six comparative methods in terms of accuracy, precision, recall, and F1-Score.

Keywords: DNS; firewall; malicious; domain; name

Citation: Wagan, A.A.; Li, Q.; Zaland, Z.; Marjan, S.; Bozdar, D.K.; Hussain, A.; Mirza, A.M.; Baryalai, M. A Unified Learning Approach for Malicious Domain Name Detection. *Axioms* **2023**, *12*, 458. https://doi.org/10.3390/axioms12050458

Academic Editors: Oscar Humberto Montiel Ross, Hsien-Chung Wu and Darjan Karabašević

Received: 23 November 2022
Revised: 10 April 2023
Accepted: 24 April 2023
Published: 9 May 2023

1. Introduction

Preserving the integrity of network security is an important consideration for all organizations. As nearly every aspect of business becomes increasingly digital, enterprise network security software can help organizations mitigate the effects of cyber-attacks, particularly by protecting against them, thereby safeguarding their operations and ensuring their competitiveness in a rapidly changing marketplace.

Intrusion detection systems (IDS) and intrusion prevention systems (IPS) have long been part of the network security toolkit to detect, monitor, and block malware and malicious traffic. IDS are designed to detect intrusions into the network and issue warnings when they detect a potential cyber-attack. However, the system itself cannot protect against attacks, and that responsibility is left to human analysts. Meanwhile, IPS work proactively to prevent successful attacks and respond according to predefined rules when an intrusion is detected. One of the most important IPS are firewalls [1]. One such firewall for IP is the DNS firewall. The majority of DNS firewalls are built on regularly updated lists of known malicious domain names. However, this method can only block known malicious communications, leaving a large number of malicious communications that cannot be blocked because they are unknown.

To predict new malicious domains over DNS communications in a timely manner with better DNS data features is important [2]. To resolve this challenge, Marques et al. [2]

proposed a machine learning-based DNS firewall built with new domain names and 34 features. However, they utilized traditional classification models for their proposed approach which have limited capabilities to learn from both numerical and textual features of the domain name at the same time.

To resolve this issue, we propose a unified deep learning model which can learn from both numerical and textual features. This model makes use of both numerical and textual information in the domain name. These pairs are entered into the model and correlating features between the pairs are extracted. Finally, the model uses the correlating features to determine if these pairs represent a malicious domain or not.

Our work makes the following three major contributions:

(1) We consider the domain name textual and numerical features unified representation learning issue in the context of malicious domain name detection tasks.
(2) A new deep learning model is proposed for malicious domain name detection.
(3) Experimental results for the detection of the malicious domain name demonstrate the high performance of the proposed method in comparison with state-of-the-art methods on the malicious domain name dataset.

The remainder of this article is organized as follows: Section 2 describes previous research on malicious domain name detection. The details of our unified learning approach are described in Section 3. In Section 4, the experimental setup is described, followed by experimental results in Section 5. Finally, Section 6 concludes this article.

2. Related Work

A number of approaches for malicious domain name detection have been proposed in the literature. In general, these approaches can be broadly classified into non-machine learning-based and machine learning-based approaches.

2.1. Non-Machine Learning Approaches

The non-machine learning-based approaches use some kind of data obtained from known malicious domain names for making malicious domain name signatures and using those signatures to filter unknown malicious domain names.

Zhang et al. [3] proposed a blocklisting system, the purpose of which is to create blocklists based on the relevance ranking scheme used by the link analysis community. The system creates blocklists for individuals who choose to submit data to a central log-sharing platform. The system evaluates the relevance of the submitter to the attacker based on the attacker's history and the submitter's recent log generation patterns. The blocklist system also incorporates extensive log prefiltering and severity metrics to understand the extent to which attacker alert patterns are consistent with common malware propagation behaviors. Prakash et al. [4] proposed PhishNet, which is based upon two components. The first component uses five heuristics to find new phishing URLs. The second component is made of an approximate matching algorithm that partitions the URL into several components, then these multiple components of the URL are matched with entries in the blocklist individually. Malicious URLs are often short-lived and are updated frequently to avoid being on the blocklist. If these malicious URLs are updated by the same adversary using domain name string manipulation, we can assume that unknown malicious URLs exist in the neighborhood of known malicious URLs. Based on this assumption, Akiyama et al. [5] proposed an effective blocklist URL generation method that uses search engines to discover unknown malicious URLs in the neighborhood of known malicious URLs. Fukushima et al. [6] proposed a blocklisting scheme in which they analyzed malicious URLs' metadata, such as domain, registrar, IP address, IP address block, and autonomous system. Furthermore, they evaluated registrars and IP address blocks associated with malicious URLs. From that evaluation, they made a blocklist with low-reputation IP address blocks and registrars often used in malicious URLs. Sun et al. [7] proposed an automatic blocklist generator (AutoBLG) that discovers new malicious URLs automatically by starting from an existing URL blocklist. The idea of their approach is to extend the space of web page searches, while decreasing the number of URLs to be analyzed, by applying some prefilters to expedite the process of creating a blocklist.

Unlike the above works, which use signatures for blocking malicious domain name, we propose a machine learning-based approach that learns from malicious and benign domain name data and classifies whether a given domain name is malicious or not.

2.2. Machine Learning Approaches

Machine learning-based approaches make a predictive model for detecting unknown malicious domain names by training on data of malicious and benign domain names.

Saxe and Berlin. [8] used character-level embedding with a convolutional neural network to retrieve associated features from file paths, registry keys, and URLs. The retrieved features are then fed into three fully connected layers to classify file paths, registry keys, and URLs as malicious or normal. Yang et al. [9] proposed a convolutional gated recurrent unit neural network for detecting malicious URLs using characters as text classification features. Their model, instead of using a pooling layer in a convolutional neural network, uses a gated recurrent unit for feature acquisition in the time dimension. Luo et al. [10] proposed an approach for identifying malicious HTTP GET requests using a novel architecture of convolutional neural network for classification; they used natural language processing-based analysis and Auto-Encoder for URL representation and extraction. Mondal et al. [11] proposed an ensemble learning approach to predict the class probabilities of benign or malicious URLs using multiple classifiers. After obtaining probabilities from multiple classifiers, a threshold filter is applied to the probabilities to finally determine whether the URL is malicious or not. Marques et al. [2] empirically investigated the impact of three feature selection methods applied to six classification models on the performance of malicious domain name detection. They conducted experiments on the malicious domains dataset, and found that the decision trees with recursive feature elimination were more suitable for the malicious domain detection task, compared with other baseline methods.

Unlike the above works, we simultaneously consider both numerical and textual feature unified learning for malicious domain name detection.

3. Methodology

3.1. Overview

In order to utilize both numerical features and textual features for detecting malicious domain names, we propose a combined deep learning-based model (hybrid feed forward network (FFN)-long short-term memory (LSTM) [12] model that preserves the advantages of both numerical and textual features. The network structure of the hybrid FFN-LSTM model is shown in Figure 1. The original dataset was split into numerical and textual features that were fed into the model as separate inputs. Individual learners were then built based on numerical and textual features, and the outputs of the individual learners were spliced together in parallel to form the input to a unified learner consisting of a fully connected neural network.

3.2. Numerical Learner

Following Marques et al. [2], we performed the same two pre-processing steps—the label encoder [13] and min-max normalization [14]—as them. The label encoder [13] transforms categorical features by simply mapping each category with an integer value ranging from 0 to n. After the label encoding [13], min-max normalization [14] was applied to scale the different features so that each feature has equal weight for the numerical feature learner. This sets the value of the data points from 0 to 1.

Pre-processed numerical features are sent to the FFN for learning numerical features. The FFN is composed of three types of network layers: an input layer, a hidden layer, and an output layer. Additionally, each of these layers contains n number of neurons. The first layer with n neurons is the input layer, and the input feature vectors are received in this layer. The hidden layer consists of a few layers. In the last output layer, the FFN outputs the results. We constructed the FFN with a depth of 5 layers. We denote these layers as $\delta = \{\mathcal{L}_n^1, \mathcal{L}_n^2, \mathcal{L}_n^3, \mathcal{L}_n^4, \mathcal{L}_n^5\}$.

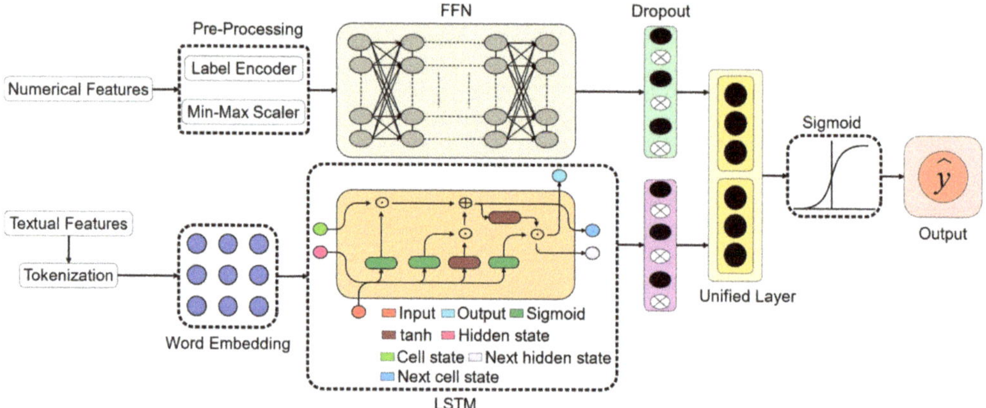

Figure 1. Our proposed malicious domain name detection model.

The FFN output layer output is forwarded to the dropout [15] layer to prevent overfitting. Owing to over-fitting, the classification ability of the FFN is limited. Dropout [15] can effectively solve this problem. Dropout [15], is a mechanism to improve the performance of the FFN by randomly setting the weights of the FFN to 0.

3.3. Text Learner

To construct an accurate malicious domain name detection model, the feature vector representation of textual features is important. Although high-level representations such as numerical features can be useful, they cannot reveal deep hidden semantics in the textual features of the domain name. LSTM [12] is a deep learning architecture, which provides a powerful representation of the textual features. It can learn semantic features automatically. Before LSTM [12] could process textual features, we needed to pre-process textual features with tokenization and represent them as an embedding matrix.

In the tokenization phase, the sequence of words is broken into multiple tokens based on the white space separator character. In this process, each word is called a token.

The embedding layer takes a token index as input and transforms it into a low-dimensional representation. Given a textual features sample tokens n which is essentially a series of words $\left[f_1, \ldots, f_{|n|} \right]$, our goal is to obtain its matrix representation $n \rightarrow \mathcal{N} \in \mathbb{R}^{|n| \times d_m}$, where $\mathcal{N}$ is a matrix consisting of a set of words $f_i \rightarrow \mathfrak{F}_i$, $i = 1, \ldots, |n|$ in the given domain name sample. Every word f_i can now be represented as an embedding vector, i.e., $\mathfrak{F}_i \in \mathbb{R}^{d_m}$, where d_m is the dimensional vector of words that appear in the textual features of the domain name. In our experiments, we randomly initialize the embedding matrix of malicious domain names textual features, and during the training process it is learned. Thus, domain name textual features matrix representation $\mathcal{N}$ having $|n|$ words sequence can be expressed as follows:

$$\mathcal{N} = \left[\mathfrak{F}_i, \ldots, \mathfrak{F}_{|n|} \right] \tag{1}$$

All domain name textual features are truncated or padded to be the same length $|n|$ for parallelization.

Lower dimensional embedding vectors are sent to an LSTM [12] network which can be regarded as a sequence of LSTM [12] units for feature learning. Let $\mathfrak{F}_1, \ldots, \mathfrak{F}_{|n|}$ be the entered words sequence (e.g., domain name textual features tokens), with having a corresponding labels sequence (e.g., the next domain name textual feature tokens). At every step t, the LSTM [12] unit takes the $\mathfrak{F}_t$ input, along with the previous hidden and cell state, and computes the next hidden state and cell state using a set of model parameters. The output is predicted using the output state (e.g., the next domain name textual feature token based upon previous ones) at each step t.

Similar to the numerical learned features, the textual learned features using LSTM [12] are fed to the dropout [15] layer to prevent overfitting.

3.4. Unified Learning

The highlight of our method is that we utilize textual and numerical features and automatically determine the optimal ratio of each feature at the fusion stage. We feed the outputs of numerical learner $\mathcal{Q}_i$ and text learner $\mathcal{T}_i$ into a new unified vector $\mathfrak{U}$ that represents both textual and numerical features in a given sample, which can be formulated as follows:

$$\mathfrak{U} = \mathcal{Q}_i \oplus \mathcal{T}_i \tag{2}$$

The new unified vector is then fed into the fully connected layer, resulting in a vector output γ as follows:

$$\gamma = \sigma(W_\gamma \cdot \mathfrak{U} + b_\gamma) \tag{3}$$

where $\cdot$ represent the dot product, the vector γ weight matrix is represented as W_γ, the bias value of the vector γ is represented as b_γ, and the activation function ReLU [16] is represented as $\sigma(\cdot)$. Finally, the γ vector is sent to an output layer.

Each numerical and textual feature is properly represented into a feature vector by previous layers. Using these feature vectors as input, a unified binary classifier is used in the output layer to predict whether these features represent a malicious domain or not. In this study, a sigmoid classifier was used, which computes a probability score for a given domain name sample as follows:

$$\hat{\mathfrak{P}}_i = sigmoid(\gamma) \tag{4}$$

3.5. Training

During the process of training, our model tries to learn the following parameters: the five dense layers of FNN, the word embedding matrices of textual features, the LSTM [12] layer, and the weights and bias of the fully connected output layer. After these parameters are learned, the malicious domain name can be detected. These model parameters are learned by minimizing the following binary cross-entropy loss function:

$$BCE = -\frac{1}{\mathfrak{N}} \sum_{i=0}^{\mathfrak{N}} \mathfrak{G}_i \cdot \log(\hat{\mathfrak{P}}_i) + (1 - \mathfrak{G}_i) \cdot \log(1 - \hat{\mathfrak{P}}_i) \tag{5}$$

where all samples in the dataset are denoted as $\mathfrak{N}$, and the output layer probability score is denoted as $\hat{\mathfrak{P}}_i$, defined by Equation (4), $\mathfrak{G}_i = \{0, 1\}$ indicates whether the i-th sample is a malicious domain or not. As it has previously been shown that Adam [17] is less memory intensive and more computationally efficient than other optimization methods, we decided to minimize the objective function using the Adam [17] optimizer. In order to efficiently calculate parameter updates during the learning process, we use backpropagation [18], a simple implementation of the chain rule of partial derivatives.

4. Experiment Setup

4.1. Dataset

We use a public dataset in this study that has been collected and processed by Marques et al. [19]. The dataset contains approximately 90,000 malicious and non-malicious domain name samples of equal size. Each sample contains 34 features, such as IP, geolocation, open ports, domain name entropy, etc.

4.2. Performance Evaluation

There are four possible outcomes of our model prediction: true positive (TP) indicates that a malicious domain name is predicted as a malicious domain name, true negative (TN) indicates that a non-malicious domain name is predicted as a non-malicious domain name, false negative (FN) indicates that a malicious domain name is predicted as a non-malicious domain name, and false positive (FP) indicates that a non-malicious domain name is predicted as a malicious domain name. Based on these outcomes, the performance evaluation metrics accuracy, recall, precision, and F1-Score can be calculated as given below.

Accuracy is perhaps the most intuitive way of measuring the performance of any binary classification model. The accuracy metric can be interpreted as the percentage of samples correctly classified by the model. Based on the notation introduced earlier, the accuracy is defined in Equation 6 as follows:

$$Accuracy = \frac{TP + TN}{TP + TN + FP + FN} \tag{6}$$

As an alternative measure of classifier performance, precision can be interpreted as the accuracy of positive predictions. Precision is defined by the following equation:

$$Precision = \frac{TP}{TP + FP} \tag{7}$$

Precision is often used in conjunction with a metric called recall, as precision measurement tends to look very high for models that predict few positives. Recall indicates the proportion of positive instances that are correctly detected by the classifier, as defined by the following equation:

$$Recall = \frac{TP}{TP + FN} \tag{8}$$

The F1-Score is the harmonic mean of precision and recall, as defined by Equation (9). A large F1-Score can only be obtained if both recall and precision are high.

$$F1 - Score = \frac{2 \times Precision \times Recall}{Precision + Recall} \tag{9}$$

5. Results

RQ.1 What is the effectiveness of our approach against the state-of-the-art baseline methods?

Our goal is to provide a method that can automatically classify malicious and non-malicious domain names. However, one challenge to the usefulness of this approach is to determine how much its performance is improved over the baseline approaches. By answering this research question, it would become clear how far ahead our method is in detecting malicious domain names compared to state-of-the-art methods. To compare the performance of our method, we chose six baseline methods, which are listed below.

Linear discriminant analysis (LDA) [20–23]: The purpose of the discriminant analysis is to find the linear function of the data that best separates the two data points. Each data point is classified into one of its two groups. When considering the data, the between-group variance is maximized, and the within-group variance is minimized. In LDA [20–23], it is assumed that the data can be represented linearly. However, this often does not reflect realistic relationships between the data, and discriminant analysis is limited in its ability to best classify data points. In simple terms, LDA [20–23] attempts to find a separable subspace of data points whose dimensionality is lower than that of the original data sample. This is achieved by finding a hyperplane that maximizes the mean and minimizes the variance between the two classes.

Support vector machine (SVM) [24]: The SVM [24] method works by finding the hyperplane that divides feature space between two regions. In this way, on one side of the plane are all data points belonging to class one, and all points that belong to class two are on the opposite part of the plane. The aim is to maximize the margins so that the optimal linearly separable hyperplane can be determined among multiple alternatives. Margin optimization means maximizing the distance between the closest data points (also called support vectors) on either side of the hyperplane. This results in a hyperplane that is as far away as possible from the closest data points (support vectors). In other words, the support vectors can be viewed as a hyperplane parallel to the main hyperplane, helping to create the most efficient hyperplane.

K-nearest neighbor (KNN) [25–27]: To classify a point in the feature space, KNN [25–27] examines the k nearest neighbors of the point and makes predictions based on majority voting from those nearest neighbors. The input variable k is used to determine the number of nearest neighbors to use. This k number is usually an odd number, to avoid situations where new data points are equally distributed in two classes. KNN [25–27], as its name implies, calculates the distance between different points in the feature space. In some situations, it may be useful to weigh the "votes" of adjacent points according to distance, so that nearby points contribute more to the classification than distant points.

Logistic regression (LR) [28]: LR [28] uses optimization methods such as gradient ascent or the more efficient stochastic gradient ascent to find the optimal parameters of a nonlinear function, called a sigmoid, which is very suitable for binary problems. The advantage of using stochastic gradient ascent as an optimization algorithm is that it can learn from new data by several batches and iterations.

Naive Bayes (NB) [29]: The NB [29] is a classifier that is based upon Bayes' theorem of statistical probability. It assumes every single predictor is equally important and independent of one another. That is, given a class variable, it assumes that the absence or presence of a particular feature is independent of the absence or presence of other features. Rather than a simple classification, the NB [29] will report probabilities of an instance belonging to an individual class. The class with the highest posterior probability in our case is the prediction of whether the domain name is malicious or not.

Decision tree (DT) [30]: DTs [30] are essentially a set of questions designed to arrive at a classification decision. As its name suggests, a DT [30] is a tree-like structure, which is composed of several parts, such as root nodes, internal nodes, leaf nodes, and branches. The root node represents the top-level decision node. In other words, it is where the classification tree begins to be traversed. A leaf node is a node that is not split into more nodes; it is where a class is assigned by majority vote. DTs [30] are constructed via an algorithmic approach that makes a classification decision for a given data point by recursively partitioning the available data. In other words, the algorithm partitions the data recursively into subsets with rules that maximize information gain.

Figures 2–5 show the results of our approach in comparison with other baseline approaches in terms of precision, recall, F1-Score, and accuracy. In terms of precision, our approach obtains 0.989%.

The average precision value improvement by our approach over all baseline approaches is 0.051%, 0.049%, 0.019%, 0.04%, 0.034%, and 0.016% compared with LDA [2], [20–23], SVM [2,24], KNN [2,25–27], LR [2,28], NB [2,29], and DT [2,30], respectively. In terms of recall, our approach obtains 0.988%. The average recall value improvement by our approach over all baseline approaches is 0.119%, 0.093%, 0.045%, 0.11%, 0.134%, and 0.036% compared with LDA [2,31], SVM [2,24], KNN [2,25–27], LR [2,28], NB [2,29], and DT [2,30], respectively. In terms of F1-Score, our approach obtains 0.988%. The average F1-Score value improvement by our approach over all baseline approaches is 0.093%, 0.079%, 0.035%, 0.083%, 0.097%, and 0.029% compared with LDA [2,31], SVM [2,24], KNN [2,25–27], LR [2,28], NB [2,29], and DT [2,30], respectively. In terms of accuracy, our approach obtains 0.988%. The average accuracy value improvement by our approach over all baseline approaches is 0.081%, 0.069%, 0.031%, 0.072%, 0.08%, and 0.026% compared with LDA [2,31], SVM [2,24], KNN [2,25–27], LR [2,28], NB [2,29], and DT [2,30], respectively. All of these results indicate that the unified learning approach is more effective compared with the baseline approaches.

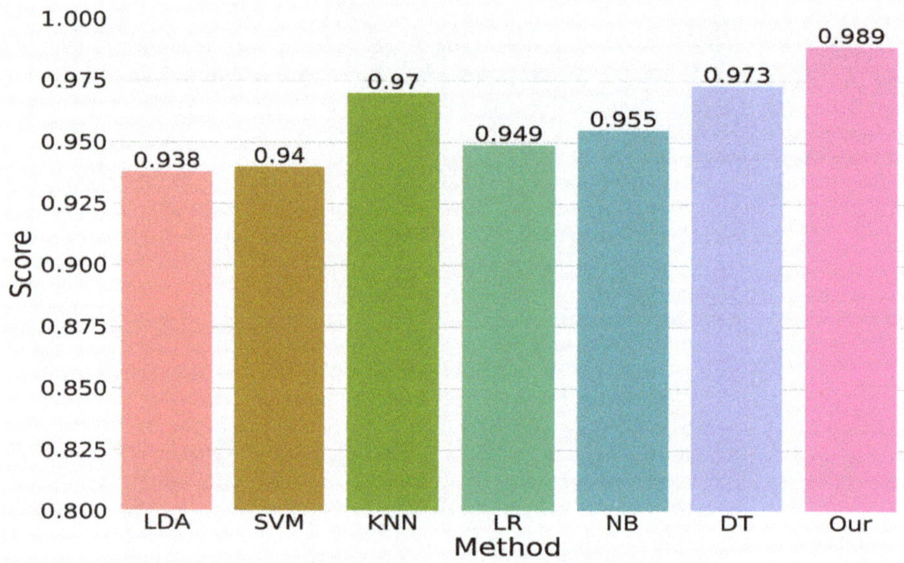

Figure 2. Comparison with baseline approaches on the precision metric.

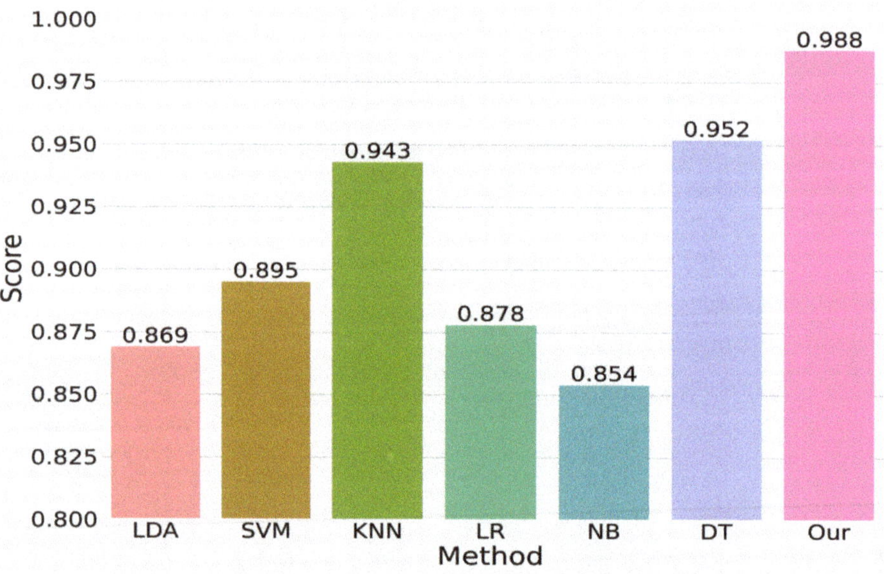

Figure 3. Comparison with baseline approaches on the recall metric.

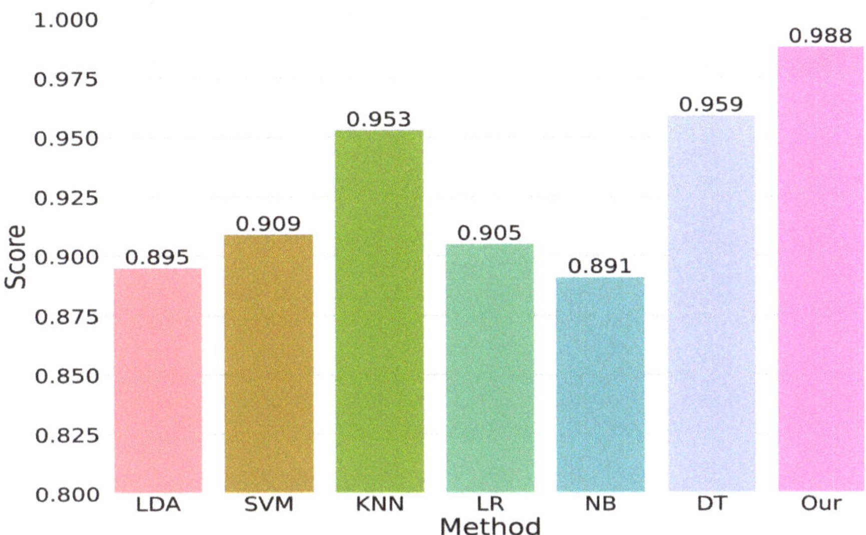

Figure 4. Comparison with baseline approaches on the F1-Score metric.

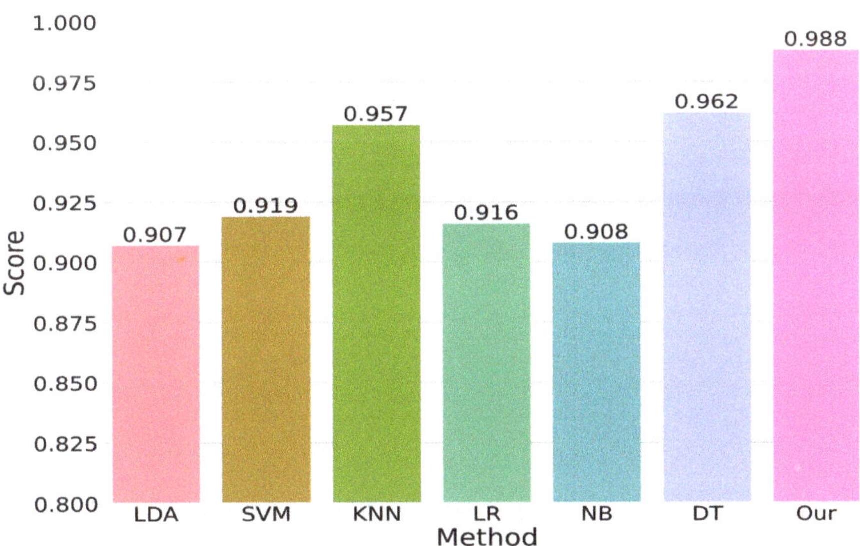

Figure 5. Comparison with baseline approaches on the accuracy metric.

RQ.2 How effective is our unified learning approach in comparison with individual features learning?

Numerical or textual features each provide separate useful information to distinguish whether the domain name is malicious or not. To investigate the numerical or textual features' impact on performance individually, we remove the numerical or textual features from the input and its respective modules in the model, and perform comparative experiments while keeping other conditions unchanged.

We evaluate performance through the accuracy, recall, precision, and F1-Score, and Figure 6 shows the experiment results. From Figure 6 it can be observed that, after removing the numerical or textual features, the performance on all the metrics decreased. In terms of accuracy, the decrease is 0.159% and 0.194% using numerical and textual features, respectively. In terms of precision, the decrease is 0.023% and 0.248% using numerical and textual features, respectively. In terms of recall, the decrease is 0.307% and 0.067% using numerical and textual features, respectively. In terms of F1-Score, the decrease is 0.19% and 0.169% using numerical and textual features, respectively. We notice that, after removing numerical or textual features, although the performance has declined, this decline does not reach 0%. The reason for this might be that numerical or textual features only account for a small fraction of the total features used to accurately detect malicious domain name. All of these results indicate that the unified learning approach is able to fully exploit its ability to capture local numerical and textual features and extract deep relationships between them. This shows that it is useful to feed numerical features along with textual features to the model for the accurate detection of malicious domain names.

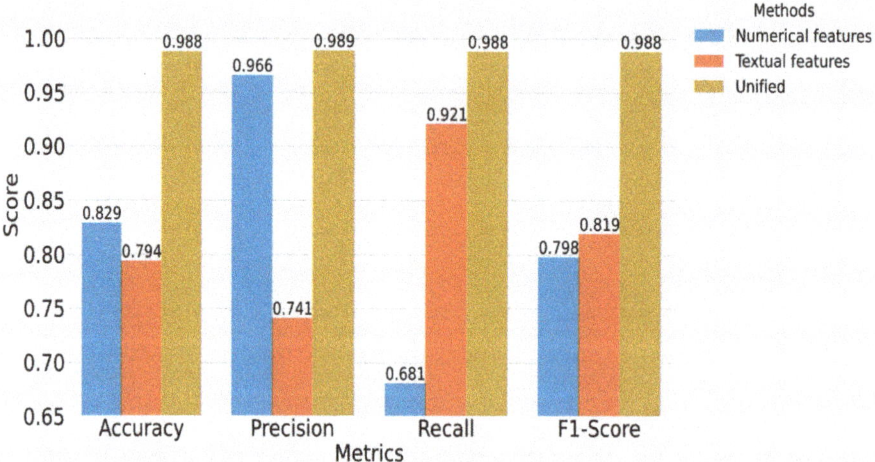

Figure 6. Individual and unified features comparison.

6. Conclusions

In this article, we proposed a novel approach that utilizes numerical and textual features of a domain name for malicious domain name detection. These domain name pairs are fed into a deep learning model that captures the semantic relationship between numerical and textual features. Then, these association features are used to classify whether a domain name pair is a malicious domain or not. We investigated the performance of our approach using a public malicious domain name dataset. We evaluated our approach's performance against the state-of-the-art machine learning-based approaches for malicious domain detection, and found our approach performs better for malicious domain name detection. We made a comparative study by removing either numerical or textual features, and performed experiments without changing other conditions. We found that it is more effective to feed both numerical and textual features to the deep learning model. This indicates that our proposed numerical and textual features pair representation is effective. In the future, we will investigate how our approach performs on industrial projects.

Author Contributions: Conceptualization, A.A.W.; Funding acquisition, Q.L.; Investigation, A.A.W.; Methodology, A.A.W.; Project administration, Q.L.; Software, A.A.W.; Supervision, Q.L.; Validation, S.M. and A.M.M.; Visualization, A.H. and M.B.; Writing—original draft, A.A.W.; Writing—review and editing, Z.Z. and D.K.B. All authors have read and agreed to the published version of the manuscript.

Funding: This work was supported in part by "Research on the Key Technology of Endogenous Security Switches" (2020YFB1804604) of the National Key R&D Program; "New Network Equipment Based on Independent Programmable Chips" (2020YFB1804600); the 2020 Industrial Internet Innovation and Development Project from Ministry of Industry and Information Technology of China; the Fundamental Research Fund for the Central Universities (30918012204, 30920041112); the 2019 Industrial Internet Innovation and Development Project from Ministry of Industry and Information Technology of China; Jiangsu Province Modern Education Technology Research Project (84365); National Vocational Education Teacher Enterprise Practice Base "Integration of Industry and Education" Special Project (Study on Evaluation Standard of Artificial Intelligence Vocational Skilled Level); Scientific research project of Nanjing Vocational University of Industry Technology (2020SKYJ03).

Data Availability Statement: A publicly available dataset was used in this study, which can be found at https://doi.org/10.17632/623sshkdrz.5 (accessed on 17 November 2022).

Conflicts of Interest: The authors declare no conflict of interest.

References

1. Liu, A.X. *Firewall Design and Analysis*; World Scientific: Singapore, 2010. [CrossRef]
2. Marques, C.; Malta, S.; Magalhães, J. DNS Firewall Based on Machine Learning. *Future Internet* **2021**, *13*, 309. [CrossRef]
3. Zhang, J.; Porras, P.; Ullrich, J. Highly predictive blacklisting. In Proceedings of the 17th Conference on Security Symposium, San Jose, CA, USA, 28 July–1 August 2018; USENIX Association: Berkeley, CA, USA, 2008; pp. 107–122.
4. Prakash, P.; Kumar, M.; Kompella, R.R.; Gupta, M. PhishNet: Predictive Blacklisting to Detect Phishing Attacks. In Proceedings of the 2010 Proceedings IEEE INFOCOM, San Diego, CA, USA, 14–19 March 2010; pp. 1–5. [CrossRef]
5. Akiyama, M.; Yagi, T.; Itoh, M. Searching Structural Neighborhood of Malicious URLs to Improve Blacklisting. In Proceedings of the 2011 IEEE/IPSJ International Symposium on Applications and the Internet, Munich, Germany, 18–21 July 2011; pp. 1–10. [CrossRef]
6. Fukushima, Y.; Hori, Y.; Sakurai, K. Proactive Blacklisting for Malicious Web Sites by Reputation Evaluation Based on Domain and IP Address Registration. In Proceedings of the 2011 IEEE 10th International Conference on Trust, Security and Privacy in Computing and Communications, Changsha, China, 16–18 November 2011; pp. 352–361. [CrossRef]
7. Sun, B.; Akiyama, M.; Yagi, T.; Hatada, M.; Mori, T. Automating URL Blacklist Generation with Similarity Search Approach. *IEICE Trans. Inf. Syst.* **2016**, *E99.D*, 873–882. [CrossRef]
8. Saxe, J.; Berlin, K. eXpose: A Character-Level Convolutional Neural Network with Embeddings For Detecting Malicious URLs, File Paths and Registry Keys. *arXiv* **2017**, arXiv:1702.08568.
9. Yang, W.; Zuo, W.; Cui, B. Detecting Malicious URLs via a Keyword-Based Convolutional Gated-Recurrent-Unit Neural Network. *IEEE Access* **2019**, *7*, 29891–29900. [CrossRef]
10. Luo, C.; Su, S.; Sun, Y.; Tan, Q.; Han, M.; Tian, Z. A Convolution-Based System for Malicious URLs Detection. *Comput. Mater. Contin.* **2020**, *62*, 399–411. [CrossRef]
11. Mondal, D.K.; Singh, B.C.; Hu, H.; Biswas, S.; Alom, Z.; Azim, M.A. SeizeMaliciousURL: A novel learning approach to detect malicious URLs. *J. Inf. Secur. Appl.* **2021**, *62*, 102967. [CrossRef]
12. Hochreiter, S.; Schmidhuber, J. Long Short-Term Memory. *Neural Comput.* **1997**, *9*, 1735–1780. [CrossRef] [PubMed]
13. sklearn.preprocessing.LabelEncoder. Scikit-Learn. Available online: https://scikit-learn.org/stable/modules/generated/sklearn.preprocessing.LabelEncoder.html (accessed on 16 December 2022).
14. sklearn.preprocessing.MinMaxScaler. Scikit-Learn. Available online: https://scikit-learn/stable/modules/generated/sklearn.preprocessing.MinMaxScaler.html (accessed on 26 February 2022).
15. Srivastava, N.; Hinton, G.; Krizhevsky, A.; Sutskever, I.; Salakhutdinov, R. Dropout: A simple way to prevent neural networks from overfitting. *J. Mach. Learn. Res.* **2014**, *15*, 1929–1958.
16. Nair, V.; Hinton, G.E. Rectified linear units improve restricted boltzmann machines. In Proceedings of the 27th International Conference on International Conference on Machine Learning, Haifa, Israel, 21–24 June 2010; Omnipress: Madison, WI, USA, 2010; pp. 807–814.
17. Kingma, D.P.; Ba, J. Adam: A Method for Stochastic Optimization. In Proceedings of the 3rd International Conference on Learning Representations, ICLR 2015, San Diego, CA, USA, 7–9 May 2015; Bengio, Y., LeCun, Y., Eds.; 2015. Available online: http://arxiv.org/abs/1412.6980 (accessed on 20 November 2022).
18. Goodfellow, I.; Bengio, Y.; Courville, A. *Deep Learning*; MIT Press: Cambridge, MA, USA, 2016.
19. Marques, C.; Malta, S.; Magalhães, J.P. DNS dataset for malicious domains detection. *Data Brief* **2021**, *38*, 107342. [CrossRef] [PubMed]

20. Wayback Machine. 2022. Available online: https://web.archive.org/web/20220615132544/http://datajobstest.com/data-science-repo/LDA-Primer-[Balakrishnama-and-Ganapathiraju].pdf (accessed on 27 March 2023).
21. Lu, J.; Plataniotis, K.N.; Venetsanopoulos, A.N. Face recognition using LDA-based algorithms. *IEEE Trans. Neural Netw.* **2003**, *14*, 195–200. [CrossRef] [PubMed]
22. Fu, R.; Tian, Y.; Shi, P.; Bao, T. Automatic Detection of Epileptic Seizures in EEG Using Sparse CSP and Fisher Linear Discrimination Analysis Algorithm. *J. Med. Syst.* **2020**, *44*, 43. [CrossRef] [PubMed]
23. Elnasir, S.; Shamsuddin, S.M. Palm vein recognition based on 2D-discrete wavelet transform and linear discrimination analysis. *Int. J. Adv. Soft Comput. Appl.* **2014**, *6*, 43–59.
24. 1.4. Support Vector Machines. Scikit-Learn. Available online: https://scikit-learn/stable/modules/svm.html (accessed on 26 February 2022).
25. kNN Definition | DeepAI. 2022. Available online: https://web.archive.org/web/20220701054511/https://deepai.org/machine-learning-glossary-and-terms/kNN (accessed on 27 March 2023).
26. Hassanat, A.B.; Abbadi, M.A.; Altarawneh, G.A.; Alhasanat, A.A. Solving the problem of the K parameter in the KNN classifier using an ensemble learning approach. *arXiv* **2014**, arXiv:1409.0919014.
27. Sklearn.Neighbors.KNeighborsClassifier—Scikit-Learn 1.2.2 Documentation. 2023. Available online: https://web.archive.org/web/20230315064604/https://scikit-learn.org/stable/modules/generated/sklearn.neighbors.KNeighborsClassifier.html (accessed on 27 March 2023).
28. Advantages and Disadvantages of Linear Regression. 2023. Available online: https://web.archive.org/web/20230111220233/https://iq.opengenus.org/advantages-and-disadvantages-of-linear-regression/ (accessed on 27 March 2023).
29. 1.9. Naive Bayes—Scikit-Learn 1.2.1 Documentation. 2023. Available online: https://web.archive.org/web/20230307185232/https://scikit-learn.org/stable/modules/naive_bayes.html (accessed on 27 March 2023).
30. 1.10. Decision Trees—Scikit-Learn 1.2.2 Documentation. 2023. Available online: https://web.archive.org/web/20230320174546/https://scikit-learn.org/stable/modules/tree.html (accessed on 27 March 2023).
31. Blei, D.M.; Ng, A.Y.; Jordan, M.I. Latent dirichlet allocation. *J. Mach. Learn. Res.* **2003**, *3*, 993–1022.

MDPI
St. Alban-Anlage 66
4052 Basel
Switzerland
Tel. +41 61 683 77 34
Fax +41 61 302 89 18
www.mdpi.com

Axioms Editorial Office
E-mail: axioms@mdpi.com
www.mdpi.com/journal/axioms